PET Packaging Technology

Sheffield Packaging Technology

Series Editor: Geoff A. Giles, Global Pack Management, GlaxoSmithKline, London.

A series which presents the current state of the art in chosen sectors of the packaging industry. Written at professional and reference level, it is directed at packaging technologists, those involved in the design and development of packaging, users of packaging and those who purchase packaging. The series will also be of interest to manufacturers of packaging machinery.

Titles in the series:

Design and Technology of Packaging Decoration for the Consumer Market
Edited by G.A. Giles

Materials and Development of Plastics Packaging for the Consumer Market
Edited by G.A. Giles and D.R. Bain

Technology of Plastics Packaging for the Consumer Market
Edited by G.A. Giles and D.R. Bain

Canmaking for Can Fillers
T.A. Turner

PET Packaging Technology
Edited by D.W. Brooks and G.A. Giles

PET Packaging Technology

Edited by

DAVID W. BROOKS
Corporate Technologies
Crown Cork & Seal
Oxford

and

GEOFF A. GILES
Global Pack Management
GlaxoSmithKline
London

CRC Press

First published 2002
Copyright © 2002 Sheffield Academic Press

Published by
Sheffield Academic Press Ltd
Mansion House, 19 Kingfield Road
Sheffield S11 9AS, UK

ISBN 1-84127-222-1

Published in the U.S.A. and Canada (only) by
CRC Press LLC
2000 Corporate Blvd., N.W.
Boca Raton, FL 33431, U.S.A.
Orders from the U.S.A. and Canada (only) to CRC Press LLC

U.S.A. and Canada only:
ISBN 0-8493-9786-3

Printed on acid-free paper in Great Britain by
Antony Rowe Ltd, Chippenham, Wilts

British Library Cataloguing-in-Publication Data:
A catalogue record for this book is available from the British Library

Library of Congress Cataloging-in-Publication Data:
A catalog record for this book is available from the Library of Congress

Preface

PET is a polymer which is enjoying substantial growth as a packaging material—across global markets and for diverse applications. The growth of PET into packaging, as a replacement for glass, metal and other plastics materials, has been extraordinary. No other rigid plastics packaging sector has experienced the level of growth of PET bottles over the last 20 years. PET is now a commodity polymer competing directly with polyethylene, polypropylene and styrenics in the markets for food and beverage packaging, as well as other products. Its success as a packaging material is due largely to technological developments in the processes used to convert PET into flexible and rigid packs at high outputs, which is crucial for the minimisation of costs. Developments in PET technology in recent years have focused on improvements in gas barrier and heat stability—a process that will continue as new opportunities develop in the food and beverage markets.

This volume is designed to inform the reader about PET, about the development of materials and technologies to improve its barrier and heat stability, about the various process technologies for converting it into rigid and flexible packs, and about environmental and recycling issues. It is the first volume to provide a comprehensive source of reference on PET packaging technologies, and it includes the latest developments.

The historical development of PET is also covered in an introductory chapter, which brings together many aspects of the early development of PET, with special reference to the 'father' of PET bottles, Nathaniel Wyeth, and his recollection of the first time he ran PET resin to make a bottle. This chapter considers the birth of the polyester industry, with the process of stretching and the uniaxial orientation of polyester fibre, which was a precursor to the biaxial orientation of film and bottles; the demise of Lopac (the Monsanto acrylonitrile copolymer), which gave PET an invaluable lead as the polymer for rigid bottles to replace glass; the development of the preform neck finish, initially based on a glass finish so that plastic closures suitable for carbonated beverage PET bottles could be used; and the further developments in both the materials and the technologies.

PET Packaging Technology will be used by packaging technologists, packaging engineers, chemists, materials scientists and engineers, those involved in the design and development of packaging, and those responsible for specifying or purchasing packaging.

The editors have been fortunate to have the support of the most knowl-
edgeable experts in PET technology as authors for this volume. They are
busy people, and we are grateful to them for their efforts and contributions.

David Brooks
Geoff Giles

Contributors

Mr Bob Blakeborough	Synventive Molding Solutions Ltd, Unit 1 Silver Birches Business Park, Aston Road, Bromsgrove, Worcestershire B60 3EX, UK
Mr Gordon J. Bockner	Business Development Associates, 4550 Montgomery Avenue, Suite 450 North, Bethesda, MD 20814, USA
Mr David W. Brooks	Corporate Technologies, Crown Cork & Seal Company, Inc., Downsview Road, Wantage, Oxon OX12 9BP, UK
Mr Geoff A. Giles	Global Pack Management, GlaxoSmithKline, London, UK and Holmesdale Cottage, Nuffield, Henley-on-Thames RG9 5SS, UK
Dr Kenneth M. Jones	47 Aspin Lane, Knaresborough, North Yorkshire HG5 8EX, UK
Dr Michael Koch	Ferromatik Milacron Europe, Riegelerstrasse 4, D 79364 Malterdingen, Germany
Dr William A. MacDonald	DuPont Teijin Films UK Ltd, PO Box 2002, Wilton, Middlesborough TS90 8JF, UK
Dr Duncan H. MacKerron	DuPont Teijin Films UK Ltd, PO Box 2002, Wilton, Middlesborough TS90 8JF, UK
Dr Vince Matthews	35 Rossett Beck, Harrogate, North Yorkshire HG2 9NT, UK
Mr Paul Swenson	Kortec Inc., Cummings Center #128Q, Beverly, MA 01915, USA
Dr Bora Tekkanat	Consultant, 1913, Orchardview Drive, Ann Arbor, MI 48108, USA
Mr Mike Wortley	Jomar Europe Ltd, Clarkson, Upper Crabbick Lane, Waterlooville, Hampshire PO7 6HQ, UK

Contents

5   PET film and sheet   116
WILLIAM A. MACDONALD, DUNCAN H. MACKERRON and DAVID W. BROOKS

6   Injection and co-injection preform technologies   158
PAUL SWENSON

7 One-stage injection stretch blow moulding   184
BOB BLAKEBOROUGH

1   Introduction
David W. Brooks and Geoff A. Giles

1.1   Introduction

*'The first time we ran the PET resin, our little machine groaned and grunted
like it never had before. Our fear was that we not only were going to fail to
produce an acceptable bottle again, but that the experiment would damage
our test equipment. . . I remember the mix of expectation and depression I felt
when we opened our moulding machine. It was late in the evening, and there
was a little cross light hitting the mould. After months of frustration during
which we had produced nearly 10,000 bottles by hand, we had grown used
to seeing blobs of resin caked on the mould, or crude shapes that looked like
anything but a bottle. This time, at first glance, it looked as if the mould was
empty. A closer look revealed something else: a crystal clear bottle. Since
then I have seen countless truly beautiful PET bottles, but none of them will
ever be as memorable as the first'* [1].

These were the words of Nathaniel Wyeth, considered to be the 'father'
of PET (polyethylene terephthalate) biaxially oriented bottles, at the Fourth
International Conference on biaxially oriented bottles and containers in PET
and other engineering and thermoplastic resins, in Düsseldorf in October 1983.
His patent, assigned to DuPont, was issued on 15 May 1973. Amoco Container
was the first licensee. Other major bottler makers followed; and the rest, as they
say, is history. So began the evolution of the PET bottle, which has revolutionised
the global bottle market, especially in the carbonated beverage sector.

1.1.1   History

While the events noted by Nathaniel Wyeth stimulated the development of PET
resin into the global material that we know today, PET (or polyester, as it is
commonly known) developed long before.

Although polyester was first discovered in the nineteenth century, it was not
until 1941 that J. R. Whinfield and J. T. Dickson of the British company ICI,
while searching for a replacement for silk, identified polyester as a fibre-forming
polymer. After the Second World War, DuPont was licensed to manufacture the
new polyester fibres in the United States, while ICI retained the exploitation
rights for the rest of the world. The polyester fibre industry was born when
the process of stretching and uniaxially orienting polyester fibre to improve its
physical properties was perfected in the 1950s. Demand for this material from

the polyester fibre industry, leading to economies of scale in PET manufacturing plants, resulted in PET becoming more cost-competitive with other materials.

In the 1960s, the technology of 'biaxial orientation' was developed—stretching the PET sheet in one direction and then again at right angles to the original stretch. This resulted in a polyester film with enhanced mechanical properties and low gas-permeability that is now widely used for food packaging, film stock, videotapes and computer floppy disks. That side of the business has continued to grow. As PET bottles underwent development for heat stability, so did sheet material, and CPET (crystallised polyethylene terephthalate) was developed for ovenable food trays. Although now levelling out in demand, this development opened up a whole new packaging sector.

However, the story should really begin before PET, with bottles made by Monsanto from their Lopac® acrylonitrile (AN/S) resin. This was just one of a range of barrier resins Monsanto had in the late 1960s that had performance bottle potential. They made extrusion blow-moulded (EBM) and injection blow-moulded (IBM) 4 oz and 8 oz Boston Rounds for product testing. Coca Cola® became interested and the first Coke bottle was made in 1969. In 1970, 10 oz bottles produced on the Monsanto BDS-8 horizontal EBM wheel were test marketed in New Bedford, MA, US and then Providence, RI, US. They were green, with the traditional Coca Cola® shape. The go-ahead for commercial production came in 1972, for 10, 16 and 32 oz sizes. Coca Cola® were not interested in the family-size segment at that time. The first Monsanto plant started up in Windsor, CT, US in 1975, followed by others in Chicago, IL, US and Havre de Grace, MD, US. By 1976, when one-way bottle production was in full swing, Monsanto also had a refillable bottle ready for production in Chicago.

In 1977, the US Food and Drug Administration (FDA) revoked approval for the Coca Cola® 32 oz Cycle-Safe bottle, moulded by Monsanto in Lopac resin. Millions of bottles were sent to the scrap heap; reportedly more than 11 million bottles from one plant alone. The FDA claimed that the Lopac bottle failed to meet the government's new limit of 50 ppb (parts per billion) for acrylonitrile extraction. The FDA claimed that this limit was exceeded in a test when a bottle was subjected to a temperature of $48°C$ ($120°F$) for 3 months. Monsanto disagreed with this result, suggesting that under normal shelf-life conditions there was no detectable migration of monomer into the contents of the bottle. This move by the FDA, as much as any, started the PET bottle revolution.

The first Pepsi-Cola® bottles were in fact also acrylonitrile. In 1968, plastic bottle moulders, DeBell & Richardson, were working with Sohio's Barex (AN/MA), which showed promise, when Pepsi-Cola® approached them to explore plastic bottle possibilities. In 1970, Pepsi-Cola® carried out a test market in Las Vegas, with 250,000 10 oz AN/MA bottles made on a Bekum EBM machine.

DuPont approached Pepsi-Cola® with PET in 1972. Mylar film existed then but no one knew how to make a PET bottle. Pepsi-Cola® were looking for a company that could make a close-tolerance, injection-moulded preform and Broadway Inc. welcomed the challenge. After some initial development they brought along their first preform; for a one litre bottle, the preform was almost as long as the bottle.

DuPont, in the meantime, was working with Cincinnati-Milacron on the 'SOM' (slug oriented moulding) approach. An extruded tube was placed in a machine and formed with about 4,000 psi (pounds per square inch) hydraulic pressure. Somehow, without heating, bottles emerged; but the tools were destroyed after about 5–10 cycles. Another idea, namely reheat blow (RHB), was suggested. With Cincinnati, the hand-operated RHB-1 was built. By late 1974, the RHB-3 was ready. The Broadway team had shortened the preform, had an 8-cavity preform injection mould ready, and were making base cups (early bottles had hemispherical bases).

The first PET test market was held in New York State during 1975–76, with bottles made by DuPont's Christina Laboratories. Pepsi-Cola® had planned to make their own bottles, but after the test had ended in early 1976, Pepsi helped transfer the technology to others. By late 1976, Amoco Containers' first bottles were ready: 64 oz bottles with a 38 mm neck and weighing 82 g with a snap-on base cup. They were shipped to Michigan, filled and on the market in March 1977. Hoover-Universal's bottles followed a couple of weeks later.

One of the best stories arising from recollections of that time was of events that happened in the early 1970s, when Pepsi-Cola® called Goodyear for a sample of polyester resin. They wanted to try to make a bottle but, since the idea seemed far-fetched, they failed to mention this to the Goodyear salesman. He had a choice of five possible grades: three with TiO_2, one low-IV (intrinsic viscosity) resin, and one solid-state resin without TiO_2. For some reason, he picked the last grade, the only one that had any chance of making a clear preform. It is said that if the salesman had made another choice Pepsi-Cola® might have given up on the whole idea.

1.1.2 IP—patents

In 1973, DuPont issued a series of patents (filed in 1970) assigned to inventors Nathaniel Wyeth and Ronald Roseveare. '*Process and Apparatus for Producing Hollow Biaxially Oriented Articles*'. US 3,733,309, '*Biaxially Oriented Poly(ethylene terephthalate) Bottle*', and GB 1341845-8, '*An At Least Partly Biaxially Oriented Thermoplastic Bottle*' [2].

The main claims of this patent can be summarised as follows: a biaxially oriented bottle of a generally cylindrical section, with a density of about 1.331 to 1.402 $g \cdot cm^3$, prepared from polyethylene terephthalate, having an IV of at least about 0.55, has an axial tensile strength of about 5,000–30,000 psi, a hoop tensile strength of about 20,000–80,000 psi, an axial yield stress of at least 4,000 psi, a hoop yield stress of at least about 7,000 psi, and a deformation constant—equal to the slope of the log (strain rate^{-1}) *vs* strain—of at least about 0.65 at 50°C. The bottle is capable of storing a carbonated liquid with a deformation of less than 5% over 100 h at 50°C, under a pressure of 75 psig (pounds per square inch gauge).

The invention can be summarised as follows: it provides a method and apparatus for producing a hollow, biaxially oriented thermoplastic article, which has improved strength properties. The articles are oriented by stretching, typically an average of up to about four times in the axial direction and about 2.5–7 times in the hoop direction. The article has reduced permeability to carbon dioxide, oxygen and water. The bottles produced are useful for storing beverages under pressure, such as carbonated sodas and beer.

1.2   Market growth and the development of materials

The number of global PET producers has doubled in the last 10 years, in response to the growing demand for PET for rigid packaging applications. Global demand has seen steady growth through the 1980s and 1990s; in 1985 about 0.5 M tonnes of PET was consumed, in 1990 about 1.5 M tonnes, and this increased to 7 M tonnes in 2000 (see Figure 1.1). PET global capacity has seen a sharp increase in the last three to five years to match the high demand for packaging; from 1990 to 1995 this increased by an average 15% per year, but from 1995 to 2000 the growth was up to 20% per year. The prediction for the next few years is growth of about 10 per cent per annum.

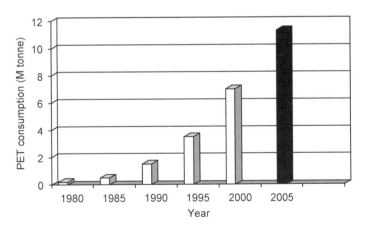

Figure 1.1 Global consumption of polyethylene terephthalate for packaging (M tonne).

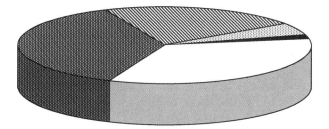

Figure 1.2 Regional supply of polyethylene terephthalate for packaging in 2000. Key: □, North America; ▦, Asia Pacific; ▨, Europe; ▩, South America; ■, Middle East/Africa.

The biggest growth has been observed in Asia, where producers have tripled their regional capacities from 1 to 3 M tonnes during 1995 to 1999 (see Figure 1.2). Asia-Pacific now has 37% of global PET capacity. In the period between 1990 and 2000, the regional split for North America has changed from 54% to 34%, for Asia-Pacific from 25% to 37%, and for Europe from 19% to 22%. In this period, the number of PET producers has doubled from about 35 to 75; mostly new companies in Asia-Pacific and South America (see Tables 1.1–1.4).

The size of PET plants has also grown in this period; a typical line in 1990 would be about 50 K tonnes, whereas a world scale line is now 200 K tonnes. The major global PET producers have also changed, largely through change of ownership. Eastman has remained the largest global PET producer, with plants in North America, Europe and South America. A consolidation of the market

Table 1.1 The major PET suppliers in North America in 2000

PET supplier	Output (K tonnes)
Eastman Chemicals	836
KoSa	475
Wellman	390
M & G	360
Nanya	200
DuPont	175

North America includes the US and Canada.

Table 1.2 The major PET suppliers in South America in 2000

PET supplier	Output (K tonnes)
Eastman Chemicals	247
Rhodiaster	177
KoSa	160

South America includes Mexico, Brazil and Argentina.

Table 1.3 The major PET suppliers in Europe in 2000

PET supplier	Output (K tonnes)
Eastman Chemicals	390
Dow	280
DuPont	240
KoSa	230
M & G	220

Table 1.4 The major PET suppliers in Asia-Pacific in 2000

PET supplier	Output (K tonnes)
Kohap	206
Shin-Kong	160
Tong-Kook	140
Far Eastern	120

Asia-Pacific includes South Korea and Taiwan.

can be expected in view of the current supply and price situation. Hoechst, the second largest supplier, has steadily dropped out of commodity chemical businesses, and its PET operations are now run by KoSa. DuPont took over ICI's PET business, and they have now merged with SASA of Turkey to form DuPontSA. In 1999, Shell decided to exit PET in the course of restructuring of its chemical business, which has led to the sell-off to M&G (Mossi & Ghisolfi Group). Wellman and Dow are the other major PET producers. For a period of time, the PET polymer business has been in oversupply; this is expected to change to a more balanced situation in the next few years. In Europe, demand may well outpace supply beyond about 2005, balanced by excess supply globally because of new capacities being constructed in the Asia-Pacific region. PET is now the most widely used plastic for bottles. The non-return bottle is established around the world; in contrast, returnable bottles have not penetrated to any great extent. The demand for PET bottles has been driven by its use in beverage applications, especially for carbonated soft drinks (CSDs), mineral waters and other beverages, which together account for 80% (while food accounts for 10%) of total PET in packaging (see Figure 1.3). The growth of PET has resulted mainly from the replacement of existing packaging; for CSDs it has replaced glass and to a lesser extent metal cans, and for waters it has replaced PVC and glass. The big question is, will beer in PET progress beyond niche markets? The projected growth for beer in PET bottles is about 10 per cent per annum over the next five years.

With technological development, PET has become a suitable packaging material for more diverse end-use applications. Developments involving the use of high gas barrier polymers are progressing with a number of companies.

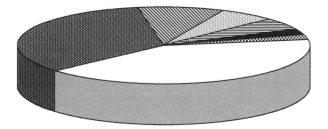

Figure 1.3 Polyethylene terephthalate packaging applications in Europe in 2000. Key: □, Carbonated soft drinks; ▨, Mineral water; ▨, Food; ▨, Film/sheet; ▤, Beer/other beverages; ■, Cosmetics; ▨, Others.

These include ethylene vinyl alcohol copolymer (EVOH), nylons in multilayer structures, coatings with organic materials, the use of polyvinylidene chloride (PVDC), epoxy-amines as external coatings on PET bottles, inorganic coatings including silica and silicon oxide, and plasma coatings of carbon and silicon oxide as both internal and external coatings on PET bottles. These developments will extend the PET market into beverage applications that require high barrier properties, the major target being beers, for which glass and metal cans predominate. There are also many food applications, especially baby food, which is at present largely packaged in glass and metal cans. Newer developments with liquid crystal polymers (LCPs), oxygen scavengers and nanocomposites will extend the barrier potential of PET. The impact of high barrier polyesters, such as polyethylene naphthalate (PEN), as an alternative to PET has so far been insignificant. PEN has better barrier and thermal properties but is more expensive, and its gas barrier properties are not as good as many multilayer and coated PET options. Modification of PET grades by copolymerisation or the incorporation of additives has allowed PET to be processed more rapidly, and thus more economically, by stretch blow moulding with better processing machinery, and can also reduce the formation of acetaldehyde (AA) for sensitive applications such as mineral waters. In hot-fill beverage and food applications, modified PET grades and new process technologies, utilising hot moulds and cryogenic cooling to increase the heat stability, are extending the use of PET.

1.2.1 PET fibre and film

In the 1930s, Carothers and Hill at DuPont produced fibre-forming polyesters that had low melting points and were unsuitable for commercial production. The fibre-forming properties of PET were first discovered in 1941 by Whinfield and Dickson in the laboratories of Calico Printers in the UK. They resulted in a patent which DuPont acquired in 1948 for the US, and which ICI was granted for the rest of the world. PET fibres became commercial in the US in 1953 and production expanded rapidly in the 1960s and 1970s. The production of textile filament yarn for use in manufacture of garments grew rapidly and, in addition, polyester soon became the principal material for reinforcement of tyres in the US. Growth in the PET fibre market continued worldwide, the enormous expansion being attributed to its great versatility and its importance to the world economy. The discovery of PET's fibre-forming properties soon led to the development of PET films at ICI. Production of polymer films by a drawn orientation process was known for polystyrene (PS) and polyamides (PA). PET was well-suited for biaxial orientation; its properties (molecular weight and melting point) were ideal for producing high-strength films. In the 1950s, the demand for fibres led to rapid expansion in the supply of PET monomers, and the economics of large-scale polymer production became more attractive. PET film manufacture benefited from this rapid growth, and soon chemical companies producing PET fibre were also producing PET films. The key suppliers of films in the US and Europe were DuPont with Mylar, ICI with Melinex and Hoechst with Hostphan, while captive producers, such as Agfa Gevaert, Eastman Kodak, 3M, Fuji and DuPont, were manufacturing for their own internal consumption.

In the 1960s, ICI developed the Merolite pack—a sausage pillow-style sachet using flexible, lightweight materials and targeting single-serve carbonated beverages. A thin board sleeve (not unlike a toilet roll inner core) surrounded the body of the 'sausage' for the purposes of decoration. This tube also provided a strengthening feature to hold the container when opening, and also to minimise the risk of squeezing the product out after opening and release of the carbonation gas. The pack was opened by means of a removable tape sealing a small hole through which the product could be dispensed. Sadly, it failed to capture a broad market interest, probably because it was difficult to handle at both the filling and retailing stage, as well as in the hands of the consumer.

However, a more widely available system is the CheerPack from Hosokawa Yoko in Japan, or their global licensee, Gualapack, in Italy. Developed during the 1980s, the polyester-laminated pouch has an injection-moulded neck attached, through which the container is filled and consumed. It is very popular for sports drinks because it is practical for use in sports bags, being lightweight and reclosable, and unbreakable in normal use. Again, merchandising issues have limited its market expansion.

1.2.2 Barrier materials

1.2.2.1 OXBAR

In the 1980s, Metal Box developed its OXBAR barrier technology, a system designed to provide a chemical barrier within the PET container wall; and specifically to absorb or scavenge oxygen migrating into the pack, preventing loss of product quality during its shelf-life. Using polymer-based chemistry, OXBAR offers a total oxygen-specific barrier independent of temperature and humidity and has a long but finite lifetime. The components of the system are a blend of PET polymer with an oxidisable polymer (nylon MXD6) at about 5% addition, and a metallic catalyst (cobalt salt >200 ppm). The active ingredients are the nylon and the cobalt salt, with cobalt catalysing the reaction of the nylon with oxygen permeating through the pack wall. The OXBAR system consists mostly of PET bottle resin and could be readily converted by standard techniques into preforms and PET barrier bottles and containers. The low levels of active ingredients do not significantly affect the processing of the PET. The important feature of this system is its ability to maintain total barrier properties under elevated temperature and humidity conditions. The OXBAR system provides a total barrier to oxygen ingress for a period of up to about two years. Oxygen transmission data on 1 litre PET bottles produced from a PET blend containing 4% nylon MXD6 and the same blend containing the OXBAR system with added cobalt give an improvement approaching 100-fold.

The same bottles were also tested under real storage conditions for two years. The bottles were filled with water, nitrogen-flushed before capping, and the headspace oxygen content was monitored. The control PET blend with nylon MXD6 showed a 10% increase in oxygen into the headspace, while the OXBAR system showed no ingress. The lifetime for the OXBAR system depends on the chemical activity of the cobalt and the level present in the system, and also the thickness of the bottle wall. A further factor is the rate of consumption, which depends on the temperature, wall thickness and the type of PET used. High temperature storage conditions and levels of cobalt below about 200 ppm reduce the lifetime of the OXBAR system. A number of pack tests were carried out with lager beer, orange juice and wine over a 6 month period. In all cases, OXBAR proved superior to PET coated with PVDC, and it was comparable to glass in taste tests.

Food contact issues were critical for a novel application involving new chemistry, and extensive extractability tests were carried out by Metal Box chemists on a range of simulants and conditions. At this time, the conclusions were that the levels of cobalt, nylon oligomers and other species were within statutory limits for intended end use. However, the OXBAR system did not become commercial and was shelved in 1990. OXBAR was revisited by Crown Cork & Seal later in the 1990s, and development continues today [3].

1.2.2.2   PEN

In the late 1980s, both Eastman Chemicals and Goodyear developed the first PEN polymers. PET did not have sufficient gas barrier or thermal resistance for oxygen-sensitive foods and beverages, hot-fill foods, juices and sauces. The very limited supply of monomer prevented a major commercial introduction, but sufficient PEN homopolymer was produced to enable trials and testing to be carried out. Eastman's PEN was called 5XO (five times better oxygen barrier then PET), while Goodyear's PEN was called HP (high performance). Compared to PET, PEN has a high melting point, higher glass transition temperature (Tg) resulting in a higher thermal resistance, and higher gas barrier properties. Bottles could be stretch blown on standard two-stage equipment using high preform temperatures. Bottle tests have confirmed the improved barrier and thermal resistance of PEN bottles compared to PET. However, the projected very high cost, uncertainty about recyclability, and the question of FDA approval prevented PEN becoming commercial. PEN reappeared again some years later. PEN is discussed in more detail in Chapter 4, Section 4.2.1.

1.2.2.3 Barrier developments
A number of barrier developments were being progressed in the early 1980s. Mitsubishi Gas Chemical (MGC) was developing nylon MXD6 in Japan, and also using a five-layer, co-injection process to produce 1 litre bottles with PET and nylon MXD6 with tie layers. The bottle had a fourfold better oxygen barrier compared to PET, and was superior to a PVDC-coated PET bottle. In Japan, Unitika together with Nissei were also evaluating a new polymer, polyarylate, called U-polymer, a clear polymer with high temperature resistance and good gas barrier properties (similar to PET). In the US, Eastman developed a high barrier polyester together with tube extrusion processing to produce a three-layer tube and preform, with the polyester having an oxygen gas barrier similar to PVDC. In Japan, Kuraray developed a five-layer co-extrusion line, and used their own EVOH barrier resin in a multilayer co-extrusion structure with PET and adhesive tie resins. Eastman were also co-extruding PET with EVOH in a five-layer structure with tie layers. They developed a new tie layer resin specifically for this application, and they were also producing a range of PET/EVOH blends. Goodyear were developing high barrier polyesters with a threefold better oxygen barrier compared to PET.

During 1981, ICI (working with Fibrenyle and Metal Box) developed a PVDC-coated PET bottle, using their Viclan PVDC coating resin. The addition of a PVDC coating improves the oxygen barrier and reduces carbonation loss, thus extending the shelf-life of PET bottles. PVDC can be applied by one of three methods: dip coating, spray coating and flow coating the emulsion onto the external surface of the bottle. Once applied, the coating is cured by drying in a hot-air tunnel at a temperature high enough to form a film, but not so high as to distort the bottle. Metal Box developed a pilot coating line, and a market trial was conducted with coated bottles from the supermarket chain, Finefare, with beer brewed by Watney Mann, which was flash pasteurised and filled by Davenports. The market trial was successful and led to the development of production coating lines by Fibrenyle and Metal Box, and the commercial launch with Jaeger Lager in 1982. Further product launches followed in 1982, with 25 brands in the UK in coated PET bottles. The coated PET bottle achieved a 4–6 month shelf-life for beers, the limiting factor being oxygen ingress, which over time changes the taste and eventually taints the beer. In the UK, a total of 40 beer brands were packed in coated PET bottles by the end of 1983, with sales of 8 M bottles rising to a predicted 30 M. The coated PET bottle in 2 and 3 litre sizes is still available today from Constar UK (Crown Cork & Seal), who manufacture bottles for breweries in the UK. Wine has been tested and filled in PET bottles and PVDC-coated PET bottles but the penetration has been low. Initial commercial trials were in coated bottles in the US. Shelf-life estimates were about 18 months in coated bottles and 8 months in uncoated bottles.

1.2.3 Competitive materials: PVC, AN/S (Lopac) and AN/MA (Barex)

In the 1960s, PVC was being considered for food packaging. In France, edible vegetable oil presented a market opportunity because it was difficult to clean returnable glass bottles. PVC was confirmed to be the best available plastic material because of its transparency, good resistance to oils, low taste, odour and cost. By the end of the 1960s, spring waters were also being packed in PVC in France, once legal approval for packaging waters was completed. Other countries in Europe followed the French example, and PVC remained the main plastic material for packing non-carbonated waters, juices, wine and edible oils for more than 20 years, until challenged by PET. Rigid PVC is an amorphous polymer with good stiffness, clarity and chemical resistance, and good gas barrier properties. Most PVC is processed by extrusion blow moulding and, with the development of high molecular weight polymers, biaxial orientation opened up markets, including packaging of carbonated waters and soft drinks.

Monsanto developed Lopac (AN/S), a copolymer of acrylonitrile and styrene (70/30) among a range of barrier polymers in the 1960s. Initially, bottles were extrusion blow moulded and injection blow moulded. Biaxial orientation was also possible; bottles could be stretch blown via a preform, processed on equipment developed by Monsanto. AN/S had excellent gas barrier properties, high stiffness, and excellent clarity and chemical resistance. Monsanto developed the Cycle-Safe oriented bottle in Lopac. The high barrier, tough, refillable bottle was later vindicated with the issue, in 1984, of an FDA regulation allowing the use of the Cycle-Safe bottle for soft drinks, as long as the container met the FDA specifications, which are only achievable using Monsanto's proprietary technology. In Monsanto's Bloomfield, Connecticut Technical Centre, they installed a commercial scale bottle line for production of 64 oz bottles at 125 bottles per minute, with an annual capacity of 50 M containers. The line consisted of an SBC100 stretch blow machine, associated injection moulding, an E-beam facility, material handling and utilities. Monsanto identified refillable soft drinks as a market opportunity for Cycle-Safe bottles because no other plastic material had the toughness and heat resistance to withstand hot caustic washing.

Monsanto offered licences for Cycle-Safe technology for refillable bottles, and they were investigating the potential for application to beer bottles. Monsanto held a number of patents for Cycle-Safe technology. The key patent [4] describes packaging materials formed of copolymers of acrylonitrile with a content in the range 55–90% by weight and a high resistance to oxygen and water vapour, providing an improved food packaging material. Other patents describe the use of hydrogen cyanide (HCN) scavengers, such as formaldehyde, which scavenge extractable HCN within the melted copolymer. Also described is the use of low-dosage E-beam radiation to remove residual nitrile monomer from preforms prior to stretch blow moulding, to ensure containers meet the requirements of the FDA and to prevent distortion of the taste of bottle contents, making these containers suitable for packaging foods. Lopac was a polymer which was well-suited for packaging foods and beverages. It had a lower density than both PVC and PET, ideal for lightweight bottles. It could undergo biaxial orientation, like PVC and PET. It also had high clarity, comparable with PET and better than PVC. Additional advantages were its higher temperature resistance and higher gas barrier properties compared to both PVC and PET.

At about the same time, another acrylonitrile copolymer was being developed for food packaging in the US. Barex, a copolymer of rubber modified acrylonitrile and methyl acrylate, was developed by SOHIO (Standard Oil Company – Ohio). Rubber modification is used to enhance its impact properties, otherwise the properties are similar to AN/S copolymers, with excellent clarity, chemical resistance, high stiffness and excellent gas barrier. Barex 210 (AN/MA), a copolymer of acrylonitrile and methyl acrylate (75/25), is easily processable on conventional blow moulding machines used to mould rigid PVC. Biaxial orientation was also possible and, in 1979, 23 and 33 cl bottles were produced on a Sidel machine (Solvay/Sidel BAP) in Barex and Solvay's nitrile copolymer Soltan, for pasteurised 'Fischer' beer in France. Barex was approved for foods by the FDA but not for direct contact with beverages, and was not suitable for hot-fill applications.

There follows a consideration of the alternative polymers, PVC and acrylonitrile copolymers with PET, and the key issues for packaging. Both PVC and AN copolymers have a problem with residual monomers: vinyl chloride and acrylonitrile have been shown to have potential harmful effects and must be reduced in the polymer to very low migration levels from the package to the product. In general, residual monomer problems with PET are small and more controllable. The gas barrier properties of AN copolymers are much better than PVC and PET (see Table 1.5).

AN copolymers are not as tough as PVC and PET. Orientation enhances the physical and barrier properties and gives sufficient toughness to be acceptable for most packaging applications. All three polymers have relatively low resistance to heat and none can be considered heat resistant. AN copolymers and PVC can be processed by injection, extrusion and extrusion blow moulding. PET is

Table 1.5 Barrier properties of competitive polymers

	Oxygen	CO_2	Water
Lopac AN/S	1.0	3.0	5.0
Barex AN/MA	1.1	4.0	6.0
PET	5.0	18	2.5
PVC	10	30	2.5

Oxygen permeability: $cm^3 \cdot mil/100\,in^2 \cdot day \cdot atm.$ $20°C$.
Carbon dioxide permeability: $cm^3 \cdot mil/100\,in^2 \cdot day \cdot atm.$ $20°C$.
Moisture permeability: $g \cdot mil/100\,in^2 \cdot day.$ $38°C$.
Abbreviations: PET, polyethylene terephthalate; PVC, polyvinyl-chloride; AN, acrylonitrile; S, styrene; MA, methylacrylate.

better suited to injection; extrusion blow moulding is only possible with some difficulty, using very high molecular weight grades. All can be stretch blown using biaxial orientation via a preform. PET is very moisture sensitive during processing and will rapidly degrade if the polymer is not correctly dried. PVC and AN copolymers are less moisture sensitive but are more sensitive to heat degradation during processing. Visually, PET is transparent and has good colour and gloss, more so than PVC. AN/S copolymer has good transparency, is hard and, of the alternative polymers, has the visual and tactile perception most like glass. AN/MA copolymers include a rubber component, which decreases the clarity, colour and gloss but improves impact performance. PET has a water-white clarity; PVC and both AN copolymers have a yellow tint.

PET made the grade, especially for beverages; PVC was acceptable for a time but has now been replaced by PET. AN copolymers did not recover from the early monomer residual issues, and Lopac has disappeared, while Barex is still looking for markets.

1.3 Technology

From the early days of the development of PET bottles, two distinct technologies developed: single stage, where the preform is injection and blow moulded in one machine; and two stage, where the preform having been injection moulded is cooled down and taken to a blowing unit to be reheated and blown into the final container. In single stage, there are usually the same number of injection moulds as blow moulds; in two stage, there is no connection between injecting and blowing and so the number of cavities can be optimal. Later chapters cover these technologies in more detail.

An alternative technology existed called 'integrated two stage', which moulded the preforms with the optimum number of cavities but then carried them, still hot, to an optimum number of blow moulds all within the same machine.

In two-stage technology, names such as Krupp Corpoplast, Sidel and Cincinatti Milacron were famous. Through company mergers, these names are gradually disappearing. In single-stage technology, Nissei and Aoki have dominated. In the integrated field, Van Doorn originated but Sipa have now taken over.

1.3.1   Single-stage ISBM

While PET bottle development was proceeding in the US, Mr Katashi Aoki, Chairman of Nissei Plastic Industrial, a large manufacturer of injection moulding machines in Japan, was leading a project to develop a machine to make biaxially oriented PP (polypropylene) containers. He recognised that the prototype machine could be used to produce the new PET bottles and, in December 1975, the One-Stage ASB-150 injection stretch blow moulding machine for making the new biaxially oriented PET bottles was unveiled.

All one-stage injection stretch blow moulding machines derived from this original 'Aoki Stretch Blow' design are referred to as classic one-stage machines, as the concept has long since been extended into other PET developments (see Chapter 7). The classic one-stage machine design is extremely versatile in that the same basic machine can be used to make a wide variety of bottles and jars in all shapes and sizes. The ASB-650, which was the standard single-stage machine in the early years, had eight cavities for 1.5 litre bottles.

1.3.2   Two-stage ISBM

In the early developments, preforms were made by continuously extruding a PET tube. To make these preforms, Krupp Corpoplast developed a preform manufacturing machine (M120) that took a continuously extruded PET tube, heated and closed one end, and then heated the other and formed a thread finish by blow moulding. This process had a faster output rate, at 12,000 preforms per hour, than the early injection moulding routes of 8 and 16 cavity moulds. Being extruded, the preforms could be multilayered with barrier materials. The M120 equipment was installed by such companies as PLM in Sweden, Owens Illinois in the US and Toyo Seikan in Japan. In later years, the M120 moved from PLM to Akzo as this preform technology specialised in multilayer. The system was overtaken by injection moulded preforms as the cavitation increased to 32 and beyond. The quality of the injection moulded (IM) neck, adding for example vent slots, made the IM finish preferable. Moreover, IM technology is available from more than one company, giving customers greater technical and commercial choice.

Krupp Corpoplast also developed the B40, a 'two-stage' technology machine, with six blow moulds operating at around 4,000 bottles per hour. Subsequent mould and cooling development increased the output to 6,000 bottles per hour.

The B40 became the workhorse for the early two-stage bottle producers (see Chapter 8).

1.3.3 Integrated two-stage ISBM

In the United States, companies such as Van Doorn developed an integrated two-stage approach. Here the preforms were made by more conventional injection moulding routes (with the number of cavities optimised to match the required output) and then, while still hot, were carried to a separate blowing machine with the optimised number of blow moulds to suit the required output. This was the first 'integrated' approach to PET bottle making. Equipment developers such as ASB Nissei in Japan took the 'single-stage' approach. Here the equipment had the same number of injection cavities as blowing moulds. This was a more compact approach and proved ideal for smaller batch output (*ca* 6 million bottles per year), with excellent glossy surfaces.

1.3.4 Heat setting

The first mention of heat setting appears in the patents of Wyeth and Roseveare of DuPont [1] and Siggel *et al.* of Glanzstoff [5]. The Wyeth patents describe the fundamental aspects of PET stretch blow moulding and methods and apparatus for producing biaxially oriented PET bottles, with general ranges of preferred process conditions and bottle properties.

None of the claims in these patents refer to heat setting, although there is a section which discusses the advantages of its use for bottles intended for hot beverages and when subjected to high temperatures and pressures in a pasteurisation process. Heat treatment can be carried out at 140–220°C for a short time, sufficient to increase the material crystallinity of the finished bottle to at least 30% and up to 50%. Good results were obtained when the heat setting was carried out for 0.1–600 s. It may not be coincidental that the values quoted by Wyeth are almost identical to those given by Siggel *et al.* Siggel's patent pre-dates Wyeth, but refers specifically to vacuum deep drawing of PET sheet, with additional heat treatment of the shaped sheet to increase the degree of crystallinity. The sheet initially has 5–25% crystallinity and is shaped while at a temperature of 85–200°C. Further heat treatment of the sheet, for a period of 30–600 s while the mould surface is at 140–220°C, is designed to increase crystallinity above 25%.

In the 1970s, a patent of key significance to heat setting technology was issued by AS Haustrup Plastic [6]. It describes how a blown PET bottle is transferred to a heated mould and then reblown, while the temperature of the heated mould is maintained above 140°C. Stretch blow moulding can take place into a heated mould, or the blown bottle can be reblown into a heated mould at a later date, on a separate machine. Collins at ICI [7] describes stretch blow moulding PET into a heated mould maintained at 130–220°C, cooling the container while maintaining pressure in the mould, and then removing the container. Sabatier of Carnaud Total Interplastic [8] describes a process for reducing residual stress in PET bottles by filling with hot liquid under pressure, or by increasing the temperature of the walls of the mould to 180–200°C. Many patents, assigned to Owens, Illinois [9], PLM [10], Monsanto [11], Yoshino [12] and Rhone Poulenc [13], were issued in the 1970s. These patents extend the scope of heat setting of PET. To summarise [14], the patents describe in broad terms heat setting of bottles using heated mould systems. Heat setting can be extended to include the bottle sidewalls, neck and base. Mould temperatures of 70–250°C are quoted, with setting times of up to 8 min. A differentially heated mould can be used to heat specific sections of the bottle and neck, and oil or water can supply the mould. The bottle can be blown with high-temperature compressed air and stretched with an internally heated rod. Subsequently, the bottle can be cooled inside or out of the mould and can be held under pressure while cooling. Heat setting can increase the crystallinity of the bottle wall to 25–60%, and the bottle can be designed to resist deformation when hot filled and cooled.

1.4 Packaging

1.4.1 Bottles in the early years

The total estimated PET bottle grade capacity in Europe in 1979, when PET bottles were first manufactured in the UK, was about 25 K tonnes. The total installed bottle production capacity in the UK by the end of 1979 was 150 M bottles in 1.5 and 2 litre sizes, and the total PET resin demand was about 7 K tonnes. There were six companies manufacturing bottles, three using single-stage Nissei ASB 650 machines, two using Husky Injection/Krupp Corpoplast two-stage machines, and Carnaud GPG using equipment that they had developed themselves. PET bottle grades in Europe were available in a range of viscosities (IV 0.75–0.98) and were based on germanium or antimony catalyst. The high IV grade produced by Akzo was used by PLM for their Strongpac PET bottle, which was converted from extruded tube into preforms and then stretch blown into bottles.

As the technology of making plastic bottles for pressurised, carbonated products was in its infancy, bottle shapes had to rely on known pressure vessel profiles, so the early bottles had hemispherical shoulders and bases. In order to allow the bottles to stand up, a base cup had to be attached. These were generally made in black, high-density polyethylene for durability when chilled. Moreover, black could contain a certain amount of recycled material to reduce costs (Figure 1.4). Attempts were made to have the base cups clipped onto the bottles to ease removal. However, they soon came off as the bottles expanded and so cups had to be glued on. These were not efficient to apply and during

Figure 1.4a PET beer bottles with HDPE base cup.

Figure 1.4b PET soft drinks bottles with HDPE base cup.

very hot periods of storage or distribution the hot melt adhesive would soften and the bottles then took on a lean like the tower of Pisa!!

Some attempts were also made to have thermoformed base cups. These had the advantage of being nestable and made in PET, which meant they could be welded to the bottles. Being tapered, the base standing diameter was reduced and the sharp corner at the base made the bottles more difficult to pass along conveyors and dead plates. They also had a tendency to be brittle if a filled bottle was dropped, particularly if chilled in a fridge.

In the 1980s, developments for one-piece bottles entered the marketplace. As interest gained ground for in-plant operations, having to purchase and glue base cups onto bottles just added to the complexity of carrying out this work in-line. A one-piece bottle was ready for filling post blow moulding without secondary operations. The product developed became known as the 'petaloid' base, as it looked petal-like and had five 'feet'. A number of companies developed alternatives to avoid patents and early problems of bases cracking. Today, one-piece, footed base bottles are commonplace.

In the early 1980s, Metal Box aimed to develop a single serve (250 ml) PET bottle to service an energy drink market, where the PET bottle could increase the market beyond that achieved by glass. In order to achieve the greatest possible shelf-life, the body had to have maximum stretch all over (including the base). The neck was designed to be plastic closure only and lightweight. To enable the bottle to stand up, the base was blown to a hemisphere, giving maximum stretch, and then inverted. To prevent the inverted base from reverting when under carbonation pressure, a special PET ring was spin-welded onto the base rim. This bottle was known as the welded ring inverted base (WRIB) bottle. To provide additional shelf-life, the bottle was coated with PVDC. The shelf-life was spectacular for a 250 ml bottle in the 1980s. Even today, its performance would exceed most 250 ml PET bottles. However, it failed due to the overall cost and problems with the security of the base rim weld over time as the PET crystallised.

1.4.2 PET cans

1.4.2.1 STEPCAN

Metal Box developed STEPCAN in the 1980s at its R&D Centre, Wantage, UK. STEP is an acronym for the stretch tube extrusion process, developed to convert extruded PET tube into can bodies via a novel stretching process. Extruded PET tube was produced by Akzo Chemicals, and later also by Metal Box, using high molecular weight PET (IV ~ 1.0).

The STEP process involves reheating amorphous PET tube, and stretch blow moulding and biaxial orientation to produce a 2 m long blown tube. This tube is then constrained, heat set at a high temperature to produce a highly crystalline, thermally stable, biaxially oriented blown tube, which converts into 18 standard size (83×97 mm) can bodies. The process maximises the mechanical and barrier properties of PET and, by varying the hoop-to-axial orientation balance, can eliminate any tendency to split or delaminate in subsequent operations. Each of the processes was protected by a patent application in a number of countries.

The two key process stages are as follows. First, simultaneous biaxial orientation gives 20% strain-induced crystallisation, natural strain hardening and very uniform material distribution, both around and along the tube. There is high mechanical strength and small crystallite size, which ensures high clarity and gloss. The second stage involves controlled heat-setting of this blown tube, giving fast, high crystallisation (40%) and stabilisation of network chains. High clarity and high gloss are maintained and gas barrier properties are enhanced.

The blown stabilised tube can then be cut into can bodies, which can be flanged at both ends. This allows the can body to be subsequently double seamed, with standard metal can ends for high integrity, suitable for packaging process-able foods, pasteurisable beverages and hot-fill products without distortion.

It is worth considering the background to the development of STEP. PET was developed in the 1970s to produce a high molecular weight polymer suitable for conversion into oriented bottles by stretch blow moulding. The injection stretch blow moulding process was developed to first produce a preform, which is then biaxially stretched into a bottle. The bottle has an injection moulded neck, which is not heated or stretched and remains amorphous. The bottle may also have sections which do not have uniform thickness, do not have uniform material distribution, do not have uniform gas barrier performance, do not have uniform mechanical properties. In the 1970s, PET bottles had a hemispherical base, which helped to improve their orientation and properties. PET bottles are satisfactory for the use for which they are designed, especially for packaging carbonated beverages, but not for processed foods and other products that require high temperature treatment.

Developments at Metal Box have also involved investigation of free blowing of PET tube, initially without a mould, to determine the natural stretch ratios in the hoop and axial directions at different stretching temperatures and blow pressures. This development provided the basis for designing the tube needed to be converted into the 2 m length blown tubes. Further developments involved free blowing with additional rod stretching to control bubble propagation during the biaxial process [15].

Thermal crystallisation of amorphous PET is an alternative process to heat-set, strain-induced, crystallised PET material for generating heat stable contain-ers. However, this process results in loss of clarity and poor impact performance. It is commercially used for manufacture of ovenable trays and for crystallising necks on some hot-fill bottles.

Further thermal crystallisation of already crystallised, strain-induced oriented PET, achieved in the novel heat setting process of STEP, retains the high clarity, improves the gas barrier and mechanical performance, and maintains impact resistance. STEPCAN could be processed successfully in commercial filling processes of high quality fruit and acidified vegetables at pasteurisation temperatures of 93–104°C. Double seam integrity was confirmed by a range of physical and biological leakage tests (biotesting) on unprocessed and processed filled cans. The ultimate test was the successful seaming of STEPCAN under the commercial conditions achieved in various countries in the world. The next stage was to extend STEPCAN for low acid foods that required sterilisation temperatures of 121°C, to achieve a specification for a can that could be sterilised in steam without overpressure, to allow processing on standard canners' equip-ment. STEPCAN was introduced commercially in the UK by Marks & Spencer, with a range of high quality fruit packs (Figure 1.5). The STEPCAN also had

Figure 1.5 STEPCAN processable PET food can.

the new Carnaud (before the merger with Metal Box) easy-open, full aperture metal end (EOLE).

STEPCAN was a unique packaging concept, offering the market a glass-clear, seamless PET can body alternative to a metal can, suitable for conventional food processing systems. The limitations in the can production output and the requirements for a higher barrier for packaging foods prevented commercial success. At the time of termination, development of a multilayer can and a continuous drawn tube route were progressing. The production facility at Perrywood, UK, and development at the R&D centre in Wantage, UK, were closed in 1990 [16].

1.4.2.2 Arnican
While STEPCAN was being developed by Metal Box, Akzo Plastics was developing a similar PET can, called Arnican. This can was manufactured from a biaxially drawn PET tube. Akzo produced the extruded PET tube and blew the can bodies. Akzo had Thomassen & Drijver Verblifa as development partners. Can bodies were double seamed with metal ends, and limited filling trials were carried out. Akzo were also developing a multilayer version to improve the barrier properties of the can. Because Akzo were also close partners with Metal Box, supplying them with extruded tube for the STEPCAN development, they did not progress any further. This, as much as any other reason, prevented further commercial development of Arnican.

1.4.2.3 Thermoformed can
In 1979, Plastona in the UK made a 33 cl PET can, called Plastocan, on its own thermoforming lines, and these cans were available for commercial use in the UK in 1980. At about this time, the global market for 22–33 cl metal cans was increasing, and attention turned to the possibility of making a PET can in

these sizes for beverages. At the same time, Thermoforming S.p.a in Italy was developing a thermoformed PET can, called Thermocan. They were carrying out R&D on the whole process, involving the production of PET sheet, modifications and improvements of thermoforming equipment, and optimisation of PET properties for can manufacture. These involved biaxial orientation in the sidewalls of the can to obtain greater strength and improved barrier properties, also can design, sheet treatment to optimise thickness distribution, and techniques to control flange thickness for seaming metal ends on to the can. The possibility of improving the gas barrier properties of cans was investigated, using co-extruded multilayer sheet, with PVDC as the gas barrier layer, PE as the internal layer and PET as the external layer. The advantages stated for Thermocan included: cost reduction compared to aluminium cans, small scale production capability, local manufacture, lightness, easy handling, high mechanical performance, suitability for packing CSDs, and a shelf-life of at least 6 months. Patents were filed, which extended to the US and other countries. The thermoformed can could not hold high-carbonated pressures, as it deformed and distorted, but it was satisfactory for still and some low-carbonated products [17].

1.4.2.4   Petainer

In 1983, PLM AB in Sweden and Metal Box in the UK formed a joint venture company, Petainer SA. The ambition of this venture was to further develop the Petainer technology, which PLM invented and patented. At this time, a machinery agreement was signed with Krupp Corpoplast, which included them in a three-way technology development of a new packaging system. Coca Cola® joined this development, having initially tested the first prototype cans made by Petainer. To further develop Petainer into the US, a US partner was sought, to contribute experience with market know how. Sewell Plastics joined forces with PLM and Metal Box, and Petainer Development Company was formed in 1984. The purpose of this company was to take the process from laboratory to full-scale production. In 1985, a site for the first plant was selected in Atlanta, US and equipment was in place later that year.

The scope of the venture was to completely develop the package, and would require a major investment of resources and technology. In order to be commercially viable, it was necessary to put together a total system, including production, handling, filling, closing, packing, transportation, distribution, vending, collecting and recycling.

The background to the development of Petainer starts with the PLM joint venture with Akzo Chemicals, a major producer of PET polymers. They developed a tube technology for PET bottles, together with Krupp Corpoplast, and manufactured the first preform machine. This machine could take an open-ended, extruded, amorphous PET tube and convert it into preforms for a 1.5 litre bottle. The PLM company, Strongpac, used this machine to produce bottles from extruded tube, which had improved mechanical performance and was lightweight. This work provided the background to a new technology, the Petainer lightweight neck technology, which allows the neck finish to be oriented, and also reduces material and increases thermal stability in the neck and shoulder of the bottle.

PLM developed a novel stretching and heat stabilising technology for PET, to provide an alternative pack for beverages and foods. The technology consists of three key process stages, which utilise either extruded tube, injection-moulded or thermoformed preforms, and involves:

- preform stretching to maximise uniaxial orientation and crystallinity
- heat shrinking the stretched preform to provide heat stability
- forming containers

Each of the processes was protected by a patent application in a number of countries. The process stages involve the injection moulding of preforms in amorphous PET. The preform is then heated to about the Tg (75°C), is wall-ironed by uniaxial orientation, and a dome base is formed. The stretched preform is trimmed to provide the correct length and is then flanged and reformed into a can body. The open top can body is now trimmed to a precise length in preparation to neck-in and flange to the correct diameter to take a seamed metal end (209 diameter). The finished can is a 12 oz oriented PET necked-in beverage can with a draw wall ironed (DWI) base.

The first commercial scale production lines using modules produced by Krupp Corpoplast were installed in the Petainer plant in Atlanta, US, and the first Petainer cans were produced in 1986. Direct comparisons with an aluminium beverage can in terms of cost, barrier and recyclability were always going to be a major challenge, which eventually prevented commercial success. The equipment was subsequently sold to Yamoro Glass in Japan [18].

1.4.3 Refillable bottles

In a number of European (and some South American) countries, the major beverage container market was for returnable/refillable glass bottles. The only way for PET to gain a foothold was to also have a returnable/refillable bottle. Therefore, a 1.5 litre, 109 g bottle was developed in the 1980s to be heat stable enough to withstand the hot wash and caustic solutions used to ensure the bottles were cleaned. As PET could absorb certain chemicals and odours, special equipment had to be developed to 'smell' inside the bottles. If the 'sniffer' could not detect the smell of the original product, the bottle was rejected. Also, as the bottles would have a tendency to experience continued minor shrinkage at each washing, care had to be taken to record on each bottle the number of trips it had made. Given the continued growth of one-trip PET bottles, some markets have now found it better, and cheaper, to have returnable, one-trip PET bottles, with the deposit built into the purchase price of the filled bottle. As the returnable/refillable bottle market is static or even declining, Chapter 10, in this volume, 'Hot-fill, heat-set, pasteurization and retort technologies', focuses on the higher performance containers and new markets.

1.4.4 Preform design, neck design – BPF, vent slots

The neck finishes in the bottles test marketed in Europe in the mid 1970s were based on the finish from the 1 litre returnable glass bottle as used by Coca Cola®, who led the PET bottle evaluation in Europe at that time.

This was a 28 mm finish that could accept a metal roll-on pilfer-proof (ROPP) closure. (It should be noted here that ROPP is an industry-recognised definition of this type of closure; no packaging should be deemed to be 'pilfer-proof' but should be tamper evident, and the packaging should carry information for the consumer to be aware of how to recognise signs of tamper evidence.)

As the test market moved into country and area launch, the neck finish became established. As, indeed, did the 1.5 litre PET bottle having the same diameter and height as the 1 litre glass returnable bottle (because it was easier to test market a new bottle which fitted the same filling line change parts and distribution system). These dimensions exist to the present day.

As the UK came into the PET bottle era in 1979 (Able Developments, part of Wells Soft Drinks, were one of the first and went in-plant with Krupp Corpoplast B40 two-stage machines), more attention was given to the neck finish. The 28 mm ROPP finish was essentially the MCA 2 neck finish and PET bottles had to accept the metal ROPP closure. Problems arose as the PET finish was so accurately made (certainly more concentric and tighter toleranced than the glass equivalent) that the closures could be removed from the bottle faster than the gas inside the bottle (containing carbonated soft drinks) could be released to atmospheric pressure. On certain rare occasions, this caused the cap to 'missile' off the bottle, which could cause injuries; such an event was referred to by the wonderful term 'tail end blow off' or TEBO. Another factor was that the rolled-on thread finish on the cap could vary, subject to the condition of the capping chuck. If the threads were not rolled into the neck finish sufficiently, then the cap could also disengage from the bottle, which was referred to as 'premature release' and could cause injuries (several court cases occurred in the UK).

In the late 1970s and early 1980s, the UK bottle and closure industry put a great deal of effort into resolving these issues. The British Plastics Federation (BPF) and the Metal Closures Association (MCA) worked together to overcome these problems. Companies like Metal Box (now Crown Cork & Seal), who had the resources of PET bottle production, both metal and plastic closure manufacture, and capping machine supply and service as well as extensive R&D capabilities, were well qualified to participate on both the BPF and MCA bodies as well as to test and evaluate the resultant proposals. Before this time, the assessment of TEBO was rather subjective; as a result, a number of devices were constructed to ensure that consistent and repeatable testing could take place, and industrial standards were set. The outcome was the BPF C neck finish, which has remained in production on preforms and bottles for many years and is only now being superseded by the PCO (plastic closures only) neck finishes. PCO neck finishes are lighter weight finishes made possible as the industry has largely moved away from metal closures on carbonated beverage bottles. The BPF C finish superseded the A and B variants. It had vent slots that were more like channels instead of the early 'vee' slots, and had a special thread profile more suited to the manufacture of the neck 'splits' needed to mould the finish. By spark eroding the start of the thread, it was possible to increase the effective thread length and make the cap engage with the neck for more turns. The new profile of vent slots increased the rate of gas pressure release and, linked to the increased turns of thread, reduced the incidence of TEBO. Incidences of 'premature release' were reduced by better understanding of the maintenance of well-defined thread definition on RO (roll on) and ROPP caps. The launch of plastic prethreaded caps in the early 1980s also meant that consistent thread engagement and closure performance was assured.

1.4.5   Beer spheres

In the late 1970s, Johnson Enterprises of Rockford, Illinois, US developed a large PET container made in the shape of a sphere for the take-home beer market, giving rise to the registered name of 'beer sphere' (Figure 1.6). This was a 20 litre container weighing 280 g. Made on a Cincinnatti Milacron RHB IX, the preform was heated by radio frequency because of its very thick sidewalls (7 mm). This container was used by a number of regional breweries, who still

Figure 1.6 Beer sphere PET beer container.

had access to the old cask-filling machinery. By 1982, Metal Box Company took a licence for European production and distribution into Europe. In response to the UK demand, they developed a 30 litre container, in conjunction with Johnson Enterprises, utilising the Metal Box PET knowledge and the 280 g preform. This larger container was used by several UK breweries for draft beer in small UK outlets and for export to the Far East, Australia and the US. Returnable, refillable kegs are expensive for such long-haul business. Some customers developed the market opportunity into draft wine and cider. Developments in the dispense equipment added to the market opportunity for this novel container.

Johnson Enterprises continue to produce the beer sphere and in the UK the licence has been taken up by Green Sphere. Several of the larger US breweries are producing 20 litre beer spheres. To the best of the authors' knowledge the 30 litre beer sphere remains the world's largest PET container for carbonated beverages.

References

1. Bakker, M. (1994) *Bottlemaking Technology and Market News*, 3, 1994.
2. UK Patents 1341845-8. Wyeth and Roseveare, DuPont.
3. Folland, R. *The OXBAR Super Barrier System: a Total Oxygen Barrier System for PET Packaging.* CMB Technology, Europak, 1989.
4. US Patent 3451538. The Monsanto patent for acrylonite copolymers: packaging films with improved properties. Expired in 1986.
5. US Patent 3496143. Siggel *et al.*, Glanzstoff.
6. UK Patent 1523146. AS Haustrup Plastic.
7. UK Patent 1474044. Collins, ICI.
8. UK Patent 1530305. Sabatier, Carnaud Total Interplastic.
9. US Patents 4154920, 4233022 and 4439394. Owens, Illinois.
10. US Patents 4264558 and 4387815. Jacobsen, PLM.
11. US Patent 4318882. Agrawal, Monsanto.
12. US Patents 4164298 and 4375442 and Japan Patents 54066967, 54068384/5 and 56080428/9. Yoshino.
13. EP Patents F2389478 and BE866692. Bonnebat, Rhone Poulenc.
14. Brooks, D.W. *Heat Setting PET Bottles Patents Review*. Metal Box internal technical record, 1985.
15. UK Patent GB 2,145,027. Tubular Articles of Biaxially Oriented Polymers. D.W. Brooks, Metal Box.
16. Dick, D.A. *STEPCAN: the choice is clear for a long-life, processed food PET container*. CMB Technology, Europak, 1989.
17. Bocchi, L. *Experience and developments in thermoformed PET containers for carbonated beverages*. Thermoforming S.p.a. Fourth International Conference on Bi-axial Oriented Bottles and Containers and other Engineering and Plastic Resins. Dusseldorf, 1983.
18. Emilson, L. (PLM) and Folland, R. (Metal Box). *Petainer PET Can Technology*. BevPack, 1986.

2 Commercial considerations

Geoff A. Giles and Gordon J. Bockner

2.1 Introduction

There is little question that the growth of PET packaging has been the big story in packaging in the 1990s. It now appears that this story will continue to dominate our attention well into the future. PET packaging has demonstrated an unusually strong connection with consumers. For whatever reasons—portability, lightweight, convenience, safety—PET packaging is helping brand owners to sell product.

However, because PET packaging is used in a wide range of market categories, it is important at the outset of this section to make a basic distinction between these markets. Commercial considerations will also be viewed from several angles: the converter (making the container), the packer/filler and the brand owner (who may not actually handle packing and filling). This gives rise to a complex matrix of considerations when assessing the benefits of PET packaging.

The world of PET packaging can be divided into two parts. The larger part—represented, for example, by nearly 25 billion containers in the US—is the huge but maturing and increasingly commodity-driven world of carbonated soft drinks (CSDs), where merchant suppliers are (to be kind) profit-challenged. The soft drink giants keep a tight rein on pricing. Coca-Cola® self manufactures nearly 90% of its containers through its bottler cooperatives in the US. Having grown significantly over the last 20 years, the future CSD market will be driven by more customised packs which will add consumer value.

The second part, the so-called 'custom' segment of PET container consumption, represented mainly by beverage products, water, juices, teas and, to a lesser extent, a host of other food products, is often lumped together under the 'new age' label. It is younger and smaller but faster growing. The overall custom market is projected to grow from approximately 11 billion containers in 1998 to approaching 20 billion by the early 2000s. A fairly large portion of this market, especially that represented by water, which converted to PET many years ago, also provides smaller returns for container manufacturers. This will drive the 'custom' segment to seek new opportunities beyond beverages and water.

In recent years, it has been emphasised that some demanding food and beverage products cannot be packed in PET until improved performance containers are available. Furthermore, the commercialisation of such containers is dependent upon the development of cost-effective, performance-enhancing, container technologies to manufacture them. Whilst we appear to be waiting for these technologies, we read in the trade press about new food and beverage products moving into PET. This is a result of shelf-life criteria being changed to accommodate PET and of efforts to optimise the performance characteristics of 'standard' PET packaging. These are often food or beverage products that are switching from other existing packaging materials. This is not to say that these alternative materials will be totally replaced by PET. Some products will remain in glass, cans or other plastic containers for some time to come. In most instances, however, the growth of PET will come at the expense of these other packaging materials. In some cases, these will be other polymers, such as polyvinyl chloride (PVC), which are still used to some extent in Europe for water and edible oil, or multilayer, olefin-based plastics. However, the material that will be impacted most by the ongoing development of PET packaging will be glass, used in its traditional markets of food and beverages.

There are many considerations for a packer to make when selecting a particular package for their product. Three primary considerations are:

- functionality or package performance
- cost
- market preference

There are also a series of secondary considerations affecting decisions on package selection. Some of these considerations could include:

- availability of the package
- distribution requirements
- environmental or recycling issues
- traditional packaging formats already used

When viewed from the perspective of the package supplier, packer/filler or brand owner, the factors to be considered when determining package functionality or performance can be either 'internal' or 'external'. Internal factors are those within the control of the particular business making the considerations (i.e. converter, packer/filler or brand owner), including material and process and product developments, utilisation of equipment innovations, and economies of scale. In other words, factors which the particular businesses themselves can sponsor or implement over time, given the development and commercialisation of new technologies and products, and availability of capital. The PET packaging industry continues to benefit from a multitude of 'internal' developments. Conversely, the glass industry does not benefit to the same extent as it is a 'mature' technology.

External factors are those that are beyond the direct control of a particular business, although it may be able to influence their impact. Some of these external factors which can, for example, impact the converter, include new food or beverage processing techniques, introduction of a new beverage or food product, modified shelf-life requirements or distribution channels, changing consumer preferences or habits, and environmental issues. Even international competition from drinks cartons is an external factor. Resin price and price instability are also factors that can impact development and conversion to PET.

An external factor to the converter, but internal to the packer/filler, is the acceptance of the PET shelf-life. Packer/fillers who have traditionally accepted and ultimately demanded the total oxygen barrier, infinite shelf-life and product neutrality of glass have gradually considered and implemented alternative processing and marketing methods which have enabled them to convert to PET packaging.

The net influence of these internal and external factors, across the supply chain, has been to create an effective parity and competitive equivalence between glass and PET for an increasing number of end-use markets.

Products to be packed fall into three categories:

- products which are 'safe' for PET
- products which are 'safe' for glass or other traditional packaging
- products which are converting from one to the other

Generally, the 'safe' PET markets are growing in their own right, while the 'safe' glass and traditional markets are declining, and the trend for the 'converting' products appears to be in favour of PET.

There will be 'blips' along the way, for example: raw material price increases based on temporary supply/demand imbalance may delay change and even induce significant structural changes to the industry; short-term equipment or container supply shortages may distort trend lines; introduction of new products will continue to favour glass at first but will ultimately yield to consumer preference.

The critical question does not appear to be whether PET will grow at the expense of traditional packaging but how soon and in which end-use markets.

Container manufacturing technology is not the only driver of new PET product introductions. In fact, four drivers must be considered:

- container technologies
- product processing and filling requirements
- packer/filler shelf-life requirements
- cost/performance characteristics

These drivers are considered in the following sections of this chapter.

2.2 Container technologies

In the last decade we saw the first significant conversions of hard-to-pack foods and beverages switching into PET containers. These included, in particular, oxygen-sensitive products such as tomato ketchup being filled into narrow-necked containers, and the launch of single-serve beer bottles. These conversions were the result of two trends which, it is thought, will govern the growth of standard and customised PET container markets over the next several years.

The first trend is the growing market's readiness to accept plastic containers, reducing or modifying their shelf-life requirements, introducing innovative food processing techniques, and the improving cost/performance characteristics of customised PET containers.

The second trend is the development of the following two distinctive approaches to technological development:

- The aggressive and creative utilisation of 'existing' manufacturing and material technologies with continued R&D initiatives. This has resulted in a commitment among producers and end users of PET containers to optimise the performance of existing PET container manufacturing concepts.
- The continued development of 'innovative' container process and product technologies in order to provide food and beverage packers with custom containers with significantly improved performance characteristics on a competitive basis.

2.3 Product processing and filling requirements

In general, there are three main product groups:

- Carbonated soft drinks (CSDs), which are carbonated to around four volumes of carbonation and supplied originally in 2, 1.5 and 1 litre multi-serve sizes. However, use of the smaller, single-serve, sizes is increasing. In the US, the fastest growing market is for the 20 oz size (approx 63 cl). Internationally, the 50 cl size has been developing a strong market and is currently the smallest suitable size produced from standard PET technology.
- Hot-filled (85°C/185°F), single-serve containers (typically 500 ml), with thermal stability and improved gas barrier, have the fastest growing market for beverage PET containers. In the US, the 16 oz glass juice bottle market has succumbed to PET. Hot-fill ready-to-drink tea is next.
- Finally, a range of food products require hot fill at >90°C/194°F and demand gas barrier properties at least 3–4 times greater than that which is generally characteristic of commercially available, standard oriented PET containers. These include tomato-based products, spaghetti sauces, jams, preserves and pickles. The successful development of these markets will depend on the availability and commercialisation of new filling technologies, closures and improved container barriers over the next few years. Current hot-fill containers are available in the larger multiserve

(family) sizes with better surface-to-volume ratios and are more suited to the cash-and-carry and club warehouse markets. The smaller sized, single-serve containers will require the extra barrier to deal with their higher surface-to-volume ratios.

A fourth product group is developing comprising future food products with process and shelf-life requirements even more rigorous than those noted above. These are the retorted and/or pasteurised products, which are either extremely oxygen sensitive or require the higher barrier to deal with the higher surface-to-volume ratio prevalent in the smaller sized containers. This category includes baby foods, pet foods, meats and beer (which is a growing category for the PET bottle market). It should be noted that work is underway to develop appropriate PET bottles for tunnel pasteurising beer by combining manufacturing technologies for heat stability and barrier. Today, beer is almost always cold filled into PET containers.

2.4   Packer/filler shelf-life requirements

Packer/fillers (and brand owners) are examining their current practices in order to move more products into PET to increase market share. While there may be little flexibility in processing or product formulation for some food and beverage products, it is believed there is significant flexibility regarding shelf-life requirements when converting to PET. A range of factors can impact the shelf-life performance of the food or beverage product, in particular gas permeation through the sidewalls and closure, and UV light penetration through the sidewalls.

Frequently the first introductory packs produced when switching to PET have to match the shelf-life requirements (and process and filling conditions) of the 'old' package (usually glass or metal). This requires a high performance specification from the PET container. Once the launch has proved successful, ongoing development of the food or beverage product being packed, as well as the processing and filling conditions, usually leads to a reduction in the performance required from the pack and the packed costs. This is a commercial consideration worth making when starting development of a new food or beverage product. Priority is often given as to how these developments can be commercialised with minimum additional capital either at the container manufacturer or packer/filler stage. An assessment of the real, commercial, shelf-life requirements might allow some developments to be brought forward and implemented to avoid any re-tooling later on (e.g. lightweighting, or some might say 'right' weighting!). This is also important if capital is required to produce an extended shelf-life package, which is subsequently not required before the capital has been depreciated.

Oxygen scavengers are being developed which will mop up oxygen (contained in the headspace, entrained in the product, or having entered the container by sidewall permeability) to extend the shelf-life. These can overcome the need to develop special high-performance, costly PET containers.

When carbon dioxide loss is an issue, it is important to consider the mechanisms that lead to loss. First of all, a pressurised PET container will relax immediately after filling and capping. Then there is 'sorption', when the CO_2 migrates into the sidewalls of the container before reaching the outside and when steady state permeation occurs. Careful container design and 'right' weighting can help to reduce these losses. The permeation rate can be increased if the container is stored at elevated temperatures. The shelf-life can be halved for every $10°C/50°F$ above $25°C/77°F$. This is an important consideration if carbonated products are to be shipped in regions of continuously elevated temperatures or warehoused in high ambient temperatures.

Another factor to consider is the impact of UV light. As the beer sector for PET bottles grows, it will be mainly in the single-serve market and UV will play a major part on the resultant shelf-life. This will require developments in UV light barriers (in the 350–450 nm wavelength band), therefore we see mainly amber coloured PET bottles for beer. In Japan, however, all coloured PET will be eliminated from 1 April 2002! Thus, we will see developments in the area of increasing UV barrier for clear PET containers in the coming years.

2.5 Cost and performance considerations

2.5.1 Material pricing

For the packer/filler or brand owner switching to PET, a key concern is the potential for container price fluctuation due to variable resin prices. When resin availability decreases prices are raised; generally, this is a cyclic phenomenon. When new resin capacity comes on stream the additional capacity generally weakens or, at best, holds prices. As the volume of the PET container market grows, there comes a point when container demand matches resin capacity. Beyond this point the resin prices tighten until new capacity is brought on stream and the whole cycle repeats itself.

Being a polyester, there are valuable end-use markets for PET resin other than just bottle grade, namely, fibre grade. After a poor cotton crop in a major producing country, there is increased demand for polyester fibre grade resin, thereby squeezing out some capacity from the bottle grade resin, and so increasing pressure on bottle grade prices. It is this swing between different PET resin markets which can have a greater effect on bottle grade price stability than pure oil pricing, which impacts all of the resin industry. This is a point worth bearing in mind when considering long-term price forecasting.

2.5.2   In-house manufacture of PET containers for packer/fillers

A further consideration for a packer/filler, from a cost and performance perspective is whether to go in-house with container production.

The economic entry barriers for the manufacture of PET containers are much lower than for glass containers, particularly for the reheat, two-stage, stretch blow moulding systems, thereby making self manufacture of PET containers a reasonable possibility. This fact is unlikely to be a significant driver at the time of converting from, say, glass to PET. However, it will be an opportunity to be evaluated once conversion has taken place and the volume of bottles being used is sufficient to justify the investment in equipment, and will typically be carried out as part of a cost-reduction initiative. Direct container cost comparisons almost always favour glass, but an expanded systems cost analysis, including distribution and handling costs (inbound and outbound), will reduce or eliminate this differential.

When analysing the cost benefits for making a change in container materials (e.g. when switching from glass to PET) and outsource container supply to in-house supply, a number of cost factors, in addition to base container cost, must be included. These are called system costs and include such as factors as:

- inbound container freight
- warehousing costs of the finished product
- outbound freight
- elimination of glass breakages

Some of these factors lend themselves to fairly straightforward quantitative analysis (e.g. outbound freight), while others are very difficult to quantify and might be deemed 'soft' benefits, such as ease of handling lightweight, unbreakable containers, and reduction in noise when handling on a packing/filling line.

2.5.2.1   Inbound container freight
Inbound container freight of (outsourced) empty containers is unlikely to play a significant part, as the weight consideration will probably not reduce the vehicle movements. However, this factor will become significant when considering in-house container production, where the incoming freight would then be carrying preforms instead of full-sized containers.

2.5.2.2   Finished product warehousing costs
This is an area where it is difficult to be quantitative. The top load resistance for PET containers filled with a still product is lower than that for glass and can lead to higher warehousing costs, depending on how high the finished product is stacked. Most modern warehouses are equipped with racking; if this is not the case through the complete supply chain, then the system cost will generally favour glass or cans.

However, PET stack strength for cold-filled bottles can be (and is) increased by liquid nitrogen dosing just before capping. This assumes that the bottle design can take pressure without significant distortion.

2.5.2.3 Outbound freight
Outbound freight costs of filled containers are likely to show a significant improvement with use of PET compared to glass. Based on the weight of filled PET containers, which are less 'bulky' than glass, it is possible to show that the total vehicle movements can be 25–30% less than the equivalent glass containers, producing a significant cost reduction. That is, trucks carrying filled PET containers 'cube-out' before they 'weigh-out'. Reduced weight loads also permit use of trucks with lower cost specification.

2.5.2.4 Breakages in-plant
Breakages in-plant have improved considerably over recent years as manufacturers have developed methods to handle containers more gently. However, past studies have shown that in-plant breakage costs can amount to 1% of filling and packing costs, excluding the cost of the lost product.

2.5.2.5 Breakages in distribution
Breakages in distribution have been shown to be capable of running at 0.5% or higher. This is variable subject to the distribution infrastructure. The more handling and trade channels serviced, the higher the potential breakage levels. Operator and consumer safety may also be a point to be considered.

2.6   Recycling issues

When considering the impact of recycling on the selection of alternative forms of packaging and, in particular, on the specific recyclability of alternative materials, one must consider distinguishing between what can be done, what is in fact being done, and the public perception of what is being done.

Glass, aluminium cans and PET are all recyclable. That is, if collected, sorted, cleaned and appropriately handled, these materials can be reused in a manner that extends their respective life cycles, conserves material and avoids consigning containers to a landfill. However, plastic (and specifically PET) and glass each present distinct issues, particularly when considering 'closed loop' bottle-to-bottle recycling, and each has a different public perception barrier to overcome.

Other, more added value, end-use opportunities need to be developed beyond 'bottle-to-bottle' recycling, as raising the recyclate quality to food grade is costly. Recycling is discussed in more detail in Chapter 11 of this volume.

2.7   End-use market penetration

Precise unit volume numbers for individual categories of beverage and food products are very difficult to obtain. Food and beverage companies guard their own product volumes very closely, while industry trade associations and government agencies, which used to track industry-wide numbers, have increasingly cut back on these activities. While current market data may be difficult to find, the trends in the contest between PET and glass containers are very clear. In recent years, the use of PET CSD bottles has continued to grow strongly, while use of glass has decreased by a corresponding amount. In the US, CSD glass containers have virtually disappeared, except for a few small restaurant sizes.

PET has now established a strong base in a number of non-hot-fill food applications (i.e. edible oil, some ketchups, peanut butter, pourable dressings). Further market penetration has been achieved in hot-fill juice, teas and isotonics, as well as applications beyond food packaging in the personal care, household and do-it-yourself markets. As is always the case with new packaging materials, the obvious and easy applications are achieved first. Then, building on these successes and given a high level of consumer acceptance (which PET has), packaged goods companies look for other areas where the new material will fit and increase their market share. PET has moved into this phase.

The new applications will require higher thermal stability and improved barrier to gas and UV light. The resin, equipment and container manufacturers are all improving their products to satisfy new market expansion opportunities. The hot-fill market for PET containers continues to grow. Multilayer and coated, high-barrier bottles continue to be developed for single-serve size PET bottles requiring improved shelf-life due to the increased surface-to-volume ratio or for packaging more oxygen-sensitive products. The opportunity to manufacture clean bottles, to keep them clean and clean fill soon after bottle manufacture, will lead to new market developments, including cold aseptic filling. This will further increase the switch from glass and metal cans.

The perceived quality of products packed in glass compared to those packed in PET has narrowed, and in some markets PET has overtaken glass. This will further facilitate the conversion to PET and will be worth serious consideration when assessing future packaging needs for packer/fillers and brand owners. PET has become the material of choice for so many markets that it will continue to go from strength to strength for many years to come.

References

The following reports are produced by Business Development Associates, Inc., Bethesda, MA, USA.

1. Technology, Markets and Prospects: PET Packaging in Foods and Beverages in the US, 1994.
2. High Performance Polyester Packaging for Foods and Beverages, 1996.

3. Improved Gas Barrier and Thermal Properties for PET-based Containers: Polyester Packaging, The Critical Path Ahead, May 1997.
4. Custom PET Markets—and the Technologies Which Will Drive Them—PET Strategies, 1998.
5. Custom PET Packaging for Foods and Beverages, 1998.
6. Beer in PET (parts 1–3), 1999–2000.

3 PET materials and applications

Kenneth M. Jones

3.1 Introduction

This chapter describes polyethylene terephthalate materials specifically designed and made for the packaging industry, including their manufacture, properties and performance. The development and properties of copolymers produced to meet specific processing needs and/or end-use requirements is considered, including injection stretch blow moulding, extrusion and forming, film manufacture, and materials with improved gas barrier and environmental stress crack performance. Over the past 25 years, synthetic polymeric materials have played a significant role in enhancing the ability to package materials—especially foodstuffs—in a safe and convenient manner that is appreciated by both the retailer and the consumer. PET polymers have played a major role in this area, with significant developments obtained in high-barrier containers for carbonated soft drinks (CSDs), high strength films, excellent gas barrier packaging for sensitive foodstuffs, and materials of high clarity.

PET-based polymers have shown great versatility in their applications from the time they were first developed in the early 1940s by Whinfield and Dickson [1, 2]. In those early days, the polymer was used specifically in synthetic fibre applications—this is still one of its major uses [3, 4]. These products were based on the uniaxial orientation of the polymer chain. Several polyester polymer variants have now been developed to meet this sophisticated end-use market. During the latter part of the 1950s, both ICI and Dupont led in the development of second-generation products based on the biaxial orientation of extruded amorphous film. These developments produced highly oriented film products, to compete initially against the cellulose-based products in applications such as X-ray films, insulating films, audiotapes and food-packaging films. A highly intricate drawing and heat setting process technology was developed and is used to produce a quality film of high integrity to meet these highly demanding market applications.

Over the past quarter century, several new film products have been developed by various polyester film producers to meet many challenging new markets. In the mid 1970s, Dupont technical group developed and patented technology that led to the production of third generation materials that had a three-dimensional oriented structure. The work, led by Wyeth [5] at the Dupont experimental station, demonstrated that bottles and containers produced by this new technique of injection stretch blow moulding (ISBM) had exceptional strength and excellent gas barrier properties. This development again revolutionised the polyester industry and allowed the polymer to be used as a replacement for glass in bottle manufacture.

In all three developments, the major polymer backbone component is PET, which is the resin originally developed by Whinfield and Dickson. This is a polycondensation polymer based on the reaction of terephthalic acid and monoethylene glycol. The polymerisation reaction is represented in simple form by the following equation:

$$n \, HOOCC_6H_4COOH + (n + 1)HOCH_2CH_2OH$$
$$= HOCH_2CH_2O(OCC_6H_4COOCH_2CH_2O)_nH \qquad (1)$$

The initial polyester produced by Whinfield and Dickson was made using the precursors dimethyl terephthalate and monoethylene glycol, respectively; however, the majority of today's newer polymer production plants use pure grade terephthalic acid (PTA) and monoethylene glycol as the raw materials. To meet more stringent downstream processing requirements and product end-use properties, the majority of today's polyester polymers used in the packaging industry are based on novel copolymers.

3.2   Polymerisation and manufacturing processes

PET is produced by a polycondensation reaction, using one of the following series of initial chemical reactions: the ester interchange of dimethyl tereph-thalate with monoethylene glycol, the direct esterification of terephthalic acid with monoethylene glycol, or the reaction of ethylene oxide with terephthalic acid to form, initially, the monomeric and low molecular weight oligomeric precursors with various monoethylene glycol to acid ratios depending upon the route used. In the former two processes, the oligomeric mixture is polymerised without further purification, while in the latter process the monomeric species, bis hydroxyethylene terephthalate (BHET), is extracted and purified from the resulting reaction products. The need for this additional purification stage has made this process uneconomical and it is not used in practice.

The ester interchange reaction is carried out in the presence of a catalyst, usually a salt of a transition metal, such as manganese, cobalt or zinc, or an alkaline earth metal, such as calcium. The reaction is allowed to proceed until more than 99% of methanol is released, using a high mole ratio (2:1) of monoethylene glycol to ester and a rising temperature profile, starting at a temperature of around 160°C and rising to around 230°C to allow the reaction to proceed to completion. The ester interchange catalyst systems are potential degradants at high temperatures, and phosphorus-based stabilisers, such as phosphoric acid, are added prior to polycondensation. The reaction equation is shown below:

$$2HOCH_2CH_2OH + CH_3OOCC_6H_4COOCH_3$$
$$= HOCH_2CH_2OOCC_6H_4COOCH_2CH_2OH + 2CH_3OH \qquad (2)$$

The direct esterification reaction is usually carried out in the absence of a catalyst, although metal alkoxide compounds of titanium, tin and antimony are known to catalyse the reaction. The solubility of terephthalic acid in monoethylene glycol is very low and the reaction is carried out at temperatures in excess of $240°C$ and usually in the range $260–290°C$, using a mole ratio of monoethylene glycol to acid of around 1.2 to 1 [6]. This gives acceptable reaction rates allowing the process to be completed (degree of esterification of around 93%) within acceptable times:

$$1.2HOCH_2CH_2OH + HOOCC_6H_4COOH$$
$$= HOCH_2CH_2O(OCC_6H_4COOCH_2CH_2O)_n$$
$$OCC_6H_4COOH + 2 H_2O \qquad (3)$$

where, n is the average degree of polymerisation for the oligomeric reaction product and is usually in the range 3–10.

The high acidity of the terephthalic acid catalyses the formation of diethylene glycol and this must be controlled at an acceptable level to give consistent polymer properties. Both alkali and alkaline earth metals as well as quaternary ammonium compounds can be used to bring about a significant reduction in the level of diethylene glycol formed. This can be formulated as follows:

$$HOCH_2CH_2OH + HOCH_2CH_2OH$$
$$= HOCH_2CH_2OCH_2CH_2OH + H_2O \qquad (4)$$

Polycondensation reactions are equilibrium reactions; therefore, to allow the polymerisation reaction to proceed, the level of free monoethylene glycol and water must be reduced quite significantly. The equilibrium constant for this reaction favours the low molecular weight oligomeric species, and thus, in the most conventional manufacturing plants, the reaction takes place in the melt phase at relatively high temperatures under vacuum. The use of a poly-merisation catalyst is essential to allow the reaction to proceed to acceptable molecular weights. The most common catalysts in current use are: antimony compounds, such as antimony triacetate, antimony glycoloxide and antimony trioxide; germanium compounds, such as amorphous germanium dioxide and germanium glycoloxide; and titanium compounds, such as titanium alkoxides. High acid ended oligomeric product from the reaction of monoethylene glycol and terephthalic acid can be polymerised to moderate molecular weights without the use of a catalyst; however, the reaction time is quite long. It is also usual to add small quantities of melt stabilisers—such as compounds based on phosphorus, including phosphoric acid and its esters—to reduce thermal degradation and colour formation. Typical conditions in current use on a continuous manufacturing plant are: melt temperatures in the range 270–300°C and a pressure of less than 5 millibars in the final reactor. In a batch process, it is essential to achieve lower final reactor pressures to obtain adequate molecular weight build up. In this case, the final pressure should be less than 2 millibars. Polymers can be produced to various molecular weights using either the continuous or the batch process. In practice, the molecular weight will be targeted to end-product requirements. The reaction is represented by the following equation:

$$HOCH_2CH_2O(OCC_6H_4COOCH_2CH_2O)_nH$$
$$= HOCH_2CH_2O(OCC_6H_4COOCH_2CH_2O)_mH + HOCH_2CH_2OH \quad (5)$$

where, m is the average degree of polymerisation for the polymer species.

However, in the melt process a small quantity (less than 50 ppm) of acetaldehyde is produced from the degradation of the monoethylene glycol entity. This can give rise to taint of some sensitive foodstuffs, such as mineral water and CSDs, and must be reduced or removed prior to use in the packaging of these materials. For this reason, polymer used in these applications is produced in the solid state at significantly lower processing temperatures. In the solid-state polymerisation (SSP) method, the low molecular weight polymer precursor is produced by the melt phase process, as described previously [7–9]. Pellets of the amorphous polymer, with a number averaged molecular weight (Mn) in the range 15,000–25,000, which are produced by the melt phase process, are first heated at temperatures of around 160°C using good agitation under a positive dry gas stream, usually nitrogen, to develop primary crystallinity. The pellets are then gradually heated with good agitation and under a positive stream of dry nitrogen gas (dew point less than −40°C) to around 210°C to anneal and develop a higher level of crystallinity (around 48%). Good agitation is critical during these two stages to eliminate the tendency for the amorphous and the non-annealed crystalline pellets to stick and sinter together. The crystalline pellets are held at this or slightly higher temperatures under a positive stream of dry nitrogen gas (dew point less than −60°C) to allow the polycondensation reaction to proceed to the required molecular weight. The solid-state polycondensation reaction can yield a very pure grade of PET with very low levels of acetaldehyde (less than 2 ppm). For conventional injection stretch blow moulding and sheet extrusion applications, polymer is produced to Mn in the range 25,000–33,000. The reaction can also produce polymer to a significantly higher molecular weight than the melt phase process, which is suitable for extrusion blow moulding applications. For this type of application polymer with Mn greater than 35,000 is required. On achieving the required Mn, the pellets are cooled to temperatures below 100°C under a positive stream of dry nitrogen gas to minimise hydrolysis and degradation. The polymerisation reaction can be summarised by the following equation:

$$HOCH_2CH_2O(OCC_6H_4COOCH_2CH_2O)_mH$$
$$= HOCH_2CH_2O(OCC_6H_4COOCH_2CH_2O)_pH + HOCH_2CH_2OH \quad (6)$$

where, p is the degree of polymerisation of the SSP polymer.

Recently, Dupont have developed an improved process for the production of PET for packaging applications [10]. In this patented process, the low molecular weight oligomeric compound produced by the reaction of terephthalic acid and monoethylene glycol is pelletised on a heated rotoformer (temperature around 150°C) and allowed to develop a highly crystalline structure. This structure is highly robust in character and allows the pellets to be fed directly to a solid-state reactor at temperatures in excess of 200°C without the risk of sintering. The polyester products arising from this process have similar properties and characteristics to conventional solid-state polyester products.

3.2.1 Manufacturing plants

As indicated in Section 3.1, the monomer can be produced from pure dimethyl terephthalate (DMT) or pure terephthalic acid (PTA); however, for most packaging applications, polymers based on terephthalic acid are preferred. Most manufacturing plants today use highly sophisticated continuous processing routes to produce PET polymer for the packaging industry. A typical modern continuous plant will have a three or four vessel reactor, depending upon the technology used [11].

In a typical four-vessel plant [12], the first vessel is a primary esterifier operating under atmospheric or slight positive pressure (1–2 bar) and temperatures in the range 260–280°C. A slurry mixture of monoethylene glycol and terephthalic acid at a mole ratio of around 1.2 to 1 is fed into this vessel continuously. The reactor is designed as a stirred tank and steady state conditions are controlled to give a degree of esterification of around 90%. The reaction product is fed from the primary esterifier to a second vessel that is again designed as a stirred tank to give complete mixing. The product is reacted at temperatures in the range 270–290°C under atmospheric or slightly reduced pressures (600 millibars), and steady state conditions are controlled to give a degree of esterification greater than 94%. The product from the second vessel is then fed to a third vessel, whose primary function is to take off excess monoethylene glycol and to allow the initial polymerisation reaction to occur. The conditions in this vessel are set to maximise the build-up in molecular weight without excessive carry-over of solids with the liberated monoethylene glycol. Most production plants operate at temperatures around 280°C and pressures in the range 15–30 millibars. Steady state conditions are controlled to give a degree of polymerisation of around 30% and a molecular weight of around 6,000.

The fourth reactor is referred to as the high polymeriser and is designed to give the polymer maximum surface area to allow the release of monoethylene glycol and other volatiles to occur readily. There are several patented reactor designs in use consisting of a horizontal cylindrical vessel with a purpose designed horizontal agitator that moves the polymer in a plug flow with some localised back mixing through the unit. This design allows the molecular weight to build up gradually across the vessel. The polymerisation reaction is carried out at temperatures in the range 275–290°C and pressures of less than 5 millibars. The molecular weight is controlled at around 15,000–20,000 depending on end-use requirements. Higher molecular weight products can be produced, up to around 33,000, at lower throughputs or using purpose-designed plants. Polycondensation catalysts and other additives (such as stabilisers) are added to the transfer line between vessel one and two or between vessels two and three.

In a typical three vessel plant patented by Dupont [13], the first esterification vessel consists of a unique design that allows the monomer to circulate at high rates without mechanical agitation, using differential pressures from the initial high gas release at the feed end to siphon the product round the unit. A slurry mixture of monoethylene glycol and terephthalic acid at a mole ratio of around 2.0 to 1 is fed directly into the vessel immediately below the heating calandria. Both water from the esterification reaction and excess monoethylene glycol are vaporised and the relative density is reduced to force circulation within the unit. The reaction conditions are controlled at atmospheric pressure and temperatures in the range 270–290°C to give a steady state product with a degree of esterification of about 90%. The second vessel is referred to as the 'up-flow pre-polymeriser' (UFPP) unit and is designed to take the excess monoethylene glycol and volatiles off efficiently without too much solid carry-over, while allowing good molecular weight build-up to occur. The reaction product from the esterifier is treated with additional monoethylene glycol and preheated to a temperature of about 280°C prior to entry. The pressure at the top of the reactor is reduced to around 20 millibars to enhance the removal of excess monoethylene glycol, water and other volatiles and to allow molecular weight build-up to occur.

The product exiting the reactor has a degree of polymerisation of around 30 and a molecular weight of about 6000. Again, the reaction product is fed to a horizontal polymerisation vessel, as described above. However, in the Dupont finisher, the horizontal agitator has extra wiper blades that smear the surface wall. This gives additional surface area to enhance molecular weight build-up and eliminates the accumulation of degradation by-products on the reactor wall. Again, the polymerisation reaction is carried out at temperatures of around 280–290°C and pressures of less than 5 millibars, and the molecular weight is controlled at around 15,000–20,000 depending upon end-product requirements.

Higher molecular weight products can be made at lower throughputs or in a designated purpose-built plant. Catalysts, stabilisers and other additives are added with the raw materials or into the transfer line between the esterifier and the UFPP reactor. On all continuous plants, patented melt viscosity measurement devices are placed in the extrusion manifold to measure and control the molecular weight to the required specification.

Copolymers using other diacids, such as isophthalic acid or 2,6 naphthalene dicarboxylic acid, other glycols, such as diethylene glycol or cyclohexane dimethanol (CHDM), or a mixture of both acids and glycols can be produced on continuous plants. These additives can be added as separate slurries or as a mixture with the terephthalic acid and monoethylene glycol. They would normally be added to the esterifier but can be added later, in smaller quantities, to the transfer line prior to the secondary esterifier or the UFPP vessel.

Batch reactors are currently only favoured for the production of speciality polyesters containing high additive levels, copolymers which have low sales volume or which are difficult to handle on continuous plants, and the highly branched polyesters. A batch plant will usually consist of two reaction vessels: an agitated esterification pressure kettle that can handle both dimethyl terephthalate and terephthalic acid as the raw materials, and a polymerisation pressure vessel that can hold vacuum down to pressures less than 1 millibar. Additives and copolymer components can be added with the raw materials or separately to either vessel in the transfer line between the esterifier kettle and the polymeriser. The polymerisation is carried out in the temperature range 270–300°C and a final reactor pressure less than 2 millibars (and preferably less than 1 millibar). The molecular weight build-up is controlled to the required level by measurement of the torque applied to the agitator.

A typical modern, continuous, solid-state polymerisation manufacturing plant will have four major reaction vessels [14]. The polyester amorphous pellets produced on a melt phase process are fed continuously into a fluidised bed reactor or a highly mechanically agitated horizontal reactor. Here they are heated, by a positive flow of dry gas or a combination of gas and surface heaters, to a temperature around 160°C, in order to initiate primary crystallisation. The residence time (around 10–15 min) is controlled to ensure that the pellets are completely crystallised, so reducing the tendency to stick. The crystalline pellets are subsequently fed to a second reactor that can be designed either as a purpose-built column reactor, such as is used in the Buhler process, or a mechanically agitated horizontal reactor, as developed by Zimmer and Bepex. In both cases, the pellets undergo gradual heating in a positive flow of dry nitrogen gas to temperatures around 210°C. They are treated at this temperature for a period, usually around 30 min residence time, to anneal and develop a more robust crystalline structure. Differential scanning calorimetry (DSC) monitors the changes that occur in the polyester morphology at this stage in the process. Typical changes are shown in Figure 3.1.

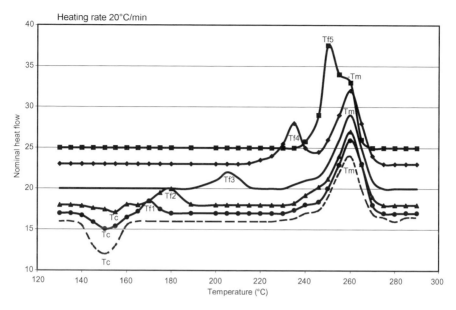

Figure 3.1 Use of DSC information to control SSP column temperatures. Heating rate 20°C/min. Abbreviations: Tc, crystallisation peak; Tf, secondary melting exotherm; Tm, primary melting point; SSP, solid-state polymerisation; DSC, differential scanning calorimetry.

The melting characteristics of the secondary crystals that are formed on annealing are closely monitored through measurement of the melting exotherm (Tf). The SSP process conditions are chosen such that Tf is always a few degrees above the temperature of the pellets. The highly crystalline pellets are transferred continuously to a vertical column reactor, heated with a positive counter current of hot dry nitrogen gas (dew point −60°C) to temperatures around 210°C to allow the polymerisation reaction to proceed. The temperature of the pellets and the gas in the column must be accurately controlled to within one degree Celsius, and kept a minimum of 5°C below the annealing temperature to eliminate any tendency to sinter. The pellets will have residence times of several hours within the reactor, depending upon molecular weight requirement and copolymer content. As indicated in Section 3.2.1, the solid-state process is highly versatile in terms of molecular weight build-up and can produce a product to a range of molecular weights ranging from 20,000 to over 80,000 by changing residence times and reaction temperatures within the column. However, it is important that the gas flow is not hindered or allowed to channel through the pellet bed, otherwise a product of non-uniform molecular weight is obtained.

In all plants, trickle samples are taken from the bottom of the column for testing in order to control the molecular weight within the required specification. The pellets are transferred from the reactor to a cooling unit that has a positive feed of cold dry nitrogen gas (dew point less than $-40°C$) to reduce tendency for moisture uptake and to prevent hydrolysis taking place. This unit can be designed as an extension to the column reactor, a separate fluid bed, or an additional horizontal vessel with mechanical agitation, such as a thermoscrew. In addition to the vessels mentioned, these plants contain sophisticated gas cleaning and purification units. The gas cleaning systems are also used to clean the pellets, removing fine dust and powder that is formed from the attrition occurring within the handling system. High levels of crystalline dust can have a deleterious influence on most downstream processing.

3.3   Structures, morphology and orientation

3.3.1   Structures

The structure of PET is based on the multiple build-up of the repeat unit mono-oxyethyleneterephthalate ($-OCH_2CH_2OOCC_6H_4CO-$) that is derived from the esterification of the terephthalic acid and monoethylene glycol entities. The combination of both the aromatic terephthaloyl and aliphatic ethylene glycol units closely defines its unique properties. In considering the physical structure of this polymer, it is best to start with the single chain and then examine the arrangement of a number of chains with respect to each other. The arrangement of a single chain is governed by the various alternative conformations of the molecule due to rotation of the many single bonds in the structure. When many molecules come together, they may form amorphous or semi-crystalline structures depending on the nature of the molecule and the thermal history to which the sample is subjected. PET exists in both amorphous and semi-crystalline forms. Part of the polyester molecule is shown in Figure 3.2, where the repeat unit is shown between the dotted lines.

The aromatic ring (paraphenylene unit) is planar and, together with the associated $-CH_2CH_2-$ on either side of the ring, it forms a single rigid structure. Rotational movements within the chain are limited to the other single bonds in the structure ($-C-O-$ and $-C-C-$ bonds). The chain repeat distance, as determined by X-ray diffraction studies, is 10.75 Å, which is slightly less than that of a fully extended, linear, zigzag chain. The average length of the molecular chains in polyester polymer used in the packaging industry lies typically between about 0.1 and 0.2 μm. However, for extrusion blow moulding polyester grades the chain length is more likely to be as high as 0.3 μm. The aromatic terephthaloyl unit in combination with contribution from some hydrogen bonding between the carbonyl oxygen and the ethylene hydrogen groupings gives the PET molecule good rigidity and high modulus. The carboxyl grouping on the ethylene $-C-C-$ bonds can take a *gauche* or *trans* conformation. In

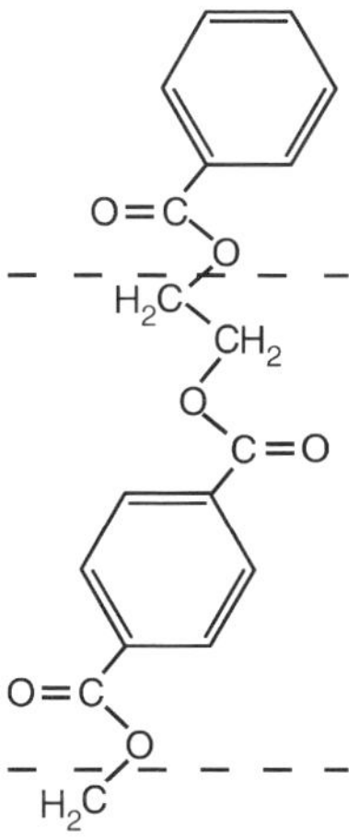

Figure 3.2 Polyethylene terephthalate: chemical repeat unit.

the amorphous state, the molecules are mainly in the *gauche* form; however, on crystallisation the molecule transforms to a *trans* configuration. The polyester molecule can also take the *trans* configuration when it is oriented.

3.3.2 Morphology

Early X-ray diffraction work indicated that highly ordered or crystalline entities with highly structured shapes (micelles) of around 100–200 Å in size were present in the polymer. However, since the molecules are significantly longer than the micelles it was concluded that the individual chains pass through a number of the highly ordered or crystalline regions, alternating with segments in the amorphous or disordered regions to give integrity to the system. The amorphous phase is assumed to form the matrix. These characteristics were the basis of the fringed-micelle model that was proposed in 1930 [15–17]. The model explains many of the characteristics associated with polyester polymer, including strength, durability, cohesion, gas barrier properties and heat setting, that are associated with crystalline regions. The non-crystalline regions are mainly responsible for extensibility, recovery, toughness and diffusion. The structural basis for the fringed-micelle model is that, on coming together, the favoured arrangement that the molecules take is one in which the free energy is at a minimum. The free energy state at high temperature, when the molecular mobility is too high for intermolecular bonding to be effective, is the random state. At lower temperatures, the molecules become less mobile and local

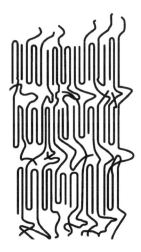

Figure 3.3 Molecular structure: modified fringed-micelle model showing crystalline/ordered domains with the chain molecule linking both amorphous and crystalline regions.

alignment of neighbouring areas of various molecules can take place to minimise free energy.

Further work, especially using single crystals [18, 19], has allowed the theory to be developed. In the late 1950s, direct evidence for folded chains was provided by electron diffraction studies on single crystals. Subsequently, folded chain structures were shown to be present in melt grown crystals. To accommodate for this observation, the fringed-micelle model was modified to show that a single molecule could fold back and forth within the micelle, as shown in Figure 3.3, as well as passing through both crystalline and amorphous regions. Although extended chain crystals have the lowest free energy, the chains prefer to crystallise by folding, since this allows the crystallisation to proceed rapidly. This indicates that the considerations are kinetic in nature and are not in the steady state of equilibrium. Consequently, the lamellar crystals are not in their most stable form and will always be liable to change when suitable conditions prevail, such as on subsequent heating. This is of fundamental importance to understanding the changes that can occur in PET on heating and is a prime cause for the changes observed in the pellet morphology during SSP. It is also important to the understanding of the mechanism as applied to heat setting (as discussed in detail in Chapter 10). In the latter case, the final structure of the article is a function of its entire thermomechanical history. However, the alternative paracrystallinity model proposed by Houseman [20], which assumes the amorphous regions to be small defects in the crystalline structure, best describes properties such as creep and barrier. This model shows the presence of voids and motions due to the dislocations.

Injection-moulded articles (such as preforms) and extruded sheet are essentially amorphous in character and various models have been proposed for their structure. All these models recognise the presence of an entangled molecular network. However, some researchers believe that the structure is not homogeneous, and that in addition to the network some crystal nuclei or extended chain molecules are present. There is evidence from work carried out by Windle and co-workers at Cambridge University, Cambridge, UK (personal communication) that some alignment occurs in the amorphous state prior to crystallisation in PET cast film, and especially in the presence of the more rigid naphthalene 2,6 dicarboxylate copolymer entity. These structures can have a significant influence on the performance of the amorphous article in any subsequent stretching process. If one considers how the network responds to stretching in various industrial processes where the drawing stresses are relatively high, it is clear that the entanglements will hinder the movement of molecules during the stretching process, and as the stretch ratio increases the ease of movement becomes more difficult.

The influence and benefit of these entanglements are clearly shown in the load-extension curves shown in Figure 3.4 for PET. In the initial stages, the system obeys the normal laws of physics, with the stress increasing sharply with strain and the polymer matrix acting as an elastic; however, as the stress

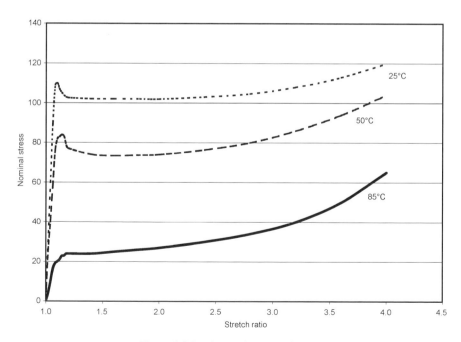

Figure 3.4 Load-extension curves for PET.

is increased beyond a critical level (the yield point) the polymer yields and the stress falls back slightly. At higher strain levels, the stress is again observed to increase sharply as the entanglements restrain movement and strain hardening sets in prior to rupture. As the stress increases, the molecules develop significant levels of orientation along the stretch direction, which contribute to high strength products in industrial processing. At the higher stretch ratios, the chain movement contributes to localised heating and some additional crystallisation occurs. These crystals can give some degree of thermal stability to the system and will influence the level of shrinkage occurring at higher temperatures; however, the level of crystallisation is not sufficient to give dimensional stability above the glass transition temperature (Tg).

3.3.3 Orientation

In practice, the stretching performance of PET is dependent on: temperature, molecular weight, strain rate, crystallisation, moisture, and copolymer type and composition [21–29]. The influence of temperature is shown by the stress/strain curves in Figures 3.4 and 3.5. At low temperatures, the stress increases sharply up to the yield point, and then falls back as the polyester molecules yield. Under these conditions, the work involved in straining the molecules generates heat and a rise in temperature of several degrees is observed at the yield point. Invariably, necking occurs under these conditions leading to a localised thinning and non-uniform stretch. As the temperature is increased to around the Tg, a much

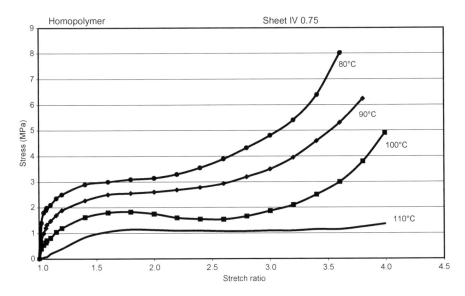

Figure 3.5 Uniaxial stretch of PET sheet at various temperatures. Abbreviation: IV, intrinsic viscosity.

Table 3.1 Properties of uniaxially stretched sheet

Draw ratio	Draw temp. °C	Density kg/m³	Crystallinity %	% Shrinkage (BW)	Tensile strength MPa	Extension at break %
4.5	100	1371.9	30.7	1.99	158	14.3
4.5	110	1371.6	30.5	0.51	112	45.0

Homopolymer: IV 0.78. Abbreviations: IV, intrinsic viscosity; BW, boiling water.

more uniform and controlled yield is obtained. At higher strain levels, the stress increases sharply again as strain hardening sets in prior to rupture. To produce both strength and rigidity (modulus), as are required for both films and bottles, it is necessary to develop a degree of hardening (high orientation) at the stretching stage. As the resin temperature is increased to above 110°C, the molecules flow more readily and a much higher stretch is required to achieve strain hardening. Test results (see Table 3.1) indicate that the level of orientation attained at the higher stretch under these conditions is less than that obtained at the same stretch ratio at the lower temperature. This is reflected in a lower tensile strength, higher extensibility at break and lower shrinkage. The optimum processing conditions for homopolyester are seen to be 85–105°C. These conditions will change if the Tg is affected by copolymerisation.

Molecular weight also has a significant influence on the stretching characteristics of PET. The molecular entanglement associated with the higher molecular weight resin has a more restraining influence on the molecules. In this case, the higher molecular weight polymer (see Figure 3.6) has a higher stress yield point, whilst the strain hardening occurs at a lower stretch ratio. For good process control it is important, therefore, to control the molecular weight within a tight specification.

As anticipated from the entanglement theory, strain rate can also have a significant influence on the properties of the final product. At high strain rates, there is considerable molecular resistance to chain disentanglement and movement. The stress/strain curves (see Figures 3.7 and 3.8) show a significant increase in the stress yield point, and strain hardening occurs at a lower stretch ratio as the strain rate is increased. At very high strain rates, one sees a significant increase in temperatures at strain hardening and a higher level of crystallinity is obtained in the product [22].

The influence of crystallinity is to contribute towards the resistance in the molecular movement during the stretching process. As the level of crystallinity increases, the stress level to yield increases (higher modulus) as the product has become more rigid and brittle. Similarly, the stretch ratios show a decrease with increased crystallinity (see Figure 3.9). It is important, therefore, in both film and injection moulding processes to minimise the level of crystallinity build-up during sheet casting and moulding, as well as at the product reheating stage prior to stretching.

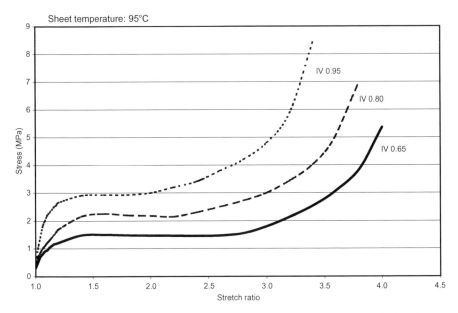

Figure 3.6 Uniaxial stretch of PET sheet: influence of IV. Abbreviation: IV, intrinsic viscosity.

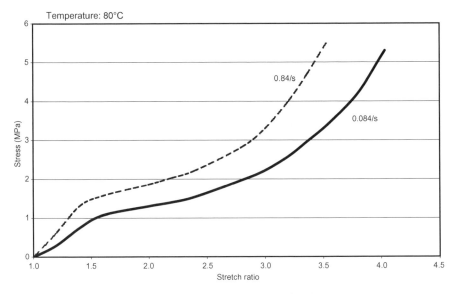

Figure 3.7 Uniaxial stretch: influence of strain rate.

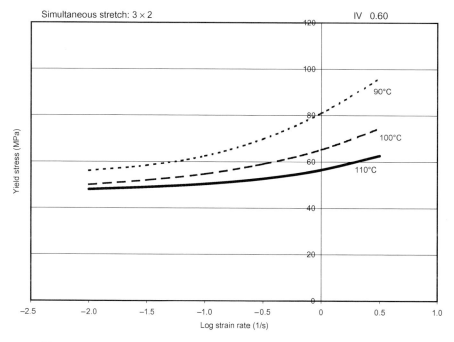

Figure 3.8 Variation of yield stress with strain rate. Abbreviation: IV, intrinsic viscosity.

Water is a good plasticiser for PET and as such it can lower the Tg quite effectively [27]. At high relative humidities (RH greater than 90%), the Tg is reduced by several degrees Celsius (see Figure 3.10). The level of moisture in the amorphous products prior to stretching can therefore have a pronounced influence on the film or container properties. The resultant effect is similar to stretching at higher temperatures: the molecules will be more mobile at lower temperature and the resin will flow rather than strain harden at the designed stretch ratio.

Most polyesters used today in the packaging industry are copolymers, and even the homopolymers that are available on the market are not true homopolymers, since they all contain some generated or added diethylene glycol moieties. The influence of copolymerisation is dependent on the type of additive and on the level added. Aliphatic, longer chain glycols or aliphatic diacids tend to reduce the Tg and will therefore act in a very similar manner to plasticisers, such as water. In this case, as described above for water, the molecules are more mobile at lower temperatures and the resin will flow rather than strain harden under the standard homopolymer stretching conditions. When an aromatic copolymer additive is used in copolymer manufacture, its influence will be dependent upon its molecular structure. Molecules such as naphthalene 2,6 dicarboxylic acid will

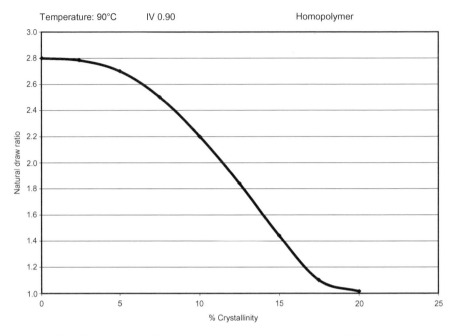

Figure 3.9 Influence of crystallisation on natural stretch ratio. Abbreviation: IV, intrinsic viscosity.

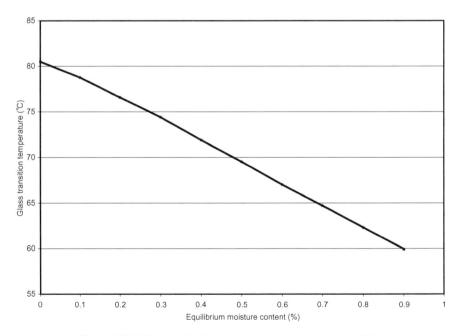

Figure 3.10 Influence of moisture on glass transition temperature (Tg).

increase the rigidity within the molecular chain, which will increase the apparent Tg and contribute towards reducing the mobility within the polymer matrix. In this case, higher temperatures are required to give satisfactory stretching. For molecules such as isophthalic acid, which act as chain disruptors, it is most likely that the overall crystallisation reaction occurring during the latter stages of stretching is hindered and thus the molecules have a greater freedom to flow. In these cases, it is found that strain hardening is delayed and that higher stretch ratios are required to obtain good physical properties.

In the more complex processes used to manufacture film or bottles [30], it is possible to have sequential biaxial stretching, simultaneous biaxial stretching, or an intermediate situation as in preform blowing. Although the same principles apply, there is a significant difference in the mode of application. In this case, the polymer is more restrained in each direction during the draw down. This can have a bearing on the stresses involved; that is, higher stresses are observed during stretching. This effect is more pronounced under simultaneous biaxial stretching as well as for the sequential sideways draw. Typical stress/strain curves obtained under both sequential and simultaneous stretching of PET polymer are shown in Figures 3.11–3.14.

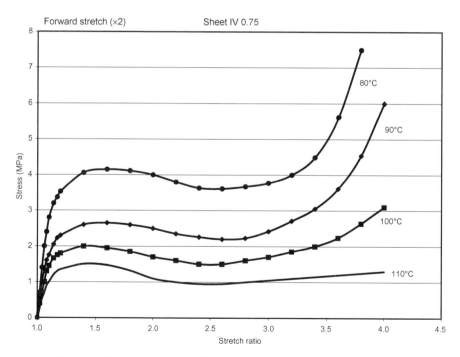

Figure 3.11 Sequential stretch of PET sheet. Abbreviation: IV, intrinsic viscosity.

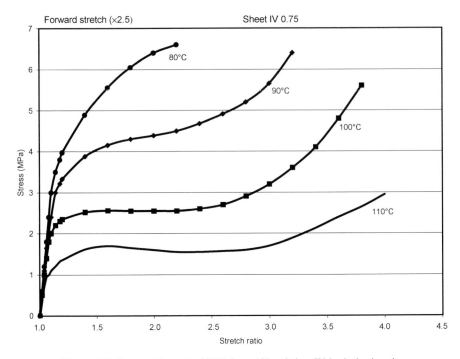

Figure 3.12 Sequential stretch of PET sheet. Abbreviation: IV, intrinsic viscosity.

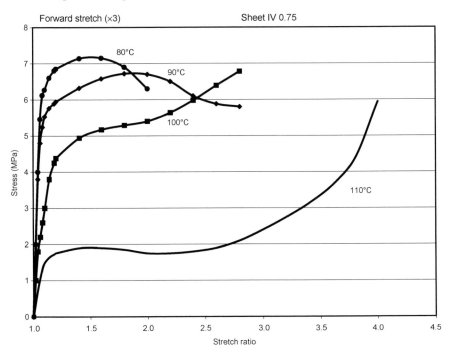

Figure 3.13 Sequential sideways stretch of PET sheet. Abbreviation: IV, intrinsic viscosity.

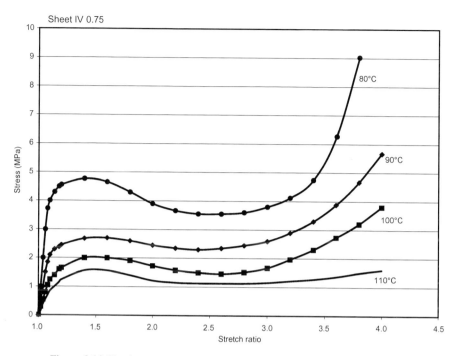

Figure 3.14 Simultaneous stretch of PET sheet. Abbreviation: IV, intrinsic viscosity.

As mentioned previously in this section, stretching the polyester beyond its natural draw can induce a degree of crystallisation in the polymer [28, 29]. This can have a significant influence on both the secondary draw in a sequential stretch process and the degree of setting achieved. The magnitude of the stress in the secondary draw is dependent upon the level of crystallisation developed during the forward draw, and therefore, to some extent, on the time delay between the forward draw and the start of the sideways draw. In the bottle blowing process, the delay is very short—less than 0.1 s—and thus the influence may not be as pronounced as that observed in the film process. However, the behaviour of the material in the sideways draw will be dependent upon the extent to which the polymer is oriented in the forward draw process. With a forward stretch ratio of 2.0 or less, a pronounced plateau is seen in the stress/strain characteristics of the sideways draw (see Figure 3.11). As the draw ratio is increased to 2.5 in the forward draw, the yield stress increases and the plateau disappears for stretch temperatures below 100°C (see Figure 3.12). At an initial stretch ratio of 3.0, higher stresses are observed for the sideways draw and the plateau is absent at temperatures below 110°C (see Figure 3.13). These results (see Table 3.2) again show that the effect of increasing the temperature is to reduce the level of orientation obtained for a given stretch ratio, increase the density and thus the degree of crystallisation, and reduce the shrinkage.

Table 3.2 Properties of biaxially stretched sheet (ratio 2.0 × 3.5)

Draw temperature °C	Density kg/m³	Crystallinity %	Shrinkage		Tensile strength (MPa)		Extension at break	
			F	S	F	S	F	S
80	1352.3	14.4	20.4	20.4				
90	1355.9	17.4	12.0	9.7				
100	1360.9	21.6	6.0	4.4	73	120	5.4	7.0
110	1367.0	26.7	1.6	0.9	62	86	5.9	

Homopolymer: IV 0.78. Abbreviations: IV, intrinsic viscosity; F, forward result; S, sideways result.

Table 3.3 Properties of biaxially stretched sheet

Draw ratio	Tensile strength (MPa)		Extension at break (%)	
	F	S	F	S
Sequential stretch 2.6 × 4.0	217	327	258.4	109.8
Simultaneous stretch 2.6 × 4.0	173	215	175.1	65.3

Homopolymer: IV 0.78. Stretch temperature: 90°C. Abbreviations: IV, intrinsic viscosity; F, forward result; S, sideways result.

In the simultaneous stretching of the polyester polymer in both directions (see Figure 3.14), the stress rises more slowly than in the secondary sideways draw and a yield point is observed in the stress/strain curves at all temperatures. The curves are of similar character to those observed in the sequential process at low initial forward stretch ratio (less than 2.0). For a given draw, the sideways stretch needs to be greater to avoid uneven stretching than is the case with sequential stretching. For similar stretch ratios, the tensile properties of sheet are lower than those achieved by sequential stretching (see Table 3.3). The influence of molecular weight on both sequential and simultaneous draw is as anticipated. The higher molecular weight samples require a greater stress to stretch and reach the strain harden regime at a lower extension than lower molecular weight material. In the sequential sideways draw, it can be seen that the higher molecular weight sample draws at a slightly lower stress (see Figure 3.15). This is because a lower level of crystallinity is formed during the forward draw.

In practice, the influence that molecular weight can have on the uniformity of stretching can be summarised using area expansion product ($\lambda_L \lambda_T$) of the natural stretch ratios in the longitudinal and transverse direction against molecular weight—in this case represented by intrinsic viscosity (IV) (see Figure 3.16). The influence of moisture, crystallisation and strain rate are also as anticipated and mentioned previously in this section. However, the effect of copolymerisation will be dependent upon the influence that the copolymer component has on the level of crystallisation occurring during strain hardening. If the additive retards the crystallisation, then the sideways draw will require less stress to

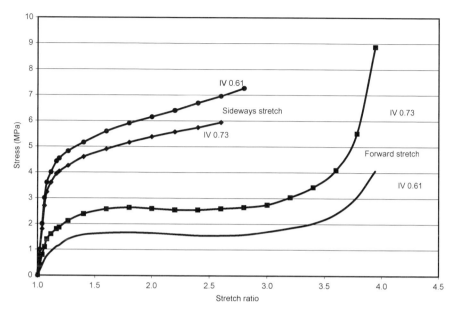

Figure 3.15 Sequential stretch: influence of molecular weight. Abbreviation: IV, intrinsic viscosity.

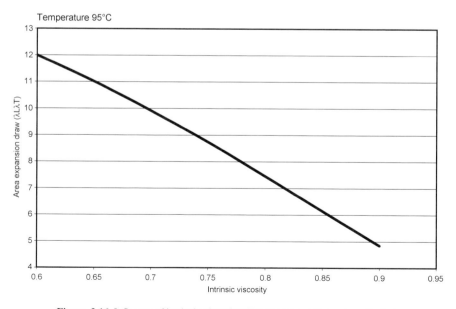

Figure 3.16 Influence of intrinsic viscosity (IV) on area expansion draw ($\lambda_L\lambda_T$).

stretch; however, if the additive enhances crystallisation, then the sideways draw will require higher stress and the product is likely to show brittle failure at high stress levels.

In addition to ensuring that good tensile properties are achieved, an understanding of the morphology of the polymer is important in ensuring that the integrity of the manufactured articles meets the customers' requirements. This is true of cast sheet, heat set film, injection-moulded preforms and containers, and blow-moulded bottles. If one considers and applies the fringed-micelle type model to these various applications, it is possible to envisage possible problem areas and probable remedies. In injection moulding, the temperature difference between the cold mould wall and the hot molten resin can give rise to different stress levels within the injection moulded article, which will cause the molecular chain to be stretched and become more oriented. In practice, these are observed as highly ordered laminar layers that in thick preform moulding can lead to stress crystallisation occurring within the preform (see Figure 3.17).

A similar effect is also observed in cast sheet and especially in thick sheet, where the temperature difference between the calendering rolls and the hot

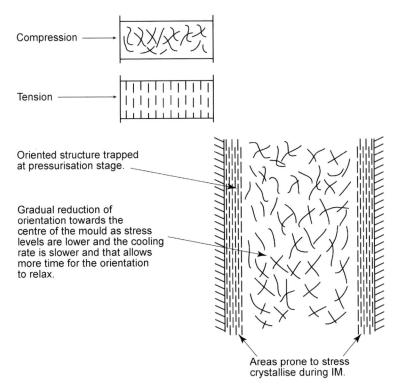

Figure 3.17 A) Influence of shear on injection moulding.

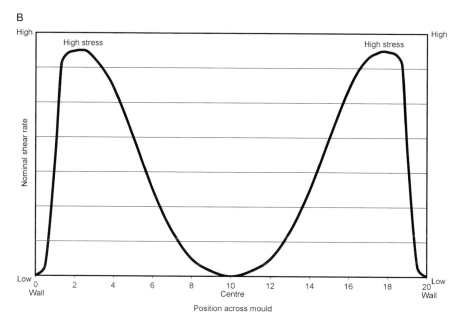

Figure 3.17 B) Shear rate profile across the mould.

extruded polymer can build up different stress levels across the sheet. In both cases, higher haze levels are observed within the samples and, on subsequent stretching, the product can delaminate. In the manufacture of bottles by ISBM, the heated preform is blown into a cold mould. If the preform has high crystallinity levels in the base or the molecular weight is not uniform, overstretching can occur in the petaloid base area leading to uniaxial orientation in critical stress areas of the container when under pressure (see Figure 3.18). In practice, this can enhance the susceptibility of the base to environmental stress cracking. All of these problems can be readily overcome by choosing the correct resin and tight process control.

The morphology of the final product is also important in obtaining good gas barrier properties, in reducing creep, and in establishing hot-fill properties in the final product. These are discussed later under the appropriate headings.

3.3.4 Creep

Creep is another important characteristic of polymers. This is particularly true of polymers under continuous stress or strain in use, as is the case for pressurised containers. In the CSD bottle, the stress levels are high and are applied for an extended period of time. As shown by Bonnebat *et al.* [26], oriented PET is quite resistant to creep at low temperatures; however, at temperatures approaching

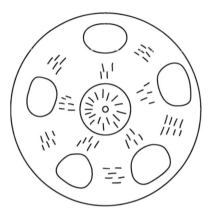

Figure 3.18 PET bottle: petaloid base. Potential areas of high stress that have partially uniaxial oriented structures are shown. These areas are likely to be prone to stress cracking in use.

Tg the creep compliance (elongation/tensile stress) increases sharply as the modulus falls and the molecules have more freedom to move. Creep is associated with movement of the molecules in the amorphous phase, and an increase in crystallisation level and/or orientation can reduce the level quite significantly. The mechanism is similar to stretching at very slow speeds, where there is a lower tendency to strain harden and therefore no resistance to chain movement. The polymer will show an increase in creep as the stress levels approach the yield point.

3.4 Properties

The properties of PET and its copolymers are determined basically by their chemical composition and molecular structures. The combination of both the aromatic terephthaloyl entity and the flexible monoethylene glycol gives PET its unique characteristics, first recognised by Whinfield and Dickson. It is a strong, tough, flexible thermoplastic that readily crystallises and can be oriented into highly ordered structures. The oriented structures can also be heat set to give increased dimensional stability at high temperatures (above 80°C). The polyesters have been used in a variety of applications because of their versatility and excellent physical properties. Their main characteristics are summarised as follows:

- General characteristics—cooled to give clear amorphous state, crystallisable, high melting temperature, melt processable, moderate softening point (amorphous Tg *ca* 80°C), worldwide food approval, excellent clarity, colourless, excellent strength, good creep resistance especially when crystallised, good barrier properties, and excellent chemical resistance.
- Characteristics of stretched products (oriented)—excellent strength and stiffness (high modulus), good creep resistance, excellent clarity,

colourless, good water vapour barrier, adequate CO_2 barrier, low taint, outstanding impact resistance, excellent chemical resistance, worldwide food approval, and favourable environmental impact.

3.4.1 Molecular weight and intrinsic viscosity

Molecular weight and molecular weight distribution are fundamental properties that determine end-use applications. The polymers are produced to various molecular weights and are available in amorphous, semi-crystalline or highly crystalline states. In the polyester industry, molecular weights have been characterised by IV measurements in dilute solutions. However, PET is not soluble in the common solvents and, over the years, resin manufacturers have used several different highly polar solvents to characterise the polymer. In more recent times, the production and use of highly crystalline solid-state materials, which are more difficult to dissolve, have exacerbated the situation. The solvents most commonly in use are the following: ortho-chlorophenol, hexafluoroisopropanol, 60:40 part mixture of phenol and tetrachloroethane, meta-cresol, trichloroacetic acid, trifluoroacetic acid, and 25:75 part mixture of ortho-chlorophenol and chloroform.

The technique is based on the concept that the hydrodynamic volume of a linear polymer is related to its molecular size and, therefore, to particular solution viscosity properties. However, each solvent will have different solvation effects on the polymer molecules and this can have a significant influence on the apparent molecule size in that solvent. For these reasons, all the solvents measure slightly different IVs for identical molecular weight materials. The difference observed is greatest with higher molecular weight products and when hexafluoroisopropanol is used as the solvent. Within the industry, both manufacturers and end users have attempted to normalise the measurements by adjusting the Huggins constants. This works well over a short molecular weight range but is not to be recommended in practice. In general, the ASTM IV testing procedure using the solvent mixture of 60:40 phenol and tetrachloroethane is used as standard for the industry.

In addition to the solution techniques, ICI has developed a modified melt flow index (MFI) testing procedure that measures the melt viscosity of the polymer at low shear rates and converts the values to IV. This instrument is manufactured and sold by Lloyds Instrument under the name Daventest melt viscometer. This test procedure is employed extensively to measure the IV of highly crystalline polymers that are used in injection moulding and sheet extrusion. For linear polymers, the IV measurement is correlated to molecular weight by the well-known empirical expression first proposed by Mark [31] and Houwink [32]:

$$[\eta] = K * M^{\alpha} \tag{7}$$

where, η is the IV, M is the molecular weight, and K and α are constants.

The values of the constants, K and α, are solvent dependent and, as expected, there are several different relationships established and published in the literature for PET. The value of the molecular weight can be expressed as Mn, Mw and Mv, which are the number average molecular weight, weight average molecular weight and the viscosity average molecular weight, respectively, depending on how the correlations are established. Although there is no preferred relationship, the ones most used by the author are as follows:

(a) $$[\eta] = 1.7 * 10^{-4} * Mn^{0.83}$$ (8)

which was established by Ravens and Ward [33] using end group analysis. IV was measured in ortho-chlorophenol at 25°C.

$$[\eta] = 1.47 * 10^{-4} * Mw^{0.768}$$ (9)

which was established using gel permeation chromatography (GPC) data for molecular weights determined in ortho-chlorophenol, and holds for polydispersity (Pd) of less than 3. With Pd above 3, the polyester shows severe branching or abnormal molecular weight distributions, such as observed in high molecular weight SSP (IV > 1.5) products.

(b) $$[\eta] = 4.68 * 10^{-4} * Mw^{0.68}$$ (10)

which was established by Moore [34] using GPC. IV was measured in 60:40 phenol/tetrachloroethane at 25°C.

However, the overall processing behaviour of the resin is established by an understanding of the molecular weight distribution. The basic reason is that some properties, such as tensile, impact strength and crystallisation, are specifically governed by the short molecules, whilst other parameters, such as solution viscosity and low shear melt flow, are influenced by the middle range molecules. The longer chains have significant influence on melt elasticity, stretching properties at strain hardening, and polymer delamination. To assess the overall performance of the resin, it is best to obtain its molecular weight distribution. This is best achieved today by high-performance liquid chromatography (HPLC)-GPC techniques, using ortho-chlorophenol as solvent. However, since this technique uses polystyrene standards for calibration purposes, the values obtained are not necessarily absolute and care must be taken in the interpretation of the data. A typical molecular weight distribution curve for the solid-state product is shown in Figure 3.19.

Typical values of Mn and Mw measured for various PET polymers are given in Table 3.4. The results show the linear polyester samples to have a Pd (i.e. the ratio of Mw to Mn) in the range 2.3–2.7 [35, 36]. Branched polyester products will have Pd values greater than 3 and usually around 4–5, depending on the

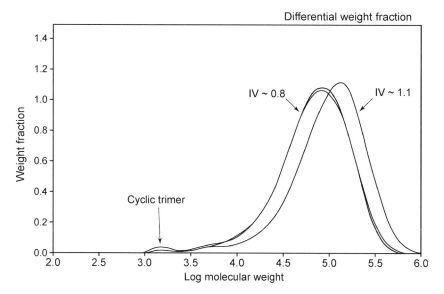

Figure 3.19 PET: molecular weight distribution curves. Abbreviation: IV, intrinsic viscosity.

Table 3.4 Molecular weight properties of various polyesters

Polymer type	IV	Mn	Mw	Pd
Melt	0.584	18232	49226	2.70
Melt	0.650	20572	56780	2.76
Melt	0.753	24765	77515	3.13
SSP	0.746	24488	68567	2.80
SSP	0.786	26079	68849	2.64
SSP	1.074	37989	108651	2.86
SSP	1.100	39100	124732	3.19
SSP	1.750	68417	233989	3.42
Bottle wall	0.796	26479	76790	2.90
Sheet	0.894	30408	82709	2.72
Branched	1.030	36122	142323	3.94
Branched	1.380	51388	227136	4.42

Both molecular weight (GPC) and IV are measured in ortho-chlorophenol. Abbreviations: IV, intrinsic viscosity; Mn, number averaged molecular weight; Mw, weight averaged molecular weight; Pd, polydispersity; SSP, solid-state polymerisation; GPC, gel permeation chromatography.

degree and type of branching occurring in the polymer matrix. The molecular weight distribution curves for PET always show a peak at the low molecular weight end that is characteristic of cyclic trimer (see Figure 3.19). The level of cyclic trimer present is dependent upon the manufacturing route used to produce the polymer [36–40]. In the melt process, there is an equilibrium value of around 1.2 wt% of cyclic trimer present; higher cyclic components are also present at significantly lower levels (less than 0.5 wt% in total). However, in the solid-state process, the high crystallinity and long residence times allows the level of cyclic trimer to be significantly reduced. On melt processing, the level will increase again and approach equilibrium values, especially if the melt residence times are long. The level is process dependent, with levels found to be in the range 0.3–0.5 wt%. The cyclic trimer has a melting point of around 319°C and is not too volatile. However, it has been known to deposit on calender rolls and within the injection moulds.

3.4.2 End group

PET-based polymers will have both hydroxyl and carboxyl end groups. Establishing an understanding and control of the end-group balance is important in controlling both the manufacturing process, especially the solid-state reaction, and downstream processing. The carboxyl end is known to catalyse both the polymerisation and the hydrolysis reactions. The level of each is process dependent, with the ester interchange process tending to give the lowest carboxyl level in the final product. However, in practice, the level of carboxyl ends is critically controlled to give optimum reaction kinetics to the polyesterification reaction without over-sensitising the hydrolysis reaction in subsequent drying and melt processing. The level of end group in polyester is traditionally quoted as 'g mole end group per 10^6 g of polymer'. The level of carboxyl in the melt process is controlled at around 30–40% of the total end-group concentration. The level is further reduced during subsequent solid-state reaction due to the preferential kinetics of the esterification reaction and the ease of water removal. In SSP products, the level of carboxyl ends is controlled at around 25–35% of the total end-group concentration.

The carboxyl end group can catalyse the hydrolysis reaction at temperatures above Tg. For products that have to operate at high temperatures, it is essential to keep the carboxyl level in the final product low and preferably below 10 g mole of carboxyl end group per 10^6 g of polymer. It is also important to control the level of carboxyl end build-up in both extrusion and injection moulding processes. At high levels (greater than 50) it can cause localised unzipping of the molecule in the presence of moisture to give terephthalic acid and low oligomeric white deposition on the calender rolls and within the mould [41]. In this case, the use of recycled material that by its nature will contain higher level of carboxyl ends can exacerbate the problem. These deposits can lead to blemishes on the sheet surface and to weakness in the container wall.

However, high levels of carboxyl ends are beneficial to nucleation and crystallisation. In the presence of critical levels of metal ions, especially sodium ion, the carboxyl ends congregate around the metal ion and impart structured sites for crystallisation. The use of sodium ion and other nucleants is discussed in more detail in Section 3.11 on crystallisation.

3.4.3 Thermal properties

The thermal properties of PET are important to defining its versatility in use. The important parameters are: the glass transition, crystallisation and melting temperatures, as well as the heats of both fusion and crystallisation and the rates of crystallisation. Differential scanning calorimetry is the technique most frequently used to measure these properties. DSC operates by measurement of the amount of heat required to maintain the temperature at the value given by the temperature programme. The measurement is made by determining the power input into the sample container and subtracting from it the power input made to a similar empty container. This is in contrast to differential thermal analysis, in which the temperature of a sample and reference are compared. The instrument is calibrated using pure compounds with accurately known thermal properties, such as tin. With these techniques, it is important to recognise that the temperatures and the magnitude of the measured transitions are affected by sample size as well as the heating and cooling rates.

In practice, the initial heating cycle shows the behaviour of the polyester as the result of its previous thermal history. The polyester will usually show glass transition, crystallisation and melting temperatures. However, to obtain true transition temperatures, it is important to heat the sample rapidly under dry inert gas to temperatures above the melt temperature, usually around 300°C, and to hold for 1 min to remove crystalline memory prior to cooling rapidly to temperatures below Tg. This technique produces unstrained amorphous samples. Typical DSC curves for various polyester polymers are shown in Figure 3.20.

These results show the strength of this test procedure in the characterisation of polyester products. The Tg is usually measured as the half-height of the step change observed in the baseline. The sample of amorphous pellet shows a Tg of about 80°C, a maximum crystallisation temperature at 160°C and a peak melting temperature at about 250°C. In contrast, the crystalline SSP pellet shows neither Tg nor crystallisation temperatures but has a very pronounced melting exotherm, with a peak at about 248°C. As shown previously (see Figure 3.1), the DSC can be used to monitor morphological changes that occur during the SSP process. In this case, a secondary melting peak is observed, which continuously moves to higher temperatures as the crystal structure matures during the heating cycle and eventually takes over from the original melt. In this case, the single peak defines a highly crystalline state that will allow the pellet to be processed through the hot SSP column without sticking to a neighbouring pellet. It can also be used to detect pellets that have been held up in the reactor or have seen abnormally high temperatures [42–46]. In this situation, the polymer matrix matures into a higher degree of crystallinity and crystal perfection and the melting peak increases to temperatures above 260°C, and occasionally approaching 300°C. These pellets are more difficult to melt under standard downstream processing conditions and appear as un-melts in the sheet or injection-moulded products.

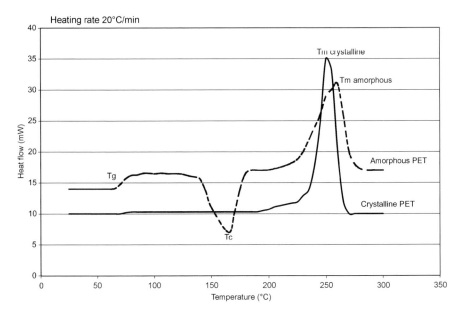

Figure 3.20 Typical DSC curves for crystalline and amorphous PET. Abbreviations: DSC, differential scanning calorimetry; Tm, melting point; Tg, glass transition temperature; Tc, crystallisation peak.

The total area under the DSC curve can be integrated to give the specific heat of the polyester over the temperature range as well as the heat of fusion and crystallisation. A summary of the thermal properties of PET is given in Table 3.5. The heat of fusion of 100 per cent crystalline PET has been determined by extrapolation to be 120 calories per gram. This value can be used in conjunction with the heat of fusion determined for the test sample to establish the level of crystallinity.

One of the important characteristics of glassy polymers is the manner in which the volume of the material changes with temperature. At the glass transition temperature, there is a moderate change in the slope of the curve of specific volume against temperature. In the glassy state, the molecules are restrained by local Van der Waals forces, while in the rubbery state a liquid-like motion of much longer molecular segments occurs. In contrast, the crystalline molecule shows an abrupt increase in volume at the melting point (see Figure 3.21). The glassy state is therefore not in equilibrium and will tend to shift to a more stable state over a period of time. This change is usually referred to as a change in 'free volume'. External environmental conditions have a pronounced influence on the degree and time scale, especially heat and high moisture levels. These changes can also shift the physical characteristics of the amorphous resin from ductile to brittle form. The DSC technique can be used to monitor changes that

Table 3.5 Summary of the thermal properties of polyethylene terephthalate

| Temperature °C | Crystalline PET | | Amorphous PET | | |
	Specific heat J/g · °C	Enthalpy J/g	Specific heat J/g · °C	Enthalpy J/g	Thermal conductivity mW/M · °C
25	1.1126	23.7	1.1194	24.1	193
45	1.1815	46.7	1.1891	47.3	198
65	1.2579	71.1	1.2626	71.8	199
85	1.4159	97.9	1.6962	100.4	175
105	1.4995	127.1	1.6932	133.9	150
135	1.6175	173.9	1.7462	185.5	122
155	1.6879	207.1	1.7228	220.5	102
175	1.7396	241.4	1.4055	252.1	97
195	1.8202	276.9	1.5277	280.1	91
215	1.8754	313.8	1.9385	315.1	92
235	2.1397	353.2	2.6274	359.9	92
245	9.1897	425.8	2.7773	388.2	92
255	2.0211	453.7	2.0051	429.8	110
265	2.0123	473.8	2.0051	429.8	120
275	2.0273	494.1	2.0227	449.9	120

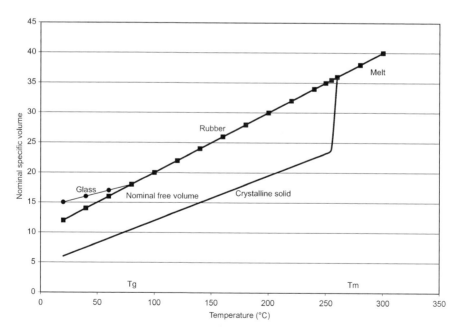

Figure 3.21 The change in specific volume–temperature curve for both glassy and crystalline polymer. Abbreviations: Tg, glass transition temperature; Tm, melting point.

are occurring over a period of time within the polymer matrix. In this case, the enthalpy changes observed in the Tg are measured with time.

The transitions occurring at low temperatures are associated with distinct molecular movement within the molecule. Measurements using dynamic mechanical analysis (DMA) show that PET has two transitions. The first transition is associated with the freezing of molecular rotational movements around the aliphatic linkages and is referred to as the α transition. This transition is strongly associated with the Tg. With increase in crystallisation level and/or orientation, the freedom of movement is decreased and this transition is shifted to higher temperatures. The second transition is associated with oscillation movements of the carbonyl group and the flipping of the paraphenylene molecule and is observed at temperatures below 0°C, with a transition peak at around −60°C. This low temperature peak has an influence on the barrier properties and gas transport through the polymer's matrix [47, 48].

The addition of copolymer entities can have a significant influence on the melt properties. In general, the melting point of the polyester is reduced by approximately 2.3°C per mole of additive. As the level of copolymer is increased, crystallisation is inhibited, and at levels above 20 moles per cent the amorphous state usually prevails. The aliphatic copolymer additives, such as diethylene glycol, polyethylene glycol and adipic acid, will give more flexibility to the molecule, and thus reduce the Tg. However, the more rigid cyclohexane derivatives, such as cyclohexane dimethanol (CHDM), have only marginal effects on the Tg. Similarly, aromatic diacids, such as isophthalic acid, also have only marginal effects on the Tg; however, diacids based on the more rigid aromatic molecules, such as naphthalene 2,6 dicarboxylic acid, will increase the Tg.

A straight linear relationship between the Tg and the mole per cent of copolymer component added is observed for the copolymers of both isophthalic acid and naphthalene 2,6 dicarboxylic acid. The addition of copolymer component can have a very pronounced influence on the crystallisation characteristics of the polymer. In most cases, the additive will hinder the tendency to crystallise from the melt; however, additives such as branching agents, which can magnify the stresses involved during cooling, can enhance the rate of crystallisation.

The DSC is used to characterise the crystallisation characteristics of the polymer. Useful data are obtained from measurements of the crystallisation enthalpy and the peak temperature of the copolymer samples at different cooling rates from the melt. As the cooling rate is increased, the molecules do not have time to align to give nucleation sites and the level of crystallisation is decreased. However, if there are nucleation sites present, a higher level of crystallisation will occur and the measured enthalpy is high [49]. This technique makes it possible to differentiate between any nucleation effects that could reduce the benefit of the copolymer additive, such as catalyst residues, to retard crystallisation. In the presence of nucleants, the enthalpy of crystallisation does not decrease as expected at high cooling rates. The technique can also be used to determine which copolymer additive is the most effective in disrupting the polymer chain and therefore hindering crystallisation. The data generated can be used to define polymer formulations for thick injection mouldings, such as the refillable bottle preform, and for thick sheet extrusions.

The difference in crystallisation characteristics can also be determined from the measurement of the isothermal crystallisation rates at various temperatures from the melt down to the Tg [50]. In this case, the rate of crystallisation is governed by the mobility of the chains of the molecule and their ability to form nuclei. As the temperature decreases, the mobility of the chain is reduced, which allows crystalline sites to form and crystal growth to occur. The rate of crystallisation increases to a maximum, and then drops off as the mobility of the molecules decreases with decreasing temperature. Again, the influence of the copolymer can be observed by a change in crystallisation rate and a shift in the temperature at which the maximum rate occurs. The latter is dependent on the copolymer additive and its influence upon the movement of the molecular chain. Typical data for both homo- and copolymer samples are shown in Figure 3.22.

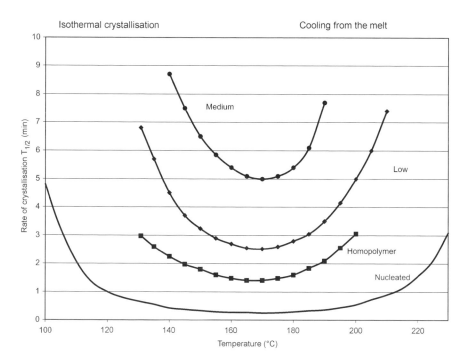

Figure 3.22 Influence of polymer formulation on the rate of crystallisation.

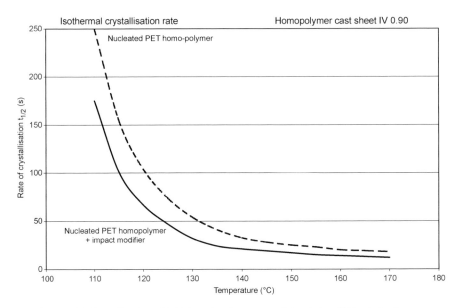

Figure 3.23 Crystallisation of PET from the glassy state. Abbreviation: IV, intrinsic viscosity.

When heating from the glassy state, one has to consider what influence the additive has had on the Tg. If the Tg has been reduced, then molecular movement will occur at lower temperatures and thus crystallisation can take place at lower temperatures. If the Tg has been increased, molecular movement will be retarded and crystallisation occurs at higher temperatures. Therefore, the influence of the aliphatic copolymer entities, such as polyethylene glycol, is to allow crystallisation to occur at temperatures below 100°C, and of additives such as naphthalene 2,6 dicarboxylic acid is to retard crystallisation to temperatures above 100°C. In heating from the glassy state, the rate of crystallisation is also influenced by the degree of stress that may have built up in the sample during cooling (see Figure 3.23). Extreme care must therefore be taken in the production of the quenched sample to ensure that the sample is stress free. The level of both moisture and carboxyl ends can have a significant influence on the results. It is therefore important that samples are dried thoroughly prior to testing and that the level of carboxyl ends is determined.

3.5 Rheology and melt viscosity

3.5.1 Melt viscosity

The melt viscosity properties and flow behaviour of polymer melts are of practical importance in both manufacturing and polymer processing. The successful melt processing of the polyester polymer is dependent, therefore, upon knowledge and understanding of the melt flow characteristics of the polyester polymer under the process operating conditions. For normal liquids, the viscosity is a material constant that is only dependent on the temperature and pressure and is independent of the rate of deformation and time. However, for polymeric materials the situation is more complex. The viscosity changes with deformation conditions and the flow of the polymeric melt is accompanied by an elastic effect; that part of the energy exerted on the system is stored in a form of recoverable energy. The polyester is viscoelastic in character and the viscosities are both time and rate dependent. This effect is observed as a die swell (Barus effect) in extruded polymer. However, the most prevalent effect of the viscoelastic phenomenon in the polymer melt is the decrease in viscosity with increasing shear rate. These effects are usually referred to as non-Newtonian properties, and are highly sensitive to the molecular weight as well as the degree of branching that may be present [51–53]. A number of empirical equations have been proposed to describe the influence of shear rate on viscosity. For PET, the author has found that an empirical equation based on a modified Ferry's relationship gives the best fit to the experimental data [54]. The relationship is given by the following equation:

$$[\eta] = \eta_0/[1 + k(\eta_0 * \psi * G_i^{-1})^n] \tag{11}$$

where, η is the viscosity (Pas s^{-1}) at the measured shear rate, η_0 is the viscosity (Pas s^{-1}) at zero shear rate, ψ is the shear rate (s^{-1}) and k, n and Gi are constants. In practice, the best empirical fit found for PET homo- and low copolymers (less than 2.0 moles per cent) is of the form:

$$[\eta] = \eta_0/[1 + 0.531(\eta_0 * \psi/1.818 * 10^5)^{0.783}] \tag{12}$$

where, η is the viscosity (Pas s^{-1}) at the measured shear rate, η_0 is the viscosity (Pas s^{-1}) at zero shear rate and ψ is the shear rate (s^{-1}).

At higher copolymer levels, the constants will change depending upon the contribution the copolymer makes to the elastic component. Typical curves for low copolymer are shown in Figure 3.24. The melt viscosity was measured on a capillary rheometer, after thoroughly drying the polymer to minimise hydrolytic degradation. Extruded samples were taken for solution IV measurements to allow correction for any molecular weight loss.

The melt viscosity at zero shear rate is correlated to molecular weight by a power function of the form:

$$[\eta_0] = K * M^n + constant + E/RT \tag{13}$$

where, η_0 is the melt viscosity at zero shear rate (Pas s^{-1}), M is the average molecular weight, E is activation energy for viscous flow, R is the gas constant and T is the temperature ($^\circ$K).

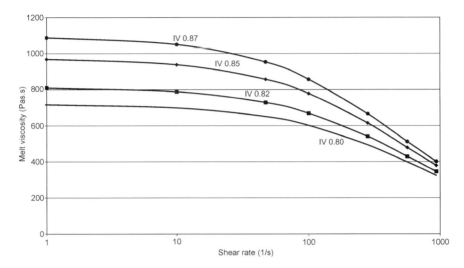

Figure 3.24 PET: melt viscosity/shear rate at 285°C. Abbreviation: IV, intrinsic viscosity.

From this equation, a correlation can be established with IV measurements. In this case, an empirical relationship can be established using the following format:

$$[\eta_0] = K' * IV^m + constant + E/RT \qquad (14)$$

where, η_0 is the melt viscosity at zero shear rate (Pas s^{-1}), IV is the intrinsic viscosity, E is the activation energy for viscous flow, R is the gas constant, T is the temperature (°K), and K' and m are constants. The relationship established for PET using ortho-chlorophenol as solvent, and used extensively by the author, is the following:

$$[\eta_0] = 0.0122 * (IV)^5 * exp. [6800/T] \qquad (15)$$

where, η_0 is the melt viscosity at zero shear rate (Pas s^{-1}), IV is the intrinsic viscosity measured in ortho-chlorophenol and T is the temperature (°K).

From these relationships, one can estimate the shear rate sensitivity of PET over a range of IVs and temperatures. In an extrusion process, the polymer melt is continuously pushed through a manifold to a die head or an injection mould. In its design, an understanding of the shear viscosity is crucial.

3.5.2 Melt flow

The melt flow characteristics of the various types of PET polymer will govern the manner to which each can be processed successfully. In an extruder, the degree of back mixing along the screw barrel and the uniformity of the product being pushed forward through the manifold, especially at any line splits, and the hot runner systems are governed by the elastic property of the polymer. Product next to the wall can have very long residence time and degradation reactions can occur that have deleterious effects on polymer properties. Excessive shear heating within an extrusion line can be minimised by adjusting the line dimension to accommodate the influence and benefit of shear viscosity. To improve consistency and to control the effects of degradation, such as acetaldehyde formation, static mixers are incorporated in the manifold.

In injection moulding applications, polymer melt is injected and immediately cooled within the mould to the desired shape. Mouldability is dependent both on the polymer properties, such as the dynamic crystallisation tendencies, and on the processing conditions [55]. In this case, the rheological and thermal properties of the polyester as well as the geometry and the temperature and pressure conditions of the mould are the most important. The flow property of the polymer is dependent not only on its melt viscosity characteristics but also on its freezing characteristics when cooled. A widely used test for mouldability is the spiral flow test [56]. In this test, the mould has a form of spiral and the polymer melt flows into the mould under pressure and freezes. The mould geometry, temperature and pressure are standardised. The test result is the length of the injected spiral, tested under standard conditions, and is expressed as a 'melt flow ratio'. The melt flow ratio is defined as the 'maximum flow path length divided by the mean thickness'. Holmes *et al.* [57] made an engineering analysis of the test and found that the ultimate length of the spiral is dependent on two sets of data, one describing the process variables and the other representing the rheological and thermal properties of the polymer.

Figure 3.25 shows the flow properties of various PET polymers made to different molecular weights. As expected, the flow characteristics are very sensitive to the melt viscosity. The influence of both pressure at injection moulding and mould temperature on the flow properties are highlighted in Figures 3.26 and 3.27, respectively. It can be seen that both have a significant influence on the flow. Table 3.6 presents the maximum melt flow ratios obtained using the DIN test procedure with a standard spiral mould for polyester made to different molecular weights. In this test, the melt temperature was 270°C, the mould temperature 40°C and the injection pressure 1000 bar. Again, the influence of molecular weight is clear, and the impact of the higher cooling efficiency in the thinner spiral product is demonstrated, with significantly lower melt flow ratios obtained. This demonstrates that the wall thickness of an injection moulding cannot be as low as is sometimes desired. The minimum wall thickness, therefore, depends on the melt flow behaviour as well as the melt flow path length and processing conditions. The injection speed and melt temperature are both of significant importance to the flow length. It is probable that the flow length will increase with increasing injection speed, since the rate of shear in the nozzle will be greater, and the relative melt viscosity could decrease. It is also possible that the melt will have a wider flow path within the

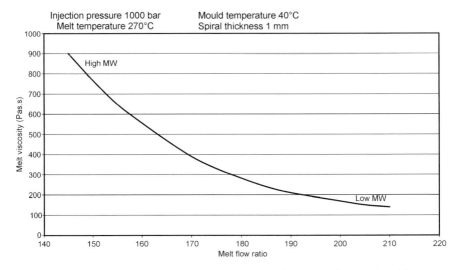

Figure 3.25 Influence of molecular weight (MW) on flow of PET into spiral mould.

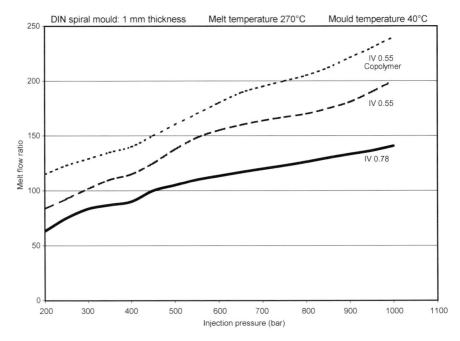

Figure 3.26 Melt flow properties of PET polymer types: influence of injection pressure. Abbreviation: IV, intrinsic viscosity.

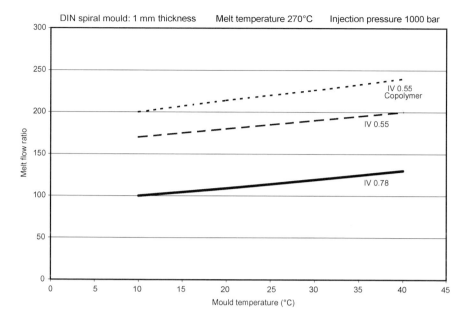

Figure 3.27 Melt flow properties of PET: effect of mould temperature. Abbreviation: IV, intrinsic viscosity.

Table 3.6 Melt flow ratio: influence of molecular weight

	Melt ratios		
		Spiral thickness	
Polyester IV	1 mm	2 mm	3 mm
0.78	145	180	220
0.75	150	205	218
0.55	200	325	>333
0.74	155	210	233
0.82	135	158	203
0.55**	240	375	>375

Melt temperature 270°C; mould temperature 40°C; injection pressure 1000 bar.
Note: **high copolymer. Abbreviation: IV, intrinsic viscosity.

cavity, since the solidified outer skin of polymer next to the mould surface is thinner.

3.5.3 Moulding shrinkage

In the injection moulding of polymeric materials such as PET, knowledge of the shrinkage characteristics of the moulding material is important to the design and integrity of the product. The understanding of the term 'shrinkage' is normally separated into 'moulding shrinkage' and 'post-moulding shrinkage'. Moulding shrinkage is defined as the difference between the dimensions of the mould cavity at $23 \pm 2°C$ and the dimensions of the moulded piece 24 h after moulding. Post-moulding shrinkage is the subsequent reduction in dimension that may occur if the part is exposed to a change in environmental conditions, such as humidity or temperature or both. The correct choice of moulding conditions, such as melt temperature, mould temperature, injection and holding pressures and the corresponding holding times, can all have a significant effect on shrinkage. The operating temperature and pressure within the mould itself can change its dimension and must also be taken into consideration.

Transparent components moulded in PET exhibit a typical linear shrinkage of about 0.45%, and the post-moulding shrinkage at temperatures below 50°C are negligible. However, for crystalline or partially crystalline moulded products, the density and thus the shrinkage of PET will change depending on the level of crystallinity attained. The level of crystallinity will be highly dependent on the presence and effectiveness of added nucleating agents, as well as the thermal history from the molten stage. It is recommended that an allowance be made at the mould design stage, to allow for minor alterations to the mould dimension following preliminary trials before the mould manufacture is completed.

3.6   Moisture uptake and polymer drying

3.6.1   Moisture level

PET polymers are highly hygroscopic in character and will absorb moisture very quickly up to saturation level. The amount of moisture absorbed is dependent on both the prevailing environmental conditions and the crystallinity level within the sample. The level of moisture in amorphous PET is found to be directly proportional to the water vapour pressure [58]. The system obeys Henry's law, as shown in Figure 3.28, and the relationship is given by the following expression:

$$S = k * p \qquad (16)$$

where, S is the solubility of water in the polymer (expressed as g/g polymer per 10 mmHg), and p is the pressure. The solubility is expressed as g/g polyester per 10 mmHg, being a more convenient unit to use in this case than the more conventional unit $(cm^3(STP)/cm^3 \cdot bar)$ normally used for expressing gas solubility.

The influence of crystallinity is to reduce the level of water present, and experiments have shown a proportional relationship between the solubility and the amorphous volume fraction at constant temperature. The level can be determined from the amorphous data by correcting for the crystallinity volume

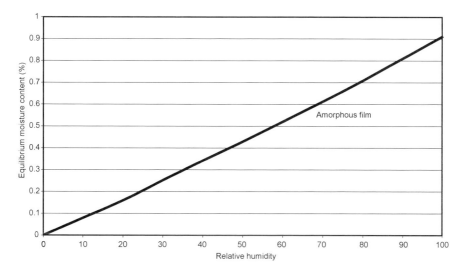

Figure 3.28 Equilibrium moisture content at various relative humidities for amorphous PET.

fraction, using the following expression:

$$S = (1 - \chi_c) * S_{amorphous} \tag{17}$$

where, S is the solubility of water in the crystalline material, $S_{amorphous}$ is the solubility of water in the amorphous state, and χ_c is the volume fraction of the crystalline structure.

The influence of temperature can be obtained from application of Van't Hoff's expression:

$$\ln[S_{sol}] = -\Delta H^\circ / RT + \Delta S^\circ / R \tag{18}$$

where, S_{sol} is the solubility of water, ΔH° is the heat of solution, ΔS° is the entropy function, T is the temperature ($^\circ$K) and R is the gas constant. For PET, the experimental data give the following equation:

$$[S_{sol}]_t = [S_{sol}]^*_{25} \, exp \, [-9900/1.98 * (1/T_t - 1/T_{25})] \tag{19}$$

where, $[S_{sol}]_t$ is the solubility (g/g of polymer per 10 mmHg pressure) at temperature, t, $[S_{sol}]_{25}$ is the solubility (g/g of polymer per 10 mmHg pressure) at 25°C, and T is the temperature in $^\circ$K. The data are summarised in Figure 3.29.

3.6.2 Polymer drying

The absorbed moisture is known to cause hydrolytic degradation of the polyester resin at high temperatures, and especially at the melt processing

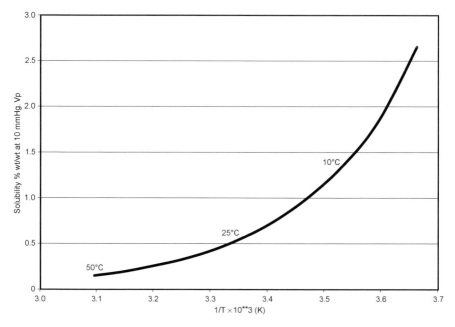

Figure 3.29 Solubility of water in amorphous PET at various temperatures.

temperatures [33, 59, 60]. The reaction can be represented by the following equation:

$$HOCH_2CH_2[OOCC_6H_4COOCH_2CH_2]_nOH +$$
$$H_2O \rightarrow HO[CH_2CH_2OOCC_6H_4COO]_pH +$$
$$HOCH_2CH_2[OOCC_6H_4COOCH_2CH_2]_qOH \qquad (20)$$

where, n, p and q are the respective degrees of polymerisation before and after hydrolysis.

To reduce the degree of hydrolytic degradation taking place, it is necessary to dry the pellets to moisture levels below 40 ppm, and preferably to less than 25 ppm, prior to melting. The effective removal of moisture is dependent upon both the level and the rate of diffusion of water through the resin. The rate of diffusion increases with temperature and, therefore, for effective drying, temperatures around 140–180°C and preferably 160°C are recommended. At temperatures above 180°C, colour forming degradation reactions are very prevalent, especially when regrind is used or more thermally unstable copolymer additives, such as polyglycols are present. There are several different drying processes currently in use, depending on the scale of operation. However, the basic operation is common to all. The pellets, if not already crystalline, will need to be carefully crystallised to reduce the tendency to stick at the higher drying temperatures. As described previously for the solid state process, the pellets are heated for a period of about 15 min, under high agitation in the presence of dehumidified air or under vacuum, to temperatures around 140°C, and allowed to crystallise. To dry, the pellets are held at the preferred drying temperature for a period of at least 4 h and preferably 6 h under vacuum or in the presence of dehumidified air with a dew-point less than −40°C, and preferably less than −60°C for high IV products. The temperature of the hot, dehumidified air should not be allowed to rise above 190°C at the air inlet to the dryer. It is recommended that the polymer pellets should be fed hot to the throat of the extruder to allow the use of milder temperature conditions, as well as to minimise the amount of shear heating required to melt the polymer in the barrel. Typical moisture levels against time for pellets dried on a column dryer are presented in Figure 3.30.

However, for highly amorphous copolyesters, such as Eastman's polyethylene terephthalate cyclohexane dimethanol copolymer (PETG) types, which do not readily crystallise, the polymer must be dried at temperatures around the Tg for periods of up to 24 h, under vacuum or using a dehumidified gas stream [61]. For these product types, care must be taken to ensure that they do not encounter high temperatures prior to entering the screw metering zone, otherwise excessive sticking will occur between the pellets and the hot surface, as well as between the pellets themselves.

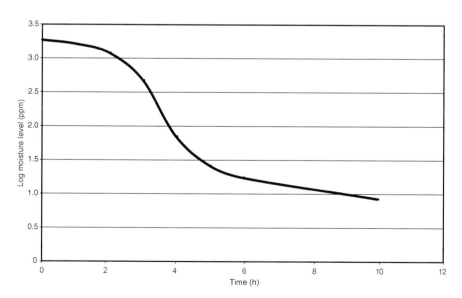

Figure 3.30 Moisture level on drying PET at 160°C.

3.7 Degradation reactions

3.7.1 Thermal and thermal oxidative degradation

The thermal degradation reactions of PET and its copolymers have been the subject of several detailed studies and reviews [62–70]. Although several schemes have been proposed, there is no conclusive evidence to support the validity of one scheme over the others. However, there are two major schools of thought regarding the nature of the mechanism involved in the degradation reaction. Marshall and Todd [65] believe that thermal degradation is initiated at chain ends, while Goodings [62] and Ritchie and co-workers [64] agree that the mechanism involves random chain scission at the ester link. In both cases, the degradation involves the methylene group. In practice, there is evidence to suggest that both mechanisms are actively involved in the degradation process. The mechanisms can be represented by the following reactions:

$$\sim OOCC_6H_4COOCH_2CH_2OH \rightarrow \ \sim OOCC_6H_4COOCH=CH_2 + H_2O$$
$$\rightarrow \ \sim OOCC_6H_4COOH + CH_3CHO$$

and/or

$$\sim OOCC_6H_4COOCH_2CH_2OH + HOCH_2CH_2OOCC_6H_4COO \sim$$
$$\downarrow$$
$$\sim OOCC_6H_4COOOCC_6H_4COO \sim +HOCH_2CH_2OCH_2CH_2OH$$
$$HOCH_2CH_2OCH_2CH_2OH \rightarrow CH_2=CHOCH_2CH_2OH + H_2O$$
$$\rightarrow 2CH_3CHO$$

Scheme 3.1

$$\sim OOCC_6H_4COOCH_2CH_2OOCC_6H_4COO \sim$$
$$\downarrow$$
$$\sim OOCC_6H_4COOCH=CH_2 + HOOCC_6H_4COO \sim$$
$$\downarrow$$
$$\sim OOCC_6H_4COOCH=CH_2 + HOCH_2CH_2OOCC_6H_4COO \sim$$
$$\downarrow$$
$$\sim OOCC_6H_4COOCH_2CH_2OOCC_6H_4COO \sim +CH_3CHO$$

Scheme 3.2

Both mechanisms can yield acetaldehyde as the main degradation product through the formation of the intermediate vinyl end entity. The presence of vinyl ends have now been confirmed in both melt and solid-state products, and this is recognised as the main precursor of acetaldehyde formation. However, the level is significantly reduced during the solid-state process, and it is almost completely eliminated from polyester copolymers used for packaging sensitive foodstuffs, such as mineral water. As the above chemical equations demonstrate, the vinyl end entity liberates acetaldehyde through reaction with either the hydroxyl end or moisture or both. There is belief in the industry that moisture has a more pronounced influence on the reaction, especially in the melt, with more released from 'wet' polymer.

The mechanism can also explain the formation of other gaseous products that are formed during extreme degradation conditions. Copolymer entities, such as diethylene glycol and polyethylene glycol, can exacerbate the situation.

Thermal-oxidative degradation is more aggressive and involves free radical mechanisms [71–73]. The oxygen reacts with the methylene groups to form peroxides, which then initiate free radical reactions. At low levels of oxygen ingress (less than 50 ppm), the gaseous by-products are identical to those obtained from the thermal degradation reaction. However, there is some evidence that the hydroxy radical is also reacting with the phenylene entity to form 2 hydroxy-terephthalic acid (N. S. Allen, Manchester Metropolitan University, UK, personal communication). This gives rise to higher fluorescence and colour. There is also an indication that, at higher temperatures, a phenylene radical is formed from which the by-product 2,4′,5 biphenyl tricarboxylic acid is produced. This is a multifunctional molecule that can give rise to localised branching. Under normal processing conditions, the level of oxygen ingress is very low (less than 5 ppm), and the level of these oxidative products is minimal (a few parts per million). In the presence of air or higher levels of oxygen, significantly more degradation occurs, giving rise to high colour as well as the release of gaseous materials, including acetaldehyde, carbon monoxide and carbon dioxide.

3.7.2 Environmental degradation

PET polymers are prone to UV degradation when left unprotected from the outside environment. The combination of UV light, moisture and air can cause severe loss of tensile properties over a relatively short period of time—depending on the prevailing environmental conditions—and the products become very weak and brittle [74–76]. For outside use, it is recommended that clear containers, sheet and film are surface coated with a UV protective layer containing a high level of UV absorber/stabiliser.

3.8 Reheat characteristics

Downstream processing of PET extruded sheet and injection-moulded preforms is dependent upon being able to successfully reheat to temperatures above Tg without imparting premature crystallinity [77–79]. The preferred heating media

is infrared radiation. PET is a strong absorber of infrared radiation in the medium
and long wave bands as shown in Figure 3.31. This strong absorption characteristic allows the polyester resin to be heated very successfully by infrared
radiation. However, in thick-walled products, this strong absorption will mask

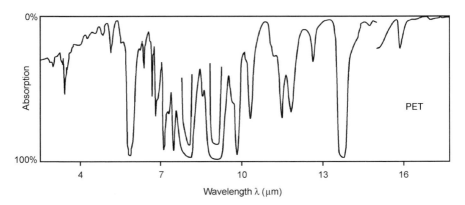

Figure 3.31 Infrared absorption spectrum of polyethylene terephthalate.

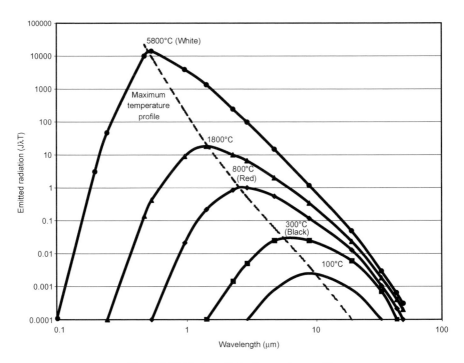

Figure 3.32 Heater radiation/temperature profile.

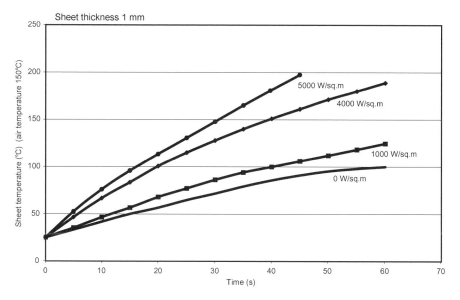

Figure 3.33 Influence of infrared heating on sheet temperature.

penetration into the core and the heating will be localised to the surface layer. The emitted radiation profiles for the various heater temperatures are shown in Figure 3.32.

The data in Figure 3.32 show that as the heater temperature increases to give white heat, the peak emission moves to a shorter wavelength. For good heating efficiency, it is best to match the emission profile with the absorption bands but, since polyester is a strong absorber of infrared radiation in the medium-to-long wavebands, care must be taken not to overheat the surface layer. It is therefore important that the temperature of the surface layer is controlled by use of air and relevant soak time, to allow the core to attain the required temperature for stretching by combination of infrared radiation and conduction. In bottle blowing technology, use is made of the short wavelength radiation (near infrared) by using quartz heaters that can penetrate more effectively into the thicker preform wall. Typical infrared heating profiles for different infrared fluxes are shown in Figure 3.33. Resin manufacturers add very effective near infrared heat absorbers to enhance the heating efficiency of their bottle resin types [80].

3.9   Gas barrier properties

In general, the gas transmission properties of polymers govern their usefulness to the packaging industry. A number of factors influence the permeability of gas through a polymer matrix. The permeability of simple gases, such as He, N_2, O_2 and Ar, in a glassy polymer is dependent upon both the solubility of the gas in the matrix and the diffusion through it. The relationship is expressed by the following equation:

$$P = S * D \tag{21}$$

where, P is the permeability of the gas, S is the solubility of the gas and D is the diffusivity of the gas in the polymer matrix.

The solubility of a gas in glassy polymers, such as PET, is sensitive to the chemical structure of the polymer and is best represented by a 'dual mode' type of sorption that involves both Henry and Langmuir sorption characteristics [81–85]. This model assigns a different mobility to the two gas populations in the gas/glassy polymer systems. In this case, one population is assumed to be molecularly dissolved and is referred to as the Henry's law mode, while the second population is absorbed into 'microvoids' and follows a Langmuir type isotherm. The solubility function is represented by the following equation:

$$S = k_D * P + C_H * bP/(1 + bP) \tag{22}$$

where, $k_D * P$ is the Henry (S_D) solubility function, and $C_H * bP/(1 + bP)$ is a Langmuir (S_H) type function, and P is the gas pressure, C_H is the Langmuir capacity constant, b is the Langmuir mode sorption affinity parameter, and k_D is a constant. As described previously for water, the solubility is temperature sensitive. The change in Henry (S_D) solubility can be represented by a van't Hoff type expression, as follows:

$$\ln[S_{sol}] = -\Delta H^\circ/RT + \Delta S^\circ/R \tag{23}$$

However, if the gas is showing strong Langmuir type characteristics, the temperature function will be more complex.

Diffusivity of simple gases depends upon the motion of the polymer molecules within the polymer matrix. For simple gases, the interaction with the polymer is weak and therefore the diffusion coefficient is independent of the concentration of the gas. The gas is seen to fill holes within the matrix, and diffusion occurs by a series of jumps from one hole to another, until the gas passes through the polymer [86]. In general, the diffusion of gases is regarded as a thermally activated process that can be expressed by an equation of the Arrhenius type:

$$D = D_0 * exp.(-E_D/RT) \tag{24}$$

where, D_0 and E_D are constants for a particular gas.

The activation energy (E_D) is the energy required by the gas molecule to jump from one hole to another. It is clear that larger molecules will require more space and thus the holes are bigger. In this case, the activation energy is larger for the diffusion of the bigger molecule and the diffusivity is smaller.

The permeability of the gas in a glassy polymer system will be influenced, therefore, both by the Henry's solubility mode and the Langmuir type solubility in the 'microvoids'. The permeability expression can be written as follows:

$$P = S_D * D_D + S_H * D_H \tag{25}$$

which can be rewritten as:

$$P = k_D * D_D[1 + F * K/(1 + bP)] \tag{26}$$

where, D_D is the Henry type diffusivity, D_H is the Langmuir type diffusivity, $K = C_H * b/k_D$, and F is the ratio of D_H to D_D.

The parameter F equals the ratio of the diffusivity of the gas in the Langmuir population, D_H, to the diffusivity in the Henry's law population, D_D. In the limit of $F = 1$, the permeability is strongly dependent on the gas pressure, whilst in the limit of $F = 0$, the permeability is independent of the gas pressure and is given by the following expression:

$$P = S_D * D_D \tag{27}$$

For PET polymer systems, the permeability function (P) in the glassy state for simple gases, such as He, O_2, N_2 and Ar, is independent of the gas pressure and follows a simple Henry's law dependence. For CO_2, on the other hand, the permeability shows a dependence on the gas pressure and the more complex 'dual mode' sorption mechanism must be applied [84, 85].

The influence of crystallinity and orientation is to reduce the level of amorphous material available for diffusion; in both cases, the highly ordered structures in these regions obstruct the movement of the molecules and increase the average length of the paths the gas has to travel [86]. In general, it is assumed that the crystalline domain is a complete barrier to the flow of gas, and as a first approximation the following relationship can be used:

$$D = D_a(1 - \chi_c) \tag{28}$$

where, D_a is the diffusivity in the amorphous polymer and χ_c the degree of crystallinity.

A similar effect is also seen on the solubility, with the highly ordered crystalline type structures reducing the solubility by a similar factor. The solubility function can be represented as follows:

$$S = S_a(1 - \chi_c) \tag{29}$$

where, S_a is the solubility in the amorphous polymer.

The permeability in both crystalline and oriented structures is therefore represented by the following expression:

$$P_{sc} = P_a(1 - \chi_c)^2 \tag{30}$$

where, P_{sc} and P_a are the permeability in the semi-crystalline/oriented and the amorphous samples, respectively.

In addition to morphological contributions, the barrier properties are also influenced by the chemical structure, and to additives that can create more tortuosity in the flow path through the polymer matrix. Copolymer additives that can increase the cohesive energy between the molecule chains, such as the chloroterephthalic acids and hydroxyterephthalic acids, improve the gas barrier properties, whilst additives such as cyclohexane dimethanol give poorer barrier properties. Additives, such as isophthalic acid, that can influence the freedom of molecular movement around the aromatic paraphenylene entity, as well as molecules, such as naphthalene 2,6 dicarboxylic acid, that give increased chain rigidity, show positive benefits [87, 89] (also, L. Monnerie, Laboratoire de Physicochimie et Macromoleculaire associe au CNRS, ESPCI, Paris, personal communication). The incorporation of additives that can have an influence on the molecular chain mobility, such as plasticisers [88] and anti-plasticisers [89] (also, L. Monnerie, personal communication), will also produce a positive or negative effect on the barrier properties depending on their influence on the chain molecules. In most cases, the addition of plasticising agents will have a negative effect, especially at high concentrations. Interestingly, it has been shown that the addition of small quantities of moisture will initially improve the barrier property, presumably through an increase in the cohesive bondage between the chain molecules. On the other hand, the addition of anti-plasticisers, such as tetrachlorophthalic dimethyl ester, dimethyl naphthalate and dimethyl terephthalate, can have a very positive effect. In all cases, the additives are suppressing the β transition as measured by dynamic mechanical analysis.

Interestingly, in dielectric relaxation measurements, the β transitions are not suppressed. Monnerie and co-workers [89] have used these two techniques in conjunction with solid state nuclear magnetic resonance (NMR) spectroscopy to resolve which chain motions are involved. They conclude that these additives hinder the rotational movement of the aromatic phenylene units, and therefore the overall freedom of the molecular chain. The results show that the effect is to reduce the diffusivity of the gas through the matrix. Yang and Jean [90] have shown a correlation between the free-volume hole size as measured by positron annihilation lifetime (PAL) technique, which supports the proposed theories on gas transport in various polyester matrixes.

Other methods of improving the barrier properties include the addition of other high barrier resins, such as MXD.6, polyesteramides and modified

Table 3.7 Gas barrier properties of PET and copolymers

Polymer type	Gas barrier properties at 30°C			
	O_2 cc·mil/100 inch^2· 24 h·Atmos	CO_2* cc·mil/100 inch^2· 24 h·Atmos	N_2 cc·mil/100 inch^2· 24 h·Atmos	H_2O g·mil/100 in^2· 24 h·Atmos
PET homo/low copolymer				
Amorphous	9	54	2.4	4
Oriented	5	40		2.5
Crystalline	4	33	1.2	2
PET medium copolymer				
Amorphous				
10% IPA	8	40		
10% PETN	8.2	53		
20% IPA	6.5	35		
20% PETN	8	47.5		
PET high copolymer				
Amorphous				
PETG	19	109	5.5	8
50% IPA	4	16		
50% PETN	6.3	39		
100% IPA	2.3	6		
100% PEN	3	18	0.4	

Note: *Values for CO_2 permeation are dependent on gas pressure: the values quoted are for low pressure ($P_{tending} \rightarrow 0$). Abbreviations: PET, polyethylene terephthalate; IPA, isophthalic acid; PETN, polyethylene terephthalate naphthalene 2,6 dicarboxylate copolymer; PETG, polyethylene terephthalate cyclohexane dimethanol copolymer; PEN, polyethylene naphthalene 2,6 dicarboxylate.

polyvinyl alcohol, the addition of platelets such as clay and liquid crystal polymer types, and the use of specific scavengers, especially for oxygen. A summary of the more relevant data is presented in Table 3.7.

3.10 Amorphous polyesters

Several different polyester resin types are available in the marketplace for use in the packaging industry. Each resin type will have been developed for specific injection moulding and/or extrusion applications, including their downstream use. The resin types are usually classed as homo, low, medium and high copolymer, and within this classification there will be other types, including branched polymers and blends.

3.10.1 Homopolymers

The term homopolymer is a misnomer for this classification, since all the polymers will contain generated and possibly added diethylene glycol (DEG) at significant levels, and should therefore be classed as copolymers. Homopolymer is the major formulation used for fibre applications and, as such, was the original formulation used for packaging applications. Polymer types with molecular weight up to 20,000 and DEG levels up to 2.5% wt/wt are still being used to make standard film products. Various formulations containing generated DEG have worldwide food approval; however, polyester made with the addition of DEG is currently undergoing the FDA approval procedures.

During the development of the ISBM process in the late 1970s, both film and fibre formulations were solid-state polymerised to an IV of around 0.75, and used successfully to produce two-piece CSD bottles. The containers had excellent tensile properties as well as a good barrier to CO_2, and were used successfully to bottle CSDs. Although this formulation has been superseded by copolymer formulations for use in the production of one-piece bottles, polymers are still being used successfully to make custom containers, such as peanut butter jars. In this application, clarity and the ability to hold container integrity during vacuum filling is critical. The resin is also used to make IM mould articles, including flat plate, as used for car number plates. However, for successful use in these applications, the molecular weight must be matched to the mould flow characteristics, as defined in the spiral mould test. Resin manufactures have developed an understanding of cleaner catalyst systems that do not cause stress crystallisation, and these are used to produce polymer grades that give good integrity and high clarity in the containers. Homopolymer is used successfully in sheet extrusion, which can be formed into thin-walled containers, display packaging and clear boxes. In the latter application, the resin has a clear advantage over competitive material in that it does not show stress whitening in the fold region. The sheet is usually co-extruded, with a copolymer layer on both surfaces containing surface modifying agents to reduce friction and improve de-nesting, as well as to improve adhesion and printing characteristics.

Homopolymers are also the most favoured resin types for heat setting and use in hot-fill applications [91]. In this case, the choice is governed by the thickness of the initial moulding. However, for mould thicknesses greater than about 4 mm, the use of copolymer is recommended to reduce the tendency for stress crystallisation to occur in the mouldings. High molecular weight (IV > 1) homopolymers, including some that are slightly branched, are used successfully in extrusion blow moulding (EBM) applications. The melt viscosity of these resins is highly sensitive to shear rate, and this enables the use of mild extrusion conditions (high shear) and high melt strength post-die (low shear). For successful use in this application area, it is essential to maintain the IV above 1.0 to reduce excessive sag in the parison prior to moulding.

3.10.2 Low copolymers

The low copolymer is the biggest market segment for PET polymer technology, and includes both the ISBM CSD and mineral water bottle applications. The move to low copolymer became essential when the market demanded one-piece bottles that have improved both bottle aesthetics and ease of recycling. To accommodate this shift in bottle design, a bottle base had to be created that could hold high pressures, in addition to keeping its shape and integrity under pressure for long periods. In this case, the standard champagne base could not hold the normal CSD bottle pressures without the use of excessive resin weight in its structure. The bottle base was redesigned into a petaloid shape that usually has five lobes and five crevasses. This base shape has areas of high stress that can give a catastrophic blow-out during manufacture, especially if the IV is low, and can lead to stress cracking during use [92–95]. To produce bottles with good base integrity, it is essential that the injection moulding and blow moulding operations be controlled to produce stress-free preforms and bottles.

During this development, the machinery manufacturers produced bigger and faster machines to handle the increased demand. To satisfy these new requirements, the resin manufacturers moved to copolymers that are more forgiving in both the injection moulding and bottle blowing operations. The level of copolymer used is 1–2.5% wt/wt, depending on the copolymer additive and overall formulation as designed specifically by the resin manufacturer to meet customers' requirements. The favoured copolymer additive is cyclohexane dimethanol or isophthalic acid. All resins for these applications are solid-state polymerised to standard IVs of 0.82 or 0.84. The higher IV resin is used to maintain molecular weight when the environmental conditions are unfavourable, such as the very warm and humid conditions that prevail in many countries during the summer months. During the reheat operation prior to blow moulding, the preforms are heated by quartz infrared heaters to temperatures around $10°C$ above Tg. It is crucial to have a uniform temperature profile across the wall to ensure that even stretching occurs during blowing in order to give maximum orientation and low creep properties in the wall [96]; also that the resin flows uniformly to the base without causing high stress areas in the crevasses.

To enhance the rate of heating and to achieve improved temperature uniformity, resin manufacturers, such as Dupont, now add specific near infrared absorbers to their resin formulation [80]. These absorbers allow thicker wall preforms (greater than 4 mm) to be used more readily in bottle manufacture, as well as giving more freedom in bottle design. The preferred absorbers are fine metallic particles and/or carbon black, and these are used at the ppm level. All the resins used satisfy worldwide food regulations. Low acetaldehyde resins are required for the manufacture of bottles for sensitive foodstuffs, such as mineral water. These resins are produced under conditions that remove the vinyl end precursor during the solid-state operation, and may also contain additional stabilisers. However, as described previously, to obtain low aldehyde levels in the bottle it is important that good drying and mild extrusion conditions are used. It is also important to give attention to the temperature profile and flow characteristics in the hot runner system, since they can both contribute to the variance seen in the acetaldehyde levels in the preforms. The use of a static mixer in the IM manifold can reduce the degree of variability. The addition of copolymer additive can have a significant influence on the stretching characteristics of the polyester polymer and blowing conditions should be adjusted accordingly. Typical data are presented in Figure 3.34.

To produce thick-walled containers, as required for the refillable CSD bottles, much thicker walled preforms are needed. To meet these requirements, a resin with a higher copolymer level is used. For this application, the copolymer additive is used at a level of around 2.3–3.0% wt/wt, depending upon the additive being used and the overall manufacturers' formulation. Again the most favoured copolymer additives are CHDM and isophthalic acid. The resin formulation must also produce bottles that are robust to the subsequent caustic wash cycle at 60°C, and must survive at least 20 trips. To survive the wash cycle without loss in integrity and especially without enhancing any subsequent environmental stress cracking, it is critical that, in the manufacture of the preform and bottle, conditions are chosen to give an adequate wall thickness to the base and not to create excessively high stress areas. To minimise stress levels in the champagne base, it is important that its design ensures that all potentially sharp angles are

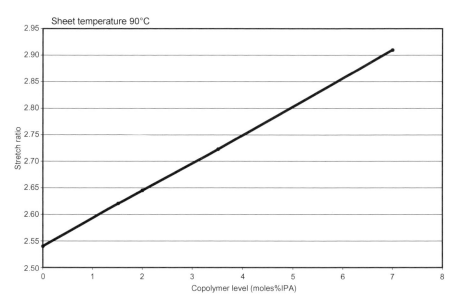

Figure 3.34 Influence of copolymer level on stretch ratio. Abbreviation: IPA, isophthalic acid.

curved. To maintain integrity in use, it is important that the IV is retained at 0.80 and that the level of carboxyl ends is kept low. To achieve this, it is essential that drying conditions are set to achieve a moisture content below 10 ppm and that mild conditions are used on both the extruder and the manifold/hot-runner system. For applications that need higher wash temperatures (around 70°C), such as containers for mineral water, a higher level of heat setting is required or additives, such as naphthalene 2,6 dicarboxylic acid, are used to enhance the Tg.

Standard grade and refillable bottle grade resins are both used for sheet extrusion, and are especially recommended for thick sheet applications. As is the case with IM, the thicker sheet is more prone to crystallise during casting and calendering. The use of copolymer reduces the rate of crystallisation during cooling and allows the production of thick sheet (up to 5 mm). Again, the IV should be maintained at 0.80 to give melt strength and adequate support during extrusion and casting.

Copolymer resins can be used successfully in many IM applications. As has been described previously, it is crucial for these applications that polymer with the optimum melt viscosity for both melt flow and product integrity, including clarity and strength, is chosen. Copolymers made to various IVs from 0.5 upwards are available for use. The spiral test data in combination with mould dimension and injection route will give a good indication of suitability for use.

3.10.3   Medium copolymers

Several PET copolymers with additive levels in the range 4.5–13% wt/wt are available as speciality resins. These polymers have a much lower tendency to crystallise and are more difficult to solid-state polymerise, as well as drying from the amorphous state. However, they do have specific benefits over standard resins, depending on the copolymer additive used. Their main benefits are seen in the following areas: improving gas barrier properties, increasing or decreasing Tg as may be required for particular end use, retarding crystallisation, increasing shear sensitivity and producing high clarity thick-walled products. Both isophthalic acid and naphthalene 2,6 dicarboxylic acid copolymers show positive improvements in gas barrier property, while naphthalene 2,6 dicarboxylic acid copolymers also show an increase in Tg. On the other hand, the use of diethylene glycol, polyethylene glycols and aliphatic diacids, such as adipic acid, will give a lower Tg and poorer gas barrier properties. While most of the copolymer additives will retard crystallisation, the most effective are CHDM and isophthalic acid. Again, isophthalic acid and CHDM are very effective in increasing the shear sensitivity, especially at low IV, and are used as polymer modifiers for IM applications. Various copolymer types are available for this use and, as has been mentioned previously, the polymers can be matched to end-use requirements through knowledge of the spiral test data, IM conditions, including mould dimension and injection route, and end-use characteristics. High clarity polymers containing very low catalyst residues are also available to produce thick-walled containers. These same copolymers can be used in extrusion to produce high clarity sheet.

The situation as regards food contact is not as clear-cut as that for a low level copolymer. In this case, it is recommended that each formulation be reviewed with relevant expertise to obtain opinion and guidance on the way forward.

3.10.4   High copolymers

The most well-known high copolymers are based on CHDM, and are referred to as PETG. These are made by Eastman and are used in both IM and extrusion applications. They are amorphous copolyesters, containing about 35% wt/wt of copolymer component. However, because of their amorphous nature, they need to be carefully dried at low temperatures (around $90°C$) for long periods (around 24 h), under vacuum or in the presence of dehumidified gas. They are made to a range of molecular weights and are renowned for high clarity. Copolyesters based on other additives, such as naphthalene 2,6 dicarboxylic acid, isophthalic acid and cyclohexane dicarboxylic acid, are available for more specific end uses, including improved gas barrier, increased Tg and high rigidity. The high copolyester grades are used as the surface layer in co-extrusion with both homopolymer and low copolymer, to give improved melt-adhesion and printing.

3.11   Crystalline polymers

High crystallinity is required in polyester products to give dimensional stability and rigidity at high temperatures. To obtain satisfactory rates of crystallisation in the mould or after forming, a suitable nucleating agent must be incorporated into the polymer. The type of additive and level must be matched to the downstream process requirements. In both extrusion and injection moulding processes, it is important that the nucleant does not cause problems during the initial melt cooling process. In both the injection and casting processes, this can affect the product integrity, with localised shear crystallisation occurring in the mould and the formation of poor quality, partially crystalline sheet during casting. As a consequence, poor impact performance is obtained as well as misshaped articles. To obtain adequate crystallinity and good product integrity, it is therefore essential to match the level of nucleant, and so the crystallisation rate, to the needs of the downstream process.

The type of nucleant will depend upon when the crystallisation should occur, such as on cooling as in the injection moulding process, or on reheating as in the forming process. For the former, the type of nucleant is usually of a particulate kind, such as talc, or a cation, such as the sodium ion [97]. In both cases, a critical mass of additive is required to produce sufficient nuclei to give a beneficial effect on the rate of crystallisation, as well as efficiency in the dispersion of the particulate to its optimum size (around 1 µm). For injection moulding applications, the optimum level is found to be around 500 ppm of sodium and around 0.2% of talc. However, these levels will vary depending on the polymer formulation and also on the level of branching that is present. To achieve the required level of crystallisation, the moulding is held for a finite time (about 20 s), in a hot mould at around 140°C. Care must be taken during cooling post-mould, to ensure that no stress areas are frozen in, which could lead to the article distorting in use. On the other hand, for extrusion into sheet form for thermoforming, it is best to use nucleating agents that are less effective in the cooling mode but very effective on heating. This type of nucleation is best achieved using a stress-induced mechanism, as is obtained when low melting point crystalline polymers, such as polyethylene and polypropylene, are finely dispersed at a level of 1–3% wt/wt into the polymer matrix. On reheating the sheet, the crystalline polymer will remelt, inducing highly localised finite stress areas, which induces crystallisation in the polyester. This is a very effective method for inducing crystallisation into thermoformed polyester trays, which are used for microwave applications. In the thermoforming process, the sheet is heated in an infrared oven to a temperature where nucleation and crystallisation are imminent, and formed as crystallisation is occurring. Again, the formed article is held in the mould for a finite time to develop maximum crystallinity (around 10 s). Care must be taken during cooling to ensure that all the moulds are cooled evenly and without build-up of localised stress areas that could affect the integrity in use.

The molecular weight of the polymer has a strong influence on the rate of crystallisation, and the level of additives must be optimised for each IV.

3.12 Polymer blends

Polymer blends can give beneficial effects that are not otherwise readily achieved. The addition of low melting point polymers enhances crystallisation from the glassy state; however, it is also observed that this creates voids and some localised orientation in the polyester, which can enhance its impact properties. The use of an elastic/rubbery polymer type, such as modified acrylic resins, as additive within the blend, at levels greater than 5% wt/wt, causes significant improvement in impact performance. Blends made with superior barrier polymers, such as polyvinyl alcohol copolymers and MXD.6, can give enhanced gas barrier properties. Some of these benefits are discussed in detail in Chapter 4.

3.13 Applications

The properties of PET that have been described have established it as an ideal polymer for use in packaging applications. It is already established as the preferred polymer for many applications, such as pressurised containers for CSD and mixer drinks, custom containers for many food and non-food applications, high quality injection-moulded containers for outlets such as cosmetic and perfumery, and sheet extrusion for use in both clear and crystalline containers, as well as in clear display packaging. PET will continue to build a market share in these areas at the expense of existing materials. It also has strength in many thin film applications, which are described in detail in Chapter 5. New and innovative developments that are currently being worked on across the whole manufacturing and processing chain, such as improved barrier and cast foam products, will allow newer end-use applications to be identified.

3.14   Trends

The interest in polyester polymers will continue across all current market sectors for the foreseeable future. The major growth and opportunities will be in the packaging sector, where it is now recognised as the preferred material for several food and non-food applications. The use of plastic materials such as PET is now acceptable to the younger generation, and it is envisaged that attempts will be made to replace more of the conventional packaging materials, such as steel, aluminium and glass, over the next 10 years. To achieve this potential growth, intrinsic properties, such as gas barrier and hot fill, must be improved. Work within the industry is focused towards achieving these goals. There is much interest in bottling beer in plastic containers, such as PET bottles. Several different routes to achieve high barrier are under investigation. Plastic bottles and containers that can retain their integrity at high temperatures are also in demand for use in hot-fill and oven pasteurisation applications. Interest in non-packaging applications, such as engineering plastics, foam and cast sheet, will allow further developments to occur, such as in blends, nucleation, filled grades, impact modified, high clarity and UV stability, to meet the end-use requirements. These developments will increase the opportunities for use in packaging applications. The consumer will expect all plastic material to be more environmentally friendly both during manufacture and in use. Although PET is acknowledged to be recyclable, a very high proportion is still not reused. This is discussed in more detail in Chapter 11. New, cleaner manufacturing processes are being developed and used, making the whole process more cost effective and more acceptable to the environment.

3.15   Global

The versatility of PET makes it a highly acceptable packaging material world-wide. The presence of polyester fibre plants and technology across all continents has allowed the material to be produced and used in newer plastic end-use applications without significant initial investment. Growth in the US and Western Europe will gradually slow down as the CSD market sector reaches maturity. However, significant growth will continue in the Eastern Bloc Countries, China and South America. Newer markets, such as the beer market, will open further opportunities worldwide and especially in the US and Western Europe.

References

1. Whinfield, R. and Dickson, J.T. UK Patent 578,079.
2. Whinfield, R. (1946) *Nature*.
3. Ludewig, H. (1971) *Polyester Fibres Chemistry and Technology*, Wiley Interscience, New York.
4. Polyester: 50 Years Achievement. The Textile Institute, Manchester, M3 5DR, UK.
5. Wyeth, N.C. and Roseveare, R.N., N.C. UK Patent 1341845-48.
6. Monsanto Corp., USA. UK Patent 1,226,154, US Patent 3,551,386.
7. Shuya Chang, Ming-Fu Sheu and Shu-May Chen (1983) *J. Appl. Polymer Sci.*, **28**, 3289.
8. Wu Rongrui, Huang Guanbao, Zhang Dashu and Li Yanli. *Proceedings of the 2nd International Conference on Man-Made Fibres*, Beijing, China, 26 November, 1987.
9. Jabarin, S.A. and Lofgren, E.A. (1986) *J. Appl. Polymer Sci.*, **32**, 5315.
10. Dupont Polyester Presentation. Polyester '99 World Congress, Zurich, Switzerland, December 1999.
11. Technical Presentations. Polyester '97 World Congress, Zurich, Switzerland, November 1997.
12. Technical Literature, Zimmer AG, Germany; Inventa AG, Germany.
13. Technical Literature, Dupont Polyester, Wilmington, Del., USA.
14. Technical Literature, Sinco Eng. SpA, Italy; Buhler Ag, Switzerland; Zimmer AG, Germany; Hosokawa Bepex Corp., USA.
15. Herman, K., Gerngross, O. and Abitz, W. (1930) *Z. physik. Chem.*, **B10**, 371.
16. van Krevelan, D.W. (1976) *Properties of Polymers and the Estimation and Correlation with Chemical Structures*. 1st Edition, Elsevier Scientific Publishing, Oxford, New York and Amsterdam, p. 16.
17. Hearle, J.W.S. and Greer, R.E.J. (1970) *Text. Progress*, **2**, 4.
18. Keller, A. (1957) *Phil. Mag.*, **8** (2), 1171.
19. Schlesinger, W. and Leeper, H.M. (1953) *J. Polymer Sci.*, **11**, 203.
20. Houseman, R. (1950) *Physik*, **128** (1), 465.
21. Jones, K.M. (1992) 'Plastic South Africa' Presentation given to the Plastic Institute, South Africa, 18.
22. Bonnebat, C., Roullet, G. and de Vries, A.J. (1979) SPE 37th Annual Conference, New Orleans, 273.
23. Jabarin, S.A. (1984) *Polymer Eng. Sci.*, **24**, 376.
24. Thompson, A.B. and Marshall, I. (1954) *J. Appl. Chem.*, **4**, 145.
25. Heffelfinger, C.J. and Schmidt, P.G. (1965) *J. Appl. Polymer Sci.*, **9**, 2661.
26. Bonnebat, C., Beautemps, J. and de Vries, A.J. (1977) *J. Polymer Sci.*, **58**, 109.
27. Jabarin, S.A. and Lofgren, E.A. (1986) *Polymer Eng. Sci.*, **26**, 620.
28. Le Bournelle, G., Monnerie, L. and Jarry, J.P. (1986) *Polymer*, **27**, 856.
29. Fuller, W., Mahendrasingam, A., Martin, C., *et al.* (2000) *Polymer*, **40** (20), 5553. See also *Polymer*, **41** (3), 1217; *Polymer*, **41** (21), 7793 and *Polymer*, **41** (21), 7803.
30. Leiger, F.P. (1985) *Plastic Eng.*, **47**.
31. Mark, H. '*Der fest Korper*' (ed. R. Sanger), Hirzel, Leipzig, Germany.
32. Houwink, R. (1940) *J. prakt. Chem.*, **15**, 157.

33. Ravens, D.A.S. and Ward, I.M. (1961) *Trans. Fara. Soc.*, **57**, 150.
34. Moore, Jr., L.D. (1960) Cleveland ACS, Meeting 4, Vol. 1.
35. Jabarin, S.A. *PET Technology: Properties and Processing*, Polymer Institute, University of Toledo, Toledo, Ill., USA, Chapter 4, p. 16.
36. Jabarin, S.A. and Balduff, D.C. (1982) *J. Liq. Chromatogr.*, **5**, 1825.
37. Jabarin, S.A. *PET Technology: Properties and Processing*, Polymer Institute, University of Toledo, Ill., USA, Chapter 4, p. 19.
38. Cooper, D.R. and Semlyen, J.A. (1975) *Polymer*, **14**, 185.
39. Pebbles, J.R., Huffman, M.W. and Ableh, C.T. (1969) *J. Polymer Sci.*, **7**, 479.
40. Goodman, I. and Nesbitt, B.F. (1960) *J. Polymer Sci.*, **48**, 423.
41. Mifume, A., Ishida, S., Kobayashi, A. and Sakajiri, S. (1962) *J. Chem. Soc. Japan (Ind. Chem. Sec.)*, **65**, 992.
42. Groenickx, G., Reynaers, H., Berghams, H. and Smets, G. (1980) *J. Polymer Sci. Polymer Physics Ed.*, **18**, 1311.
43. Groenickx, G. and Reynaers, H. (1980) *J. Polymer Sci. Polymer Physics Ed.*, **18**, 1325.
44. Alfonso, G.O., Pedemonte, E. and Ponzetti, L. (1979) *Polymer*, **20**, 104.
45. Chixing Zhou and Clough, S.B. (1988) *Polymer Eng. Sci.*, **28**, 65.
46. Sweet, G.E. and Bell, J.P. (1972) *J. Polymer Sci.*, (Pt A-2) **10**, 1273.
47. Ward, I.M., Monnerie, L. and Maxwell A.S. (1998) *Polymer*, **39** (26), 6851.
48. Davies, G.R. and Ward, I.M. (1972) *J. Polymer Sci.*, **10**, 1153.
49. Lawton, E.L. (1985) *Polymer Eng. Sci.*, **25**, 348.
50. Lin, C.C. (1983) *Polymer Eng. Sci.*, **23**, 113.
51. Pearson, G.H. and Garfield, L.J. (1978) *Polymer Eng. Sci.*, **18**, 583.
52. Axtell, F.H. (1987) A Study of the Flow Properties and Processability of Thermoplastic Polyester. Thesis, Loughborough University, Loughborough, UK.
53. White, J.L. and Yamane, H. (1985) *Pure Appl. Chem.*, **57**, 1441.
54. Ferry, J.D. (1942) *J. Am. Chem. Soc.*, **64**, 1330.
55. Lee, Y., Wang, T. and Chang, S. (1983) *J. Chinese Inst. Chem. Eng.*, **14**, 347.
56. van Krevelan, D.W. (1976) *Properties of Polymers and the Estimation and Correlation with Chemical Structures*. 1st Edition, Elsevier Scientific Publishing, Oxford, New York and Amsterdam, p. 491.
57. Holmes, D.B., Esselink, B.P. and Beek, W.S. (1966) *Processing polymers to products*. Proc. Intern. Congress Amsterdam, 't Raedthuys, Utrecht, 1967, p. 167.
58. Yasuda, H. and Stannett, Y. (1962) *J. Polymer Sci.*, **57**, 907.
59. McIntyre, J.E. (1985) *Fibre Chemistry*, (eds. M. Lewin and M. Pearce), Chapter 1, Marcell Dekker, New York.
60. McMahon, W. (1959) *J. Chem. Eng. Data.*, **4**, 57.
61. PETG Technical Literature, Eastman Chemicals Corp., Kingsport, Tn, USA.
62. Goodings, E.P. (1961) *Soc. Chem. Ind., (London) Monograph*, **13**, 211.
63. Buxbaum, L.H. (1968) *Angew. Chem. Int. Ed.*, **7**, 182.
64. Allan, R.J.P., Iengar, H.V. and Ritchie, P.D. (1957) *J. Chem. Soc.*, (London) p. 2107.
65. Marshall, I. and Todd, A. (1953) *Trans. Fara. Soc.*, **49**, 67.
66. Villain, F., Condane, J. and Vert, M. (1994) *Polymer Deg. Stability*, **43**, 431.
67. Jabarin, S.A. and Lofgren, E.A. (1984) *Polymer Eng. Sci.*, **24**, 1056.
68. Zimmerman, H. and Kim, N.T. (1980) *Polymer Eng. Sci.*, **20**, 680.
69. Aalbers, J.G.M. and van Houwelingen, G.D.B. (1983) *Fresinus Z. Anal. Chem.*, **314**, 472.
70. Khemani, K.C. (2000) *Polym. Degrad. Stab.*, **67** (1), 91.
71. Yoda, K., Tsuboi, A., Wada, N.T. and Yamadera, R. (1970) *J. App. Sci.*, **14**, 2357.
72. Nealy, D.L. and Adams, L.J. (1971) *J. Polymer Sci.*, **9**, 2063.
73. Zimmerman, H. and Becker, D. (1973) *Faserforshung u. Textil.*, **24**, 479.
74. Allen, N.S., Edge, M., Mohammadian, M. and Jones, K.M. (1991) *Eur. Poly. J.*, **27**, 1373.

75. Allen, N.S., Edge, M., Mohammadian, M. and Jones, K.M. (1991) *Textile Res. J.*, **61**, 690.

76. Allen, N.S., Edge, M., Mohammadian, M. and Jones, K.M. (1994) *Poly. Deg. Stability*, **43**, 229.

77. Rosato, Donald V. and Rosato, Dominick V. (eds.), *Blow Moulding Handbook*, J. Wiley and Sons, London, UK, 1989.

78. Technical Literature, Sidel Corp., Le Havre, France.

79. Kim, S.L. *Effect of Equilibrium Time on the Properties of Reheat Blown 2 l PET Bottles*, SPE ANTEC Conference, Boston, USA, 1983, pp. 804-908.

80. Technical Literature 'Laser[++]', Dupont Polyester, Wilton, Middlesbrough, UK.

81. van Krevelan, D.W. (1976) *Properties of Polymers and the Estimation and Correlation with Chemical Structures.* 1st Edition, Elsevier Scientific Publishing, Oxford, New York and Amsterdam, p. 403.

82. Michaels, A.S., Vieth, W.R. and Barrie, J.A. (1963) *J. Appl. Physics*, **34**, 13.

83. Vieth, W.R., Alcalay, W.R. and Frabetti, A.J. (1964) *J. Appl. Polymer Sci.*, **8**, 2125

84. Koros, W.J. and Paul, D.R. (1978) *J. Polymer Sci.*, **16**, 2171.

85. Brolly, J.B., Bower, D.I. and Ward, I.M. (1996) *J. Polymer Sci.*, **34**, 769.

86. Perkins, W. (1988) *Polymer Bull.*, **19**, 397.

87. Light, R.R. and Seymour, R.W. (1982) *Polymer Eng. Sci.*, **22**, 857.

88. Weinhold, S. Technical Presentation, Polyester '97 World Congress, Zurich, Switzerland. November 1997, Figure 1, p. 3.

89. Monnerie, L., Maxwell, A.S., Laupretre, F. and Ward, I.M. (1998) *Polymer*, **39** (26), 6835.

90. Yang, H.E. and Jean, Y.C. (1998) *Medical Plastic and Biomaterials*, January.

91. Maruhashi, Y. and Asda, T. (1992) *Polymer Eng. Sci.*, **32**, 481.

92. Tekkanat, B., McKinney, B.L. and Behin, D. (1992) *Polymer Eng. Sci.*, **32**, 393.

93. Jabarin, S.A., Lofgren, E.A. and Mukherjie, S. (1994) ANTEC conference, 3271.

94. Jabarin, S.A. and Lofgren, E.A. (1992) *Polymer Eng. Sci.*, **32**, 146.

95. Moskala, E.J. (1996) Bev-Pak Americas '96 Conference, Florida, USA, 8-1.

96. Gray, F.P. (1996) PET Oriented Containers: Creep Behaviour, presented at Bev-Pak Americas '96, Florida, USA.

97. Aharoni, S.M., Sharma, R.K., Szobota, J.S. and Vernick, D.A. (1983) *J. Appl. Polymer Sci.*, **28**, 2177.

4 Barrier materials and technology

David W. Brooks

4.1 Introduction

PET is now a commodity polymer for packaging markets—in particular for beverages and foods. It has competed successfully with traditional materials like glass and metal for containers and bottles, and is now the major material, in terms of volume, for the manufacture of bottles. PET has good gas barrier properties but, when packaging oxygen-sensitive materials and highly carbonated products, PET does not have a good enough barrier to oxygen, carbon dioxide and moisture. This is the case for products such as beers, baby foods and wine, when packed in small containers and when a long shelf-life is required. The present chapter briefly describes the different types of polyester, in particular polyethylene naphthalate (PEN), which is a high barrier polyester, and amorphous polyesters, which are clear like glass in thick mouldings, making them ideal for high value packaging.

Polymers which have better barrier properties than PET have been utilised in multilayer structures—using co-extrusion and co-injection technologies—to enhance the barrier properties of PET without detracting from its other beneficial properties. Barrier improvements can also be achieved with coatings, both internally and externally applied to bottles and containers. Now 'active' systems, such as oxygen scavengers, are being explored and, more recently, nanocomposites are being developed for use in packaging. Liquid crystal polymers (LCPs) have the highest barrier properties currently available and will be further developed for use in packaging. Whether plastics will somehow eventually merge with glass, to form plastic glass, only time will tell.

4.2 Polyesters

PET bottles produced by the stretch blow moulding process are particularly suitable for packaging products that require high clarity and gloss, a barrier against gases, resistance to creep under pressure, and toughness and impact strength. PET is a condensation polymer polymerised from two monomers, pure terephthalic acid (PTA) and a di-alcohol or diol, ethylene glycol (EG). Replacing PTA or EG with other monomers produces different polyesters, for example, isophthalic acid (IPA), which at low additions can improve the barrier properties of PET. Another widely used monomer is 1,4 cyclohexane dimethanol

(CHDM), supplied by Eastman Chemicals, which substitutes for EG in PET to produce polyethylene terephthalate glycol modified polyester (PETG) and other amorphous copolyesters. Another option is to replace EG with diethylene glycol (DEG), which doubles the EG content. It is also possible to include more than one acid or diol in the structure, to produce various copolyesters. PET copolymers used for stretch blown bottles may contain small amounts of either IPA or CHDM, which help to broaden the processing window, increase line speeds, slows crystallisation speed and reduce stress in the bottle.

Extrusion blow mouldable PET was developed some time ago but did not progress commercially because of cost and processing difficulties. This PET has a much higher molecular weight (intrinsic viscosity (IV) about 1.05) than standard, bottle-grade PET (IV about 0.8), and this increases the melt strength to help control the formation and stability of the extruded parison.

In the 1980s, Eastman Chemicals and Goodyear Tyre Co. developed a high barrier polyester called polyethylene naphthalate. PEN has a better gas barrier (five times better than PET), better UV barrier, improved temperature stability and greater stiffness than PET, but is more expensive to produce, so that it costs more than PET. PEN is being targeted at high-performance packaging but so far with limited penetration. The high price and limited availability of the monomers have restricted PEN's introduction into high volume commercial packaging applications.

4.2.1 PEN

PEN is polymerised using 2,6-dimethyl naphthalenedicarboxylate (NDC) with EG, and is being developed or studied by most polyester producers. It is a semi-crystalline polymer similar to PET, but with a double ring structure, which makes it more rigid and improves its barrier and mechanical properties. It has a high molecular weight and can be converted by injection, extrusion and biaxially stretch blow moulded into containers and bottles via injection moulded preforms. PEN is a higher melting polymer than PET, otherwise the processability is similar.

The properties of PEN make it an ideal material for high-performance packaging, with the potential to replace glass in some market segments. PEN has a gas barrier (oxygen, CO_2) about five times better than PET, moisture barrier about four times better than PET, stiffness 50% greater than PET, and a higher glass transition temperature (Tg) at 121°C compared to 75°C for PET (see Table 4.1).

Following initial objections to the use of PEN for bottles that could end up in the recycle stream mixed with PET, technology was developed to separate the two polyesters. As a result of this PEN homopolymer, PEN/PET copolymers and blends are now accepted by the Food and Drug Administration (FDA) for food contact applications. The monomers used to produce PEN are on the European

Table 4.1 Properties of polyethylene terephthalate (PET) and polyethylene naphthalate (PEN)

	PET	PEN
Tensile modulus (MPa)	1200	1800
Tensile strength (MPa)	45	60
Melting point (TM) ($^{\circ}$C)	257	270
Glass transition (Tg) ($^{\circ}$C)	75	121
Thermal shrinkage (150°C.%)	1.5	0.9
Oxygen permeability	1.5	0.3
CO_2 permeability	9	1.7
Moisture permeability	1.2	0.28

Oxygen and CO_2 permeability: $(cm^3 \cdot mm \cdot m^{-1} \cdot day^{-1} \cdot atm^{-1})$ 23°C/50% relative humidity.
Moisture permeability: $(g \cdot mm \cdot m^{-2} \cdot day^{-1})$ 38°C/90% relative humidity.

EC positive list and can be used in direct food contact packaging applications. PEN is still an expensive polymer for packaging, mainly because of the high price of the monomer, and consequently has limited capacity, which will only increase with increased demand.

These limitations have led to the development of copolymers and blends of PET/PEN that offer improved barrier compared to PET but at a lower cost. Low PEN content copolymer (with typically 10% PEN) offers the potential to achieve a pasteurisable beer bottle, while high PEN content copolymer (with 90% PEN) that would be suitable for hot-fill is similar in cost to PEN homopolymer. PEN is being targeted at food and beverage applications, where PET cannot meet the oxygen gas barrier requirements, or the heat stability for hot filling, retorting or pasteurisation. PEN is being test marketed for beers and returnable packs. Other potential markets include medical packaging (because PEN has good radiation stability, UV light barrier and low oligomer migration). It is also suitable for household and chemical packaging (because of its excellent alkaline resistance), and for cosmetics, health and beauty packaging.

At the time of writing, PEN homopolymers, copolymers and blends have not been used for any major, high-volume packaging application. The NDC monomer is now produced on a commercial scale by BP Chemicals in the US, and many polyester producers have been actively developing PEN polymers ready for packaging applications.

4.2.2 Amorphous polyesters

In the medical and health and beauty markets, different types of polyester have been developed, which are amorphous (non-crystalline) and so can be moulded into transparent, thick-section preforms and converted into glass clear components and containers. PET and PEN can be used to produce packaging for these market sectors but are more difficult to inject or injection blow mould into thick sections without crystallising. Polyesters that are amorphous contain an aromatic diol in the chain, which prevents crystallisation but retains high clarity. PETG and other copolyesters have been developed by Eastman Chemicals under the trade names Eastar, Durastar and Elegante. These include copolyesters based on acid-modified polycyclohexylene dimethylene terephthalate (PCT) with good hydrolysis resistance, excellent clarity even in thick sections, low haze and good impact performance, but poor thermal stability. They have a lower gas and moisture barrier compared to PET but have good chemical resistance. Similar copolyesters have been developed by DuPont (with Eclat) and Kanebo Gohsen (with Bellpet). The key differences between the amorphous copolyesters and crystalline, bottle-grade polyesters is their potential to achieve clear glass components, containers and bottles, even in thick sections. They offer a wider process window, higher stretch ratios and improved process cycle times. These copolyesters command a higher cost; they have FDA food contact approval and Durastar copolyesters also have Class 1 Medical Device approval.

4.2.3 Other polyesters

Polybutylene terephthalate (PBT) is similar to PET but with 1,4 butanediol replacing EG to give a polyester used for engineering applications. The number of packaging applications that have been developed for PBT is limited but includes components for medical packaging and cosmetics, health and beauty. Further development of polyesters in the future will depend on the economic costs of manufacture of high-barrier polymers, in particular naphthalate-containing polyesters. Shell has recently introduced a polyester called polytrimethylene terephthalate (PTT) for packaging films, which has a slightly better gas barrier than PET. The next stage of this development is production of the chemically similar polytrimethylene naphthalate (PTN), which has a much better gas and moisture barrier than PET or PEN. PTN has about a nine times better oxygen barrier, a four times better moisture barrier and an 18 times better carbon dioxide barrier when compared to PET.

Polyester copolymer developed by Mitsui Petrochemicals, called B010, is based on PTA, IPA, EG and special diols. Mitsui state that B010 can be blended with PET, it processes like PET and complies with FDA regulations. The gas barrier properties are about four times better than PET. Blends using 10% B010 with PET improve carbonation retention and oxygen barrier in bottles by 20%. These blends are said to have superior gas barrier when compared with PET/PEN blends.

M&G Group (ex Shell) have recently announced the development of a monolayer PET resin ideally suited for packaging beer. M&G are targeting the European 6-month shelf-life requirements for beer.

4.3   Barrier materials

PET alone does not have a good enough barrier for highly oxygen-sensitive or moisture-sensitive products. In the past, the options available to improve the barrier were to coat the outer surface of a PET bottle, film or sheet with a high-barrier polymer, such as ethylene vinyl alcohol copolymer (EVOH) or polyvinylidene chloride (PVDC). These can be applied as a thin coating on the outer surface of PET bottles to give a modest improvement in barrier and extend the shelf-life of beer in large bottles by a few weeks.

Multilayer technology to achieve barrier improvements involves the production of preforms incorporating a barrier polymer as a separate layer in a PET multilayer structure. Co-injected preforms made with three, five or more layers incorporate: PET as the main constituent, a barrier polymer, possibly an adhesive polymer to tie or bond the PET and barrier layers, and may also include post-consumer recycle (PCR) PET. Co-extrusion technology has been evaluated, and was commercial for a time for the manufacture of tube that could be converted into preforms, but this process has not been as important as co-injection technology. Over-moulding and two-layer co-injection has been developed more recently and offers further potential for incorporation of barrier layers into PET bottles and containers. The main barrier polymers are: EVOH copolymers, PVDC copolymers (less important today) and polyamides, including nylon MXD6 and similar high-barrier polyamides. In the future, it may be possible to include liquid crystal polymers (LCPs) in multilayer structures; LCPs have excellent gas barrier and moisture barrier properties (see Table 4.2).

4.3.1   EVOH

EVOH has been manufactured by and is commercially available from two major producers, both Japanese companies. Kuraray produce EVAL resins and Nippon

Table 4.2 High oxygen barrier polymers

Oxygen permeability ($cm^3 \cdot mm \cdot m^{-1} \cdot day^{-1} \cdot atm^{-1}$)		
	0% RH	100% RH
LCP	0.003	0.003
EVOH	0.003	0.25
N-MXD6	0.09	0.15
PTN	0.16	0.16
PEN	0.3	0.3
PET	1.5	1.5

Abbreviations: RH, relative humidity; LCP, liquid crystal polymers; EVOH, ethylene vinyl alcohol copolymer; N-MXD6, nylon-MXD6; PTN, polytrimethylene naphthalate; PEN, polyethylene naphthalate; PET, polyethylene terephthalate.

Gohsei produce Soarnol resins, which are both used in packaging films and for barrier layers in multilayer structures. EVOH is a random copolymer of ethylene and vinyl alcohol, it is crystalline and can be processed by extrusion and injection technologies to produce an excellent gas-barrier material, suitable for food, beverage, medical, cosmetic and household packaging. These special properties are achieved when the correct copolymerisation ratio of ethylene to vinyl alcohol is used. In addition to its unique gas-barrier properties, EVOH also has excellent oil and organic solvent resistance, so it is used in packaging for oily foods, edible oils, mineral oils and organic solvents. In addition, EVOH has good aroma and flavour barrier properties and is highly effective at retaining fragrances and preserving the aroma and flavour of the package contents. It also prevents undesirable odours penetrating the pack from the environment. Moreover, EVOH copolymers have a high gloss and low haze, which can produce good clarity in films and packs.

EVOH can be processed on conventional extrusion and injection equipment into blown or cast film, sheet, extrusion blow and co-extrusion blow moulding, extrusion coating, injection and co-injection moulding. The copolymers can be processed with a range of other polymers, including PET.

EVOH copolymer grades are characterised by their ethylene content, melt flow and gas barrier properties. Standard grades usually have an ethylene content of 32–44 mol%. The high ethylene content copolymers have a lower density, lower melting point (Tm) and glass transition temperature (Tg), and gas-barrier properties are not as good.

EVOH has excellent gas-barrier properties, superior to any other thermoplastic polymer used for packaging applications, in particular when it is used dry. The gas barrier properties will vary with ethylene content but are also affected by moisture. EVOH has —OH groups in its structure, which makes the copolymer hydroscopic and it will absorb moisture. The gas barrier properties are adversely affected by moisture absorbed, so packaging applications involving high humidities either during processing (sterilisation, retorting, pasteurisation, hot filling) or in use, need to be closely monitored or avoided. However, even under conditions of high humidity, EVOH still has excellent gas-barrier properties that will recover with time. Because EVOH is hydroscopic, its moisture barrier performance is not as good as some other polymers used in packaging. It is recommended that high ethylene content copolymers are used for applications when moisture barrier is required.

Rigid packs include bottles co-injection stretch blow moulded with PET for a wide range of foods, such as ketchups, mayonnaise, dressings, baby foods, beverages, carbonated soft drinks, juices and, more recently, for beer.

New developments in EVOH include a delamination resistant grade from EVALCA, which is reported to have better adhesion than other high barrier polymers without the need for a tie layer. The gas barrier properties have also been improved compared to earlier delamination resistant EVOH grades. This grade can be readily separated from PET in recycle systems.

4.3.2 PVDC

PVDC is a copolymer of vinylidene chloride (VDCM) with other monomers, including vinyl chloride (VCM), acrylic esters and carboxyl containing monomers. The two major producers of PVDC copolymers are Dow, with Saran grades, and Solvay, with Ixan. The content of VDCM in the polymer is usually about 70–90 mol%, and the chloride is derived from chlorine extracted from sea water, so it is inorganic and is not derived from oil-based materials.

PVDC is a semi-crystalline barrier copolymer and the volume used in packaging is low compared to commodity polymers, such as PVC. The total volume used in Europe represents less than 0.5% of the total polymers used for packaging. PVDC has excellent gas barrier and moisture barrier properties, and it can be processed by extrusion techniques, usually as a barrier layer in multilayer films and rigid packs. PVDC is now used mainly in flexible packs. It is used less today for rigid packs, such as bottles, trays and cups, having been replaced by other barrier polymers, such as EVOH and barrier nylons. The copolymer can also be used as a water-based dispersion or solvent-based coating to be applied to the external surface of containers like PET bottles to improve the gas and moisture barrier.

PVDC has benefits over other competing barrier polymers in that it has excellent gas barrier and moisture barrier properties, as well as providing a good chemical barrier to aroma and flavours and resistance to oils and fats. It does, however, have particular disadvantages in that it is more difficult to process and requires special equipment for extrusion, is not processable by injection moulding and, like PVC, is subject to criticism on environmental grounds. These issues have restricted its growth as a polymer for packaging at the expense of other barrier polymers.

The main applications of PVDC include water-based dispersion coating of PET beer bottles, extrusion and co-extrusion of multilayer films, and extrusion as a barrier layer in rigid multilayer packs for foods and beverages.

4.3.3 Polyamides (nylon)

The main polyamide or nylon used for barrier packaging is a polymer called nylon-MXD6 produced by Mitsubishi Gas Chemical (MGC). It is a semi-crystalline polymer, with high stiffness, good transparency and high gas barrier properties. This barrier nylon is produced by polycondensation of meta-xylylene diamine (MXDA) with adipic acid. It is a polymer which has some particular properties that distinguish it from other nylons, such as nylon 6 and 66. It has higher tensile strength and modulus, higher Tg, lower water absorption and excellent gas barrier properties.

N-MXD6 exhibits excellent gas barrier properties compared to other nylons and many other thermoplastics, and is comparable with EVOH and PVDC copolymers. It has a wide processing window, and so can be processed by injection, co-injection, extrusion and co-extrusion techniques, in combination with other polymers, such as PET, to produce multilayer structures as film for flexibles, and sheet, containers and bottles for rigid packaging applications. Like most other nylons, N-MXD6 is hygroscopic and absorbs moisture, and its gas barrier properties deteriorate under high humidity conditions, so it is more moisture sensitive than PVDC but less sensitive than EVOH.

N-MXD6 polymer grades are characterised by melt flow, for processing and conversion into film, sheet or containers. Its high melt temperature makes it particularly suitable for processing in multilayer structures with PET.

N-MXD6, like PET, can be biaxally stretched and thermoformed to produced oriented high barrier, multilayer, flexible films and can be injection moulded with PET into preforms for stretch blown bottles, for packaging oxygen-sensitive foods and beverages. It can also be blended with other polymers, in particular nylons and PET, to produce intermediate barrier containers. N-MXD6 is being used for the development of high gas barrier, multilayer PET bottles for beers and other beverages.

EMS Chemie in Switzerland has also developed a high barrier aromatic polyamide, Grilon BM FE4581, designed for use as a barrier layer in multilayer PET bottles and films. It is a semi-crystalline copolyamide with excellent oxygen and carbon dioxide barrier properties in conditions of low and high relative humidity. The copolymer has high transparency and is suitable for orientation processes in multilayer structures with PET for end-use applications by injection stretch blow moulding, extrusion blow moulding and film extrusion. Barrier data provided by EMS Chemie suggest that BM FE4581 has a comparable oxygen and carbon dioxide barrier to nylon MXD6, with slightly improved barrier at high relative humidities.

In Switzerland, pack tests with beer have been carried out using 0.5 litre PET multilayer bottles containing BM FE4581 as the barrier layer in a five-layer structure. The bottle has a 4-month shelf-life and is designed to be used at sports events and concerts. Typically, a barrier layer incorporated at 5–10% of the total layer thickness of a PET bottle will give a 4–6 fold improvement in gas barrier properties compared to PET alone. This could enable some oxygen-sensitive beers to achieve a shelf-life of about 4 months.

4.3.4 Liquid crystal polymers (LCPs)

LCP polymers are high aromatic polyesters with excellent gas and moisture barrier properties. They have not so far been exploited for packaging, mainly because of their high cost. LCPs are very expensive and technically difficult to process into preforms for stretch blow moulding. Blends with PET have been developed by Superex, and the bottles produced have good gas barrier properties but are opaque because of the multiphase structure. Ticona introduced its new LCP, Vectran, designed for use in packaging films. It was developed specifically to be co-processed with standard packaging resins, including PET, and using conventional processes, which include thermoforming. The material has excellent oxygen, carbon dioxide and moisture barrier, thermal resistance and chemical resistance, making it ideal for use in multilayer films and rigid packaging for retort pouches, trays and lids. The Vectran range of LCPs offers all the benefits of standard LCP resins, and has improved barrier performance compared to EVOH under conditions of high temperature and high humidity. LCPs also have excellent flavour barrier, showing almost no scalping of flavour/aroma chemicals, which is much lower than nylon 6 and EVOH. The high gas barrier properties are retained after retorting at 120°C, and when using alternative sterilisation processes. Techniques have been developed to produce Vectran LCP cast and blown films as co-extruded structures, and it can be processed on conventional extrusion and lamination equipment, produced as thin layers at 2–5 μm thickness. A clear, easily processed LCP that could be incorporated into multilayer PET structures would be a very interesting prospect for high barrier packaging.

4.4   Barrier technology

The development of a clear plastic package for oxygen-sensitive beverages and foods has long been a target for the packaging industry. PET bottles have existed for more than 20 years, and many material and technical developments have been explored to achieve a high gas barrier, clear, lightweight plastic container to replace glass and metal cans. PET alone does not have a good enough gas barrier for packing oxygen-sensitive products. PET is now widely used for packing carbonated beverages, such as colas and soft drinks, having replaced glass.

The two main approaches to achieving improved gas barrier properties in PET are: barrier coatings applied to the inside or outside of a PET bottle, or use of a barrier material as a layer in a multilayer PET structure that can be injection moulded into a preform and incorporated as the barrier layer in the structure.

Coating technologies can be divided into two types: those that use plasma or vacuum deposition of a thin layer of barrier material, like carbon or silica, onto the inside or outside surface, and those that spray a liquid organic coating material onto the outside surface of a bottle. Scavengers are 'active' chemical systems that can absorb or scavenge oxygen passing through the walls of a container, to provide additional barrier in monolayer and multilayer packages. A number of systems exist and continue to be developed. More recently, nanocomposite materials based on nanoclays have been developed, which have the potential to improve barrier systems based on their ability to provide a structure with a more gas tortuous path (see Section 4.4.4).

4.4.1 Organic coatings

External coating materials have been in use for many years. Both PVDC and EVOH emulsions have been applied to PET bottles. In the 1980s, Metal Box developed the PVDC coated 2 litre PET bottle for non-premium beers to give a two to three times improvement in shelf-life compared to PET alone.

Various coating processes have been developed. PPG offers an epoxy-amine coating material called Bairocade, which can be externally coated onto bottles to give an excellent gas barrier, clear coating suitable for packaging beer; this coating is not affected by humidity. The low temperature cure process does not affect the properties of the PET bottle and provides a tough film that can withstand handling and filling operations. The coating thickness is normally less than $10\,\mu m$ and the gas barrier properties can be up to 20 times better than a standard uncoated PET bottle. The improved gas barrier has the potential to give a 9 month shelf-life for beers. Bairocade technology has been commercialised by Containers Packaging for monolayer PET beer bottles in Australia, and by Graham Packaging in the US for juices in PET bottles.

Crown Cork & Seal is developing its coating technology, StarShield, based on the PPG Bairocade material. Dow has developed a barrier thermoplastic epoxy-polymer, BLOX (blocks oxygen), an amorphous thermoplastic epoxy resin that combines the adhesive and durability properties of epoxies with the flexibility and processability of thermoplastics. This material can be used as a barrier layer in multilayer PET structures and as a coating for bottles. Tailored comonomer design offers a range of grades from medium to high barrier. Dow believes BLOX will be more cost competitive than alternative barrier polymers in barrier structures for beverage packaging of beer and juices. The polymers have very good transparency, with low haze, excellent gloss and good colour, and strong adhesion to PET. BLOX strong adhesion barrier polymers are targeted as high barrier layer material in multilayer PET bottles. The good adhesive performance should allow complex bottle shapes to be made without the risk of delamination. Dow is working with Tetra Pak to produce a PET preform which is over injected with a layer of BLOX on the same platen. BLOX is used in TetraPak's Sealica PET container system, although it has not yet been granted FDA approval for use in the US. BLOX adhesive resins offer superior adhesion, durability and clarity, but with lower barrier properties. These resins can be used in powder coatings.

4.4.2 Inorganic coatings

The alternative inorganic coatings can be applied to the inside or outside of the PET bottle. The latest developments in plasma coatings use carbon and silica, which are applied in a high-vacuum environment. Silica technologies include an exterior coating process developed jointly by Coca Cola® working with Krones and the University of Essen, called BESTPET. A clear silica, silicon oxide (SiOx) barrier layer, 40–60 nm thick, is applied to the external surface of PET bottles using a high-vacuum plasma process patented technology based on the use of a physical vapour deposition (PVD) process. It is claimed that the technology at least doubles the gas barrier properties, with the potential to reduce the thickness of PET bottles. Following recent further development of this technology by Krones, it is now claimed that a 5–9 fold improvement in barrier is possible. PET bottles with this coating can be used for CSDs and beer. The first commercial line installed at a Coca Cola® plant in Germany runs at speeds of 20,000 bottles per hour. Krones expects the total manufacturing cost using BESTPET will add 0.5 cent per bottle.

A second process being developed by TetraPak, called Glaskin, plasma coats the inside of PET bottles with SiOx to produce a clear coating 0.2 μm thick. It is also said to improve the gas barrier properties at least two fold. The 3–4 month shelf-life achieved by silica coatings is acceptable for some beers and alcoholic drinks but does not meet the target for most European beers.

A further plasma deposition process has been developed by Sidel, called ACTIS (amorphous carbon treatment on internal surface). The amorphous carbon is deposited onto the internal surface of PET bottles to a thickness of 0.1 μm, using acetylene gas and microwave energy directed at the bottle. PET bottles can be produced with a 30 times better barrier to oxygen and 7 times better barrier to carbon dioxide, and also a sixfold reduction in acetaldehyde migration compared to PET. The coating is clear but has a yellow tint, making it suitable for beer packaging. This coating technology offers one of the best available systems in approaching gas barrier performance and economic costs compared to glass and metal cans. The ACTIS 20 production scale plasma coating system is commercial and can coat 10,000 PET bottles per hour. The ACTIS system has been approved in Europe for use in packaging, and the FDA have also granted a letter of non-objection in the US. Sidel is planning to launch ACTIS-treated PET bottles and containers for beer and other beverages, such as fruit juices, teas and soft drinks, in various countries. The ACTIS-treated PET bottle is said to be 100% recyclable.

A similar barrier coating technology developed by Kirin and Mitsubishi Shoji Plastics uses plasma vapour deposition to provide an internal coating of a thin layer of diamond-like carbon (DLC). This is said to give excellent improvement in gas barrier properties to PET bottles. The coating is said to increase the oxygen barrier 20 times and carbon dioxide barrier 7 times. Nissei ASB has built a barrier coating machine for the DLC coating system with an output of 2,000 PET bottles per hour. The DLC coating system is expected to find applications in packaging of beer, cosmetics, juices, drinks and vitamins. Because the deposited layers are brittle, it is essential that they are thin in order to prevent delamination as the bottles are subjected to creep and stresses during filling and when under pressure, and scuffing when handled and during transportation. The carbon coatings are yellow in colour and so should be acceptable for beers. The barrier performance is similar to organic coatings, giving a potential shelf-life of up to 9 months for beers.

Most plasma coating systems must be applied using a vacuum chamber, which adds cost to the bottle in terms of equipment and processing. A silica-based coating has been developed by MicroCoating Technologies (MCT), using a patented process which enables the coating to be applied to the bottle without the use of a vacuum. MicroCoating Technologies claims that it has successfully deposited transparent silica coatings onto the external surface of PET bottles and films. The company has a global license from Georgia Tech Research Corporation for its open atmosphere, flame-based combustion chemical vapour deposition (CCVD) technology. MCT uses a device called a nanomiser to form microscopic droplets, which are carried by oxygen gas to a flame where they are combusted. The polymer substrate is coated by passing its surface across this flame and the droplets react and deposit onto the surface. The coating thickness is 50–150 nanometers, and the oxygen barrier achieved is equivalent to alternative silica-based barrier coatings deposited onto PET, using plasma-enhanced chemical vapour deposition methods. The MCT process offers an alternative coating technology that may be more cost effective compared to vacuum deposition technology.

4.4.3 Scavengers

Materials that can scavenge or absorb a particular gas, chemical or material as part of a package structure act as an active packaging system, where the product, package and environment interact resulting in either an extension of the shelf-life or the attainment of a specific improvement in property. These interactions are a positive benefit to packaging systems. Oxygen scavenging or absorbing materials are one such example. They remove oxygen in the headspace of the pack, in the product or migrating into the pack, to very low levels. They are especially important for oxygen-sensitive food packaging, where oxygen ingress is an issue in determining pack shelf-life.

Oxygen scavenging systems have been developed for PET packaging over a number of years. The companies most active in this technology include BP Chemicals, Crown Cork & Seal, CPETT, Chevron Phillips Chemicals, EVALCA/Darex, MGC, Toyo Seikan and TriSeal. BP Chemicals have developed Amosorb 3000, a polyester copolymer containing an iron salt, which acts as an active barrier layer in PET bottles, is compatible with PET and does not delaminate. Crown Cork & Seal have further developed their Oxbar technology, following initial development in the 1980s by Metal Box; a blend of PET with nylon MXD6 and a cobalt salt was used as the barrier layer in

Figure 4.1 PET beer bottles courtesy of Crown Cork & Seal.

multilayer PET bottles. The Oxbar system reacts very quickly with oxygen in the bottle environment and it has a long active life, at least as long as the product shelf-life for the pack (Figure 4.1) CPETT has also developed a nylon MXD6 based system for PET bottles.

Amosorb, Oxbar and other similar systems can be incorporated into three, five or more multilayer PET systems. Other developments include, Toyo Seikan with Oxyguard and Tri-Seal with TriSO$_2$RB, using iron salt scavengers in trays and cap liners. ZapatA have developed an acetaldehyde scavenger. MGC with Ageless, and Multisorb Technologies have developed iron-based sachets which can be incorporated into rigid and flexible packs. CIBA have acquired BP Chemicals Amosorb 2000 technology, now called Shelfplus. EVAL Co. and Darex Container Products have jointly developed a range of materials, trade name DarEval, which are EVOH incorporating a proprietary oxygen scavenger. The material retains its oxygen barrier properties even at 100% relative humidity, which EVOH alone cannot achieve. Data on a three-layer co-injected PET bottle incorporating 5% DarEval indicate that it provides high oxygen barrier properties for up to 4 months. The material does not yet have FDA approval for food contact.

A new oxygen scavenging polymer, which has been developed by Chevron Phillips Chemicals, is suitable for rigid and flexible packaging. The material can be used as an active barrier layer between a passive oxygen barrier material, such as PET, EVOH or nylon, in a multilayer structure. The material is a blend of ethylene methylacrylate cyclohexene methylacrylate (EMCM) copolymer with a masterbatch containing a photoinitiator and a transition metal catalyst. EMCM is an oxidisable polymer that readily scavenges oxygen. The photoinitiator is activated by UV light, which triggers the reaction only when the package is filled.

4.4.4   Nanocomposites

Inorganic materials, like silicates, based on montmorillonites are nanoclays, which are used to produce nanocomposite materials. These nanoclays can be blended at low levels (5%) with polymers to produce high gas barrier nanocomposites. The silicates produce a lamellar structure, which extends the diffusion path in packaging materials, so improving the gas barrier properties. The nanocomposite can be a separate layer or a blend in the package structure. Nanocor is one of the leading manufactures of silicates and is developing nanocomposite materials together with a number of companies. Most of the development so far has focused on nylon-based nanocomposites, and these have been incorporated as a separate layer with nylon 6 films to improve gas barrier properties. Nanoclays are very small particles which do not interfere with light transmission through the pack, so that clarity can be maintained. The level of haze is acceptable for amber bottles but not at the moment for clear bottles.

In films, puncture resistance and stiffness is improved and the materials can be processed on existing equipment, with the potential for thinner barrier layers in both films and rigid packs. Nanocomposites offer another alternative for gas barrier improvement in PET and possibly other polymers. Eastman is working with Nanocor on nylon-based nanocomposites for use in multilayer PET bottles. Eastman has commercialised its Imperm nanocomposite barrier material, which is designed to be used as a barrier layer and is said to provide an oxygen barrier more than 50 times better when compared to PET. Imperm used in a three layer PET bottle at the 4% level can improve the oxygen barrier up to five times, and at the 10% level the barrier is about 10 times better than PET.

Honeywell International (ex AlliedSignal) has been working on nylon-based nanocomposites for some years. It now has a nylon 6 based matrix polymer, called NC1, into which a clay nanocomposite is introduced during polymerisation. The material is produced as a batch process, is cost competitive with other nylon-based systems and its barrier properties are intermediate between nylon 6 and nylon MXD6. A second material, NC2, incorporates an oxygen scavenger into the nylon-clay nanocomposite material, and provides a superior oxygen barrier performance. The scavenger is an organic material that is activated by moisture. Southern Clay (Laporte) is another major producer of nanoclays and is actively developing nanocomposites suitable for PET systems.

4.5   Oxygen barrier

The sections on barrier materials discuss the material options to improve the barrier performance of PET. Much evaluation and testing has been carried out over many years on a wide range of materials, blends, coatings and multilayer structures that include PET. This section considers PET and predicts the improvement required in oxygen barrier properties for PET based structures to make them suitable for packaging oxygen-sensitive products, including foods and beverages. The major factor that causes deterioration of foods and beverages is the presence of oxygen, either in the product when it is filled in the bottle or by permeation through the bottle wall during its shelf-life. If we consider only the oxygen that permeates into the bottle, then this level will determine what is acceptable over the shelf-life of the pack and will also depend on the type and quality of the product. For example, a 2 litre PET bottle and a non-premium beer will have an acceptable shelf-life of about 8 weeks. A PVDC-coated 2 litre PET bottle will have more then double the shelf-life at 22 weeks. When tested with premium quality beers, both PET and PVDC-coated PET bottles were unacceptable even for very short shelf-life periods.

Using this information, it is possible to estimate what level of oxygen ingress will cause deterioration of a particular product, and to predict the shelf-life for the product and the pack. The oxygen ingress into a 2 litre PVDC-coated PET bottle in 22 weeks is about 4.56 ml compared to 13.7 ml for a 2 litre PET bottle (threefold improvement), which equates to 2.28 ml/l. This level of barrier performance is acceptable for a non-premium beer and some other foods and beverages but not for a premium quality beer, baby food or wine. The barrier performance required for packing very oxygen-sensitive products needs to be higher. Discussions with end users on the topic of packing these types of products in PET bottles and containers produced a figure of less than 1 mg of oxygen ingress in 6 months per litre of contents. This equates to not more than 0.7 ml/l in 6 months.

These data provide the basis for estimating the oxygen ingress into bottles of various sizes, over a nominal 6 month shelf-life period, if we know the surface area and wall thickness of the bottle and the oxygen permeability of the PET. Consider a 0.5 litre PET bottle with a surface area of $0.036 \, m^2$, a wall thickness of 0.35 mm and an oxygen permeability for PET of $1.5 \, cm^3 \cdot mm \cdot m^{-1} \cdot atm^{-1} \cdot day^{-1}$. This will give an oxygen ingress of 11.9 ml/l in 6 months, which would clearly be unacceptable for premium beer, baby food and wine by a factor of 17 times, and for non-premium beers by a factor of five times.

Considering alternative polyesters, such as PEN, and blends, coatings and multilayer structures based on PET, a comparative analysis of various material systems can be made with a knowledge of the oxygen permeability of the polymers and of other materials being considered. If we consider two levels of barrier performance, the first would be for packaging most foods and beverages, including non-premium beer, where an oxygen ingress limit of not more than 2.28 ml/l in 6 months is required. The second is for very oxygen-sensitive products, such as premium beers, baby food and wine, where an oxygen ingress limit of not more than 0.7 ml/l in 6 months is required.

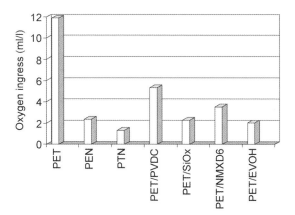

Figure 4.2 Oxygen ingress to a 0.5 litre bottle over 6 months. Abbreviations: PET, polyethylene tereph-thalate; PEN, polyethylene naphthalate; PTN, polytrimethylene naphthalate; PVDC, polyvinylidine chloride; NMXD6, nylon-MXD6; EVOH, ethylene vinyl alcohol copolymer.

4.5.1 Packaging foods and beverages

PET with an oxygen ingress of 11.9 ml/l in 6 months in a 0.5 litre bottle clearly does not achieve the target oxygen ingress of 2.28 ml/l in 6 months. PEN with an oxygen ingress of 2.38 ml/l in 6 months is very close to being acceptable, and the higher barrier PTN with an oxygen ingress of 1.3 ml/l in 6 months clearly does achieve the target. A PVDC-coated PET bottle with an oxygen ingress of 5.34 ml/l in 6 months is unacceptable, while a silicon-coated PET bottle with an oxygen ingress of 2.26 ml/l in 6 months just achieves the target. A multilayer PET bottle with a 10% EVOH barrier layer has an oxygen ingress of 2.02 ml/l in 6 months, while a PET bottle with a 10% nylon NMXD6 barrier layer only achieves an oxygen ingress of 3.5 ml/l in 6 months (see Figure 4.2).

4.5.2 Packaging oxygen-sensitive foods and beverages

PET with an oxygen ingress of 11.9 ml/l in 6 months in a 0.5 litre bottle clearly does not achieve the target oxygen ingress of 0.7 ml/l in 6 months. Both PEN and PTN are also unacceptable. An external organic-coated and plasma carbon-coated PET bottle both achieve the target oxygen ingress of 0.7 ml/l in 6 months. A multilayer PET bottle would require about 40% EVOH and more than 70% NMXD6 as the barrier layer to achieve the target oxygen ingress of 0.7 ml/l in 6 months, levels which are probably too high to be economically viable. A multilayer PET bottle with 10% LCP having an oxygen ingress of 0.45 ml/l in 6 months and an Oxbar multilayer PET bottle with an oxygen

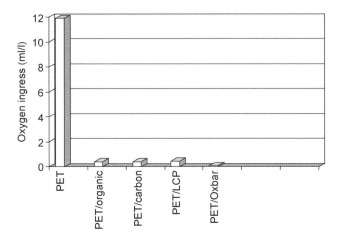

Figure 4.3 Oxygen ingress to a 0.5 litre bottle over 6 months. Abbreviations: PET, polyethylene terephthalate; LCP, liquid crystal polymers.

ingress less than 0.45 ml/l in 6 months would both achieve the target (see Figure 4.3).

4.6   Carbon dioxide barrier

PET was first developed in the 1970s for carbonated soft drinks. The orientation induced when PET is stretch blown achieves the essential stiffness and creep resistance to allow a bottle to resist deformation when filled with a highly carbonated product. PET has sufficient carbon dioxide barrier for packing CSDs, mineral waters and non-oxygen-sensitive products, especially in large volume

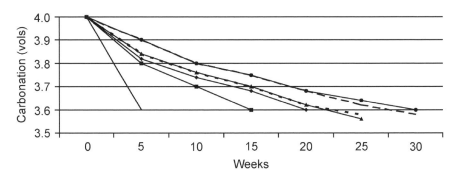

Figure 4.4 Carbonation loss for a 0.5 litre bottle. For abbreviations see legend to Figure 4.2. Key: ———, PET; —■—, PET/NMXD6; —◆—, PET/EVOH; ■ ■ ■, PET/Oxbar; —▲—, PEN; ▬ ▬, PET/organic; —●—, PET/carbon.

bottles. Products which are both carbonated and oxygen sensitive, like beers, require higher barrier polymers or structures to ensure carbonation loss is minimised. During the shelf-life of a product packed in a plastic bottle there will be carbonation loss, which is more critical in small bottles. Carbonation loss depends on a number of factors: sorption of carbon dioxide into the bottle wall, creep of the bottle, and diffusion of carbon dioxide from the product through the bottle wall.

The specification for carbonation loss for beer has been quoted by brewers to be not more than 10% loss in 6 months for a 0.5 litre bottle. PET will only achieve a shelf-life of 5 weeks, while PEN will achieve 20–25 weeks. Multilayer PET-based structures that include a high barrier polymer will have a much improved carbon dioxide barrier. A structure with 10% nylon NMXD6 will extend the shelf-life to 15 weeks, and with 10% EVOH this will be 20 weeks. The PET Oxbar system will achieve in excess of 20 weeks, while both external organic-coated and plasma carbon-coated PET bottles will meet the 6 month target shelf-life (see Figure 4.4).

4.7   Future trends

In the last 20 years, the volume of PET has grown manyfold and it now dominates the market for packing soft drinks and mineral waters, replacing glass and PVC. The growth of PET into foods, health and beauty, cosmetics and household packaging has been relatively small in comparison to beverages. The major target in recent years has been the very high volume market for beers which are in glass bottles. While PET does not have sufficient gas barrier performance, developments in barrier polymers and barrier technology are providing potential solutions for a PET-based structure for bottles, cans and containers to replace glass. PEN has not provided the solution because it does not have sufficient gas barrier performance for packing very oxygen-sensitive foods and beverages. In the future, new polyesters may be developed that could have the required gas barrier but issues of cost, recycling and environment will need to be addressed. Various options involving coatings, multilayer structures incorporating high barrier polymers, scavengers and nanocomposites will continue to be tested for oxygen-sensitive products. We are close to providing a solution that meets the requirements of food and beverage manufacturers and consumers but a single polymer material or technology may not satisfy all these requirements.

Acknowledgement

Gas permeability data and barrier calculations have kindly been provided by Mr Jim Nicholas of Crown Cork & Seal.

5 PET film and sheet

William A. MacDonald, Duncan H. MacKerron
and David W. Brooks

5.1 Introduction

Over the past four decades, biaxially oriented PET film has become one of the most important materials in the global packaging market. This chapter presents a review of the technology which produces this packaging medium. Its aim is to illustrate the wealth of science and technology which contributes to the manufacture and success of this product, and to indicate the potential that remains in polyester film for continued growth as a modern packaging material.

Biaxially drawn polyester film was developed by ICI in Europe and DuPont in the US in the 1950s, with DuPont introducing the first commercial film line in the late 1950s [1]. There was a slow increase in the number of film manufacturers throughout the 1960s and 1970s and production mushroomed in the 1980s and 1990s. From the late 1990s onwards, there has been a major consolidation in the industry, with DuPont buying the ICI Films business and then forming a joint venture with Teijin, Toray aquiring Rhone Poulenc and Chiel and joint-venturing with Saehan, and Mitsubishi acquiring Hoechst. The global capacity of PET film in 1999 was 1,250,000 tonnes. There are now about 30 producers of PET film worldwide. DuPont Teijin Films™ and Toray Saehan Inc. are the major producers, with declared capacities of about 250,000 tonnes and 220,000 tonnes, respectively. Mitsubishi and SKC form the 'second tier', with approximately half the capacity of the top two. The next in order is Kolon (Korea), with about half the capacity of Mitsubishi and SKC.

PET film's unusual balance of physical, chemical, thermal and electrical properties has made it appropriate for a wide range of specific engineering functions—from acting as a carrier to performing as an electrical, physical or thermal barrier, or as a manufacturing aid. Due to its unique balance of properties, PET film has found acceptance in many markets, including industries producing computer tapes and video tapes, motors, capacitors, flexible circuitry, graphic arts materials, release products, cards, labels and packaging. About 350,000 tonnes of all polyester film produced globally is destined for use in packaging. It is therefore understandable that the processes by which film is produced from PET and the practices which modify and extend its performance are the subject of regular interest and review. By describing the current standards in polyester film manufacture and the acknowledged underlying science and technology, this chapter is intended to provide such a review.

5.2 The film process

Biaxially oriented PET film is produced exclusively by a tenter process, in which the amorphous cast film is usually drawn in the machine direction (MD) by passing it over heated rollers and then fed into a tenter frame to achieve a draw in the transverse direction (TD). A schematic of the process is presented in Figure 5.1. The sequence of steps is normally as described above (MD-TD) but the process can be reversed (TD-MD). Moreover, a simultaneous tenter process, in which the clips are not interconnected and stretching can therefore be carried out by accelerating the clips in the MD within the diverging TD draw section, has been commercialised. The conversion of polymer into film comprises five basic stages:

1. Polymer preparation and handling
2. Extrusion and casting
3. Deformation or drawing
4. Heat setting
5. Slitting and winding

In the following sections, each stage is described and discussed in more detail.

5.2.1 Polymer preparation and handling

Polymer can be extruder fed to the drawing process or it can be directly fed from a continuous polymeriser (CP). In both cases, the virgin polymer tends to have a number averaged molecular weight (Mn) of about 20,000, although higher and lower molecular weights are filmed. In the extruder fed film lines, the polymer

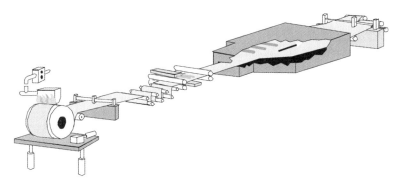

Figure 5.1 A typical film manufacturing process.

handling involves blending and drying. As discussed in Section 5.2.6, blending is a consequence of the film process never being 100% material efficient, and virgin polymer is therefore blended with polymer reclaimed from the film process. Having already been processed, the reclaim polymer is usually of lower molecular weight than the virgin polymer; therefore, the ratio of reclaim to virgin chip must be carefully controlled to ensure that the resulting mixture will yield a feedstock of adequate molecular weight to be filmed. Drying is essential in closed (single-screw) extrusion systems, as the PET is susceptible to hydrolysis resulting in a reduction in molecular weight. Less commonly, processes have evolved based on vented (twin-screw) extruders, where moisture is removed just after melting. In the drying stage, polymer is crystallised first to avoid the chip sintering during drying and is then dried for several hours at 160–180°C to reduce the moisture level to 10–30 parts per million (ppm). Close-coupled CP film lines do not have this stage and melt is pumped directly through filtration to the die.

5.2.2   Extrusion and casting

5.2.2.1   Extrusion

The blended and dried polymer is then melt extruded through a slot die. There is usually melt filtration before the die to remove degraded polymer, gels, catalyst residues and pipe deposits. The extrusion system is typically designed to deliver stable output up to about 2.6 tonnes/h over a wide range of operating conditions and throughputs. In exceptional circumstances, higher outputs are possible, up to about 3.5 tonnes/h on thick film lines using complex tandem or parallel extrusion systems. The requirements for film extruders differ from those for most other products in that it is necessary to maintain a very steady output, avoiding fluctuations that could cause thickness variation in the machine direction of the film. This high degree of output stability is often achieved by inclusion of a melt pump after the extruder. The system must also be designed to keep the polymer melt temperature low and to avoid 'hold ups', so that the polymer does not degrade causing defects during interruptions to the film process.

    The size of the extrusion system must be optimum for delivering the required amount of melt at the right temperature. This is commonly achieved by using specially designed single-screw extruders, which are able to cope with a limited range of molecular weights and the presence of fillers (typically < 3% but exceptionally up to 15%). The design of single screws to deliver uniform melt consistently at low melt temperatures becomes more complicated as the output is increased and the flexibility demands extended. To overcome this, it is common practice to feed melt via a coarse filter or screen to a melt pump, which in turns meters the polymer melt uniformly to the die. More recently, twin-screw extruders have been introduced on some film lines to widen the operating window and to provide capital-efficient high throughputs. These are able to cope with a wider range of molecular weights, improve the mixing, and they have the advantage of extruding at lower melt temperatures. Other combinations, such as tandem single screws with melt pumps, are also used to give a stable output. Parallel extrusion systems are commonly used for high output but these present the problem of ensuring homogenous melt stream blending. Whichever extrusion system is employed, its purpose is to transport a consistent flow of polymer melt to the flat film die of a stenter process. The die, which can be centre or end fed, converts the melt from a circular cross-section to a uniformly thick melt curtain of the required width. The thickness of the film is continuously measured across the web after the tenter process, giving a thickness or gauge profile. These profile data are used to make fine adjustments to flow profile at the die either through thermoviscous heating or by actuation of mechanical bolts, which physically modulate the die gap profile to achieve a uniform film thickness profile. Combinations of thermoviscous and mechanical modulation are also employed in some cases.

5.2.2.2 *Casting*

The aim of the casting is to produce a continuous uniformly thick film of non-crystalline polymer with no surface blemishes and this is achieved by drawing down the melt curtain onto a casting drum. The temperature of the polymer melt from the extrusion system will normally be 280–310°C, and to minimise crystallisation, which would increase film haze and brittleness and possibly cause a film breakage later in the filming process, the molten film has to be cooled as quickly as possible below its glass transition temperature (Tg). This is achieved by cooling the casting drum using recirculated water, which passes through a heat exchanger to control its temperature between typically 10°C and 15°C. Thin film can be satisfactorily cooled using a single drum, normally 600–900 mm in diameter, but for thicker films, where the insulating properties of the film prevent cooling through to the air (non drum) contacting side of the melt, a second drum is used to provide additional cooling.

As the casting drum rotates, air is drawn into the gap between the film of melt and the drum, affecting the contact between the two surfaces and the effectiveness of the cooling. This is avoided by electrostatically charging the film surface using a pinning wire or blade electrode stretched across the drum just below the die face. This creates an electrostatic field around the wire or blade, which induces a charge on the melt curtain surface. Since the drum is earthed, the charge forces the melt curtain onto its surface. The charged film is attracted onto the earthed casting drum evenly across its full width. The charge generated on the film surface depends on the film speed, and the position and shape of the electrode. A blade electrode gives a more focused area of high density charge than does a wire electrode, and this is important for casting at high speeds. The conductivity of the polymer can also be a factor and metal salts, such as magnesium, can be added to improve the conductivity.

As mentioned above, significant levels of crystallinity in cast film will contribute to higher levels of stress during subsequent drawing stages, which in turn create issues around profile and even web fracture. It is therefore not surprising that enormous resources have been spent to study the phenomenon of crystallisation of PET in order to understand the key factors likely to control the rate of crystallisation in the casting stage. Crystallisation comprises two principal processes: an initial event which is known traditionally as nucleation and is referred to as such in the present text, and a subsequent process of growth. The rates of both processes are sensitive to temperature but the dependency of each is different. Nucleation occurs more readily at lower temperatures, whereas the growth process passes through a maximum rate at higher temperature. In an effort to cast from the melt a PET film which is truly amorphous, the heat transfer or quenching must be sufficiently rapid to prevent both processes.

Figure 5.2 shows the dependence of the rate of isothermal crystallisation of PET on temperature. The rate is the product of the rates of nucleation and of growth, and passes through a maximum midway between glass transition temperature (Tg) and melting point (Tm). Here the diagram suggests that in order to cast PET film which is amorphous, heat must be removed from the melt on a timescale that is far shorter than that of early crystallisation. This is easily achieved for thin, clear film but more difficult for thicker cast film. However, Figure 5.2 also highlights the effect of an inorganic filler dispersed in PET on its rate of crystallisation. The enormous increase in overall rate is due entirely to faster nucleation, which is heterogeneous in nature and presents a challenge to the production of wholly amorphous film from this system. In

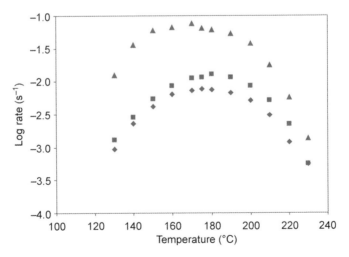

Figure 5.2 Crystallisation of PET. Key: ◆, 0.665 intrinsic viscosity (IV); ■, 0.574 IV; ▲, 0.578 IV + 0.1% filler.

practice, some crystallinity does establish in cast PET film that is either of thicker gauge or contains high loadings of inorganic additive.

Since the presence of crystallinity is known to alter the behaviour of cast film in the subsequent draw stage of the process, and in practice increases the risk of film failure [2], it is important to understand and predict the level that can establish at this early point in the process. This is achieved by combining a simple heat transfer model of the casting stage with a kinetic model for crystallisation. Examples of both are widely reported, and often a simple Avrami expression is sufficiently accurate to describe the phase transformation of the material. An Avrami model was adopted in one of the earliest published reports of PET crystallisation [3, 4]. More sophisticated kinetic models may be preferred to represent the crystallisation of complex systems based on PET but containing filler, nucleating agent or other commercial additive [5, 6].

The effort to characterise and model the crystallisation that will occur in a real film under specific casting conditions has proved successful, and most film and equipment manufacturers will possess this capability. However, it is worth noting that models based on heat transfer calculations and crystallisation kinetics measured in the laboratory still present only an ideal account of the process of change that the film undergoes on the casting drum. For example, nuclei can pre-exist in the extruded film either from large, unmelted crystals in the dried chip or as a result of shear in the die. Similarly, particulates such as catalyst residues can persist in the polymer and act to varying degrees as heterogeneous nucleants [7]. Some of the consequences of these phenomena have been well documented [8–10], and in general they can be a source of deviation from the expected crystallisation kinetics during casting.

It is virtually impossible to account quantitatively for the influence of all such parameters on film at the casting stage; however, an awareness of their presence is important and serves to emphasise the value of the underlying science to the management of the material at this stage of the process. Finally, some macroscopic considerations of the casting drum itself must also be made.

The surface of the casting drum must be of a very high standard to avoid imprinting any patterning or 'graininess' on to the cast film. The surface must also be hard in order to avoid damage and must be resistant to corrosion so that no pitting occurs. Therefore, a drum that is hard chrome-plated and highly polished is usually favoured. During operation, the casting drum must be free of vibration and must rotate smoothly to minimise any source of variation in thickness in the machine direction of the film.

5.2.3 Drawing

5.2.3.1 The forward draw preheat (FWDPH)
The preheat zone of the forward draw is worth considering separately, since any stage at which the nature of the film can change provides an opportunity to

influence and control its properties. In the preheat zone, the temperature of the cast film is raised by passage over a series of heated contact rolls until a point, usually about 15°C above its Tg, where the material can be readily stretched. Heat transfer calculations indicate that the film may be above Tg for, at longest, only a few seconds but again this can provide an opportunity for crystallisation. For example, undeveloped nuclei which are produced efficiently in the final stages of the casting process are now present to enable crystal growth to proceed immediately upon reheating. Other considerations mentioned previously can also influence the rate of crystal growth during annealing [8, 9, 11]. A further complication is that perfect thermal equilibrium cannot be achieved throughout the film. The nature of the preheat design and the thicker edges of the film, inherited from neckdown of the melt from the die, mean that the FWDPH stage will produce a film with a temperature gradient through its thickness and at its edge. These local differences will develop discrete morphologies, which will respond differently downstream; thus, it becomes critical for the preheat stage to heat film rapidly but minimise thermal fluctuation within the material. Again, simple heat transfer models can offer significant scope to optimise the process design.

5.2.3.2 The forward draw (FWD)

The forward draw stage, which physically stretches the heated polyester film between two nip roll systems, is designed to improve its tensile properties in the MD. Stretch ratios of around 3.5:1 are employed, which cause a traumatic effect on both the macroscopic and microscopic nature of the film and thereby raise its tensile modulus and strength by a factor of about three. However, detrimental effects can also arise. Additional variation in thickness of the film and further crystallisation can occur and effort must be made to limit these.

In order to achieve the desired physical and dimensional properties in film at the FWD exit, a considerable level of knowledge has been generated about the material properties and behaviour of PET. Thermal properties have been used in heat transfer models, load-deflection and viscoelastic behaviours have been studied to develop constitutive models to describe deformation at high strains, and crystallisation kinetics have been measured to predict film morphology and physical properties either later in the production process or in the finished product.

It is now possible from a knowledge of the FWD design, the process conditions and the molecular structure of the polyester, to calculate the change in the physical shape of the web [12–14] and the stress developed within it, at that stage of the process. Moreover, one can predict the accompanying evolution of both the molecular orientation [15–18] and crystallinity in the PET. This is a significant achievement since several process parameters, namely temperature, strain and strain rate, act in combination to influence the rate and extent of crystallisation in the film.

Through a programme of systematic experiments using time-resolved synchrotron X-ray diffraction methods, the very high rates of crystallisation that occur in PET, during and after rapid drawing, have recently been quantified [19]. In fact, the behaviour of PET during hot uniaxial drawing can be represented for all conditions of temperature, strain and strain rate by the two master curves shown in Figure 5.3. This work revealed two key features about the

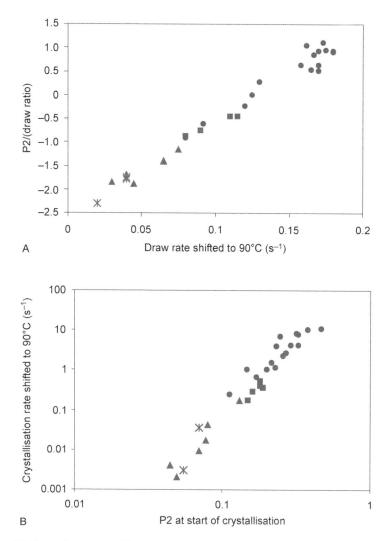

Figure 5.3 Orientation and crystallisation in PET after one-way draw. A. Orientation B. Crystallisation. Key: ●, 90°C; ■, 100°C; ▲, 110°C; ✖, 120°C.

crystallisation of PET: first, at high draw rates, the process of crystallisation commences only after the completion of the physical draw and follows first order kinetics; and second, the rate of that process is determined by the temperature and the degree of molecular orientation after stretching [20]. Thus, Figure 5.3B is a quantitative representation of the dependence of crystallisation on film temperature and molecular orientation after the draw. Figure 5.3A is critical to the description since it provides a value for the molecular orientation $\langle P_2(\cos\theta)\rangle$. This is governed by the process parameters of temperature, draw rate and macroscopic draw ratio. Inspection of the mastercurve confirms the popular understanding that low temperature, fast rates and high ratios favour efficient molecular alignment along the draw direction [21]. However, the experimental data can only be fully rationalised through an appreciation of the role of molecular relaxation. Studies have shown that segmental motion permits the oriented network to relax during and after draw and, under certain conditions, is capable of offsetting completely the effect of stretching. The interplay of molecular alignment under strain and entropically driven relaxation was quantitatively explained in terms of known modes of relaxation for polymers [22], and was shown to relate to the final morphology in the one-way drawn film [20].

Work of the sort described above illustrates the nature and extent of transformations that can occur on the molecular level and on the timescale of each stage of the film process. Clearly, similar mechanisms will operate but with different results in subsequent stages of the film production line. Tangible benefits can now be achieved from this knowledge by compiling the kinetic data with heat transfer models of the film in a real FWD environment. It becomes possible to make accurate predictions about the final levels of molecular orientation and crystallinity in PET after the first draw stage of a real film process [23]. Thus, the process engineer can make fundamental comparisons between different production lines and may even redesign the FWD stage in favour of specific attributes of the finished film product.

However, close control of the film structure in the FWD step is also critical to the subsequent stages of the production process, where a knock-on effect of creating either well-behaved or deteriorating film performance can be seen downstream. In the sideways draw (SWD), for example, serious profile loss can be caused through inconsistent morphology inherited from the FWD stage, while the phenomenon of bowing which appears in the heat set stage of the tenter is also sensitive to the same parameter.

5.2.3.3 *The sideways draw preheat (SWDPH)*
The tenter oven comprises a series of zones in which air temperature is closely controlled and which perform specific tasks on the resident film. The first zone, known as the sideways draw preheat is designed to deliver film at temperatures typically between 90°C and 110°C and with a consistent

morphology across its width and along its length, to the sideways draw stage of the process. Fundamental studies confirm that when uniaxially drawn PET is reheated above its Tg, significant structural reorganisation in the form of secondary crystallisation begins to occur [24], which is known in turn to affect its physical behaviour [25, 26]. Therefore, good temperature control across the film must also be provided in the preheat zone, such that not only is the final temperature of the web at the zone exit constant within narrow limits but also the rate of heating has been similar for each position across the film.

In order that all regions of film follow a similar thermal profile throughout the preheat zone, film thickness and local heat transfer efficiency must be identical at all times across the web. Heat transfer will, in turn, depend on local air temperature and velocity, which again must be identical in all regions of the zone. In practice, this is difficult to guarantee and instead is compensated to some extent by ensuring that thermal equilibrium is achieved ahead of the zone exit (see Figure 5.4, curve of film A). However, Figure 5.5 demonstrates that consistent levels of crystallinity and microstructure across the film will not automatically follow.

One-way drawn film crystallises as a function of its FWD history and dwell time in the preheat zone. For film 2 (see Figure 5.5), this means that areas of the web that reach thermal equilibrium earlier than others will have progressed by the preheat exit to higher levels of crystallinity, which in turn respond differently in the subsequent SWD zone. Similarly, if the final temperature of the film varies across its width at the zone exit, so too will its local modulus, its yield stress and its draw behaviour in the following SWD stage.

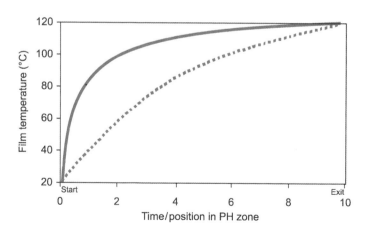

Figure 5.4 Film temperature in sideways draw preheat (SWDPH). Key: ▬▬, film A; ▪▪▪▪, film B.

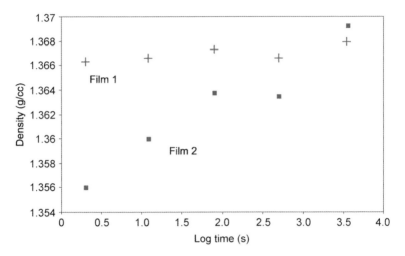

Figure 5.5 Crystallisation of one-way drawn film. Key: ■, FWDR × 3, FWDT 85°C, biref 0.114; +, FWDR × 4, FWDT 100°C, biref 0.111. Abbreviations: FWD, forward draw; R, ratio; T, temperature.

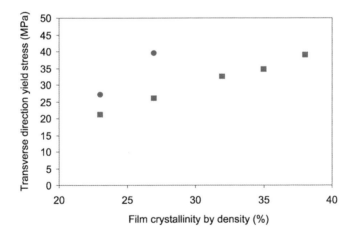

Figure 5.6 Influence of crystallinity developed in the sideways draw preheat (SWDPH) zone. Key: ●, SWD Temp 100°C; ■, SWD Temp 110°C.

Figure 5.6 shows the dependence of the yield stress of one-way drawn PET film on its crystalline content, when strained in the perpendicular direction. Clearly, if the crystallinity in a film web changed sufficiently across its width, the conditions for an even, homogeneous draw in the sideways direction would disappear, resulting in problems concerning the quality of the film. This is supported by experiment and finite element modelling. Moreover, the fracture

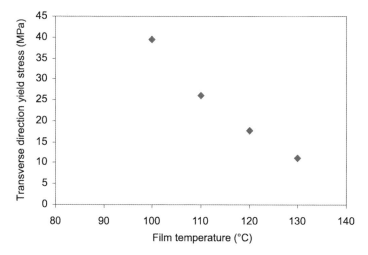

Figure 5.7 Influence of film temperature at exit of sideways draw preheat (SWDPH).

stress of one-way drawn film when strained in the TD can be around 75 MPa at 100°C; this suggests that, in the worst case, poor control of crystallinity in the preheat zone can have catastrophic consequences in the subsequent SWD zone. The yield stress described is also directly sensitive to temperature, as shown by Figure 5.7.

A similar logic therefore applies when relating temperature control across the web in the preheat zone to the consistency of draw in the SWD zone, quite independent of crystallinity. In fact, fluctuations in film temperature should be less than 10°C and preferably below 5°C at any point in the preheat zone in order to maintain a well-behaved film process. In practice, this presents a challenge since many features of the tenter process design and engineering conspire to cause real conditions in the zone to vary with position and over time.

5.2.3.4 *The sideways draw (SWD)*

In the second stage of the tenter oven, the edges of the web are led along diverging rails that cause the material to be stretched, for the second time, by a factor of three to four. The object of the SWD step is to develop the properties of the film in the TD, via orientation at the molecular level, to a point where they balance or approximately balance those measured in the MD. As in the first draw operation, the film is stretched above its Tg, which due to restricted molecular motion is now above 90°C [27, 28]. However, in contrast to the forward draw performed under conditions of constant load, the TD stretch is carried out under a constant rate of deflection. The SWD is usually continued beyond the point of initial stress hardening, which helps to minimise any added profile variation across the web [29, 30].

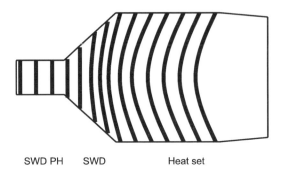

SWD PH   SWD   Heat set

Figure 5.8 Development of the macroscopic bow in film. Abbreviations: SWD, sideways draw; PH, preheat.

A more interesting complication is a Poissons ratio effect, which generates a stress in the film, at this stage in the process, that is perpendicular to the direction of draw, namely along the MD. As a result, the speed of the film at its centre momentarily falls below that at its edges and the physical retardation of the centre of the web relative to its edges remains throughout the rest of the process. The classic demonstration of this feature simply involves marking the surface of film with lines parallel to the TD prior to entry to the tenter oven, as indicated in Figure 5.8. The distortion of the film is then highlighted by the surface markings, which appear curved or 'bowed' on the web when it emerges from the tenter [31]. The film therefore experiences a combination of strain fields in the SWD zone, from simple elongation at its centre to a mixture of elongation and shear nearer its edge. Not surprisingly, the underlying microscopic structure which develops from this process history is equally complicated, yet it is clearly necessary to elucidate that structure in order to establish its relationship to properties [32–34].

One of the most powerful techniques that can be applied to the problem is the measurement of X-ray diffraction (XRD). In Figure 5.9, a series of wide angle X-ray scattering (WAXS) intensity plots are presented, which were produced using specimens collected from the centre of a PET film at various points along the SWD stage. Each trace is a plot of intensity at $2\theta = 17.7°$, which corresponds to the reflection of the (010) plane in the PET crystal lattice and is used to indicate the relative abundance of crystals at a particular orientation [35, 36]. By rotating the specimen in the beam, the so-called Chi plot is produced, which shows the distribution of orientations of the PET crystals in the plane of the film. Typically, four maxima are seen through 360°, corresponding to the preferential alignment of the two crystal populations, along the MD and TD.

From this and a knowledge that the overall level of crystallinity remains relatively constant during the sideways draw [37], and that the average crystallite size increases through the zone, it has been possible to establish the likely mechanism by which the biaxial morphology of PET film evolves. In the

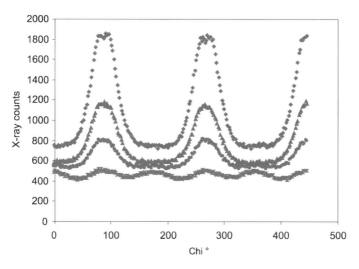

Figure 5.9 In plane orientation of crystalline fraction through the sideways draw (SWD). (Forward draw ratio of film was ×3.6). Key: ◆, sideways drawn ratio (SDR) = ×1; ▲, SDR = ×1.8; ●, SDR = ×2.5; ✗, SDR = ×3.5.

mechanical field of the SWD, large crystals present from the first FWD stage appear to survive, while smaller, less perfect crystals are destroyed. The rotation of crystallites from MD to TD does not account for the appearance of new crystallites aligned in the sideways direction. Rather, the second population is generated by a mechanism similar to that in the FWD, namely nucleation that is enhanced by an apparent increase in supercooling from orientation, and growth.

The effect of bowing by the film on its underlying microstructure can be demonstrated to similar effect using the Chi plot. Figure 5.10 shows the result recorded for a specimen of film at three positions across its width [38]. The principal alignment of the two crystal populations has clearly rotated away from the MD and TD as a result of the shear forces described previously. In practice, the extent of this rotation is seen to increase across the film from centre to edge, in accordance with the increasing angle of bow.

Although this is a qualitative description of the evolution of microstructure in the SWD, much of the mechanism has been quantified by post-analysis of film [39–41], and work continues to establish the kinetics of the transformations, using time-resolved X-ray synchrotron methods. The early results presented in Table 5.1 demonstrate that the rate constant for crystallisation in biaxial film, immediately after stretching, is less than that for uniaxially drawn film (A. Mahendrasingam and C. Martin, personal communication). However, the rate of crystallisation is significant on the timescale of the process and critical at the elevated temperatures that the system will experience in the final, heatset zone of the stenter oven.

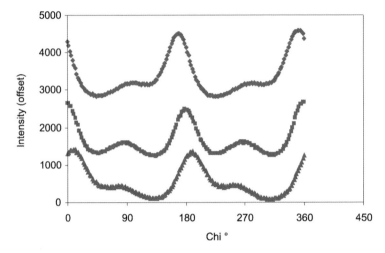

Figure 5.10 Crystallite orientation across a film. (Film draw ratios were ×3.4 and ×3.6). Key: ◆, near left edge; ■, film centre; ▲, near right edge.

Table 5.1 Crystallisation of biaxially drawn PET*

Temperature (°C)	Final draw ratio MD × TD	Rate of crystallisation (s^{-1})
90	3.3 × 1.0	8.9
90	3.4 × 2.8	8.3
90	3.5 × 3.5	7.0
120	3.7 × 1.0	18.6
120	3.0 × 2.1	17.6
120	3.3 × 3.0	10.1
120	2.7 × 3.4	9.5
120	3.4 × 3.2	8.1

*Biaxial draw process was performed simultaneously.
Abbreviations: MD, machine direction; TD, transverse direction.

Once again, knowledge of the fundamental behaviour of the polyester, its molecular alignment, molecular relaxation and crystallisation, can be combined with heat transfer and constitutive models of the film process. This offers the process engineer a complete description of the material throughout and after the SWD stage, and provides predictions of behaviour and property in support of product development.

5.2.4   Heat setting

The final stage in the stenter oven is designed to develop a crystalline morphology in the film that will retain the improved mechanical properties from the drawing stages and that is more stable over time and at temperature.

The heat set or crystallisation stage of the process comprises three or more regions of the stenter oven, each with independent temperature control and the ability to adjust the lateral dimension of the web. Thus, film can be treated to a range of thermal and strain programmes to optimise its final properties. Temperatures of the film can exceed 230°C and, although residence time may be only a few seconds, this is sufficient for density changes equivalent to a rise of 30–40% in crystallinity to occur. On the same timescale, the non-crystalline regions of the film can exhibit significant molecular relaxation, with or without accompanying macroscopic strain relaxation, and chemical reactions such as ester interchange may even occur [42].

Much has been reported about the nature and mechanism of the structural change in polyester film under conditions typical of a heat set stage, and some articles offer excellent reviews for further information [37, 43–48]. However, it should be recognised that the exact thermal history of polyester film in a commercial heat set zone is often difficult to mimic in the laboratory. A variety of treatments, such as toe-in, the strain relaxation in the TD and MD, haul-off tension and cooling rate, are superimposed on the biaxial film, and their interplay can cause complex and often subtle effects on the structure and behaviour of commercially manufactured film.

For example, curve A in Figure 5.11 is the differential scanning calorimetry (DSC) trace of PET film approximately two years after manufacture. The trace shows a pronounced endothermic peak in the region of the Tg, caused by the reduction in free volume with time. Curve B (Figure 5.11) shows the

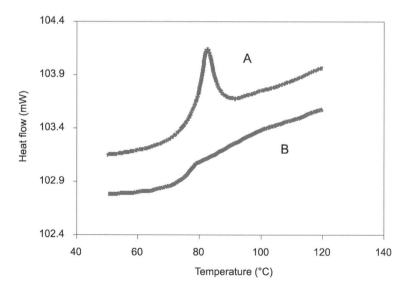

Figure 5.11 Differential scanning calorimetry (DSC) of commercial PET film.

Table 5.2 Residual shrinkage in PET film

Heat set temperature (°C)	Strain relaxation (toe-in)	Residual shrinkage at 150°C	
		MD	TD
210	None	1.90	2.30
220	None	1.78	1.59
230	None	0.92	1.39
210	MD only	−0.32	2.40
210	TD only	1.76	0.16

Abbreviations: MD, machine direction; TD, transverse direction.

corresponding response by similar film immediately after manufacture, before the ageing process has progressed [49]. From this perspective, film which emerges from the heat set stage is metastable [50]. A less subtle and often more important behaviour, which is exhibited by film as a consequence of the history of the heat set stage, is depicted in Table 5.2.

Unless all physical anisotropy can be removed from the non-crystalline fraction of biaxial PET film, the product will undergo residual shrinkage at elevated temperature. By managing both the film temperature and a relaxation of strain (known as toe-in) during the heat set stage, it is possible to achieve considerable control of this film property. However, this control cannot be managed at all temperatures during exit from the stenter oven and in practice some degree of frozen-in strain will exist in the final film. Both phenomena underline the point that the heat set of the process does not cause true structural equilibrium to be achieved in the film. A stable structure may be closely approached at the high temperatures of the stenter oven but this is not the equilibrium structure under ambient conditions. Upon exiting the heat set zone and cooling rapidly, the film material is unable to equilibrate structurally to its ideal state at room temperature and, consequently, will display properties which are both time and temperature dependent. Nevertheless, the film which exits the heat set stage of the stenter oven has completed its primary manufacturing programme and all knowledge about its structure and composition at this point is of value to predict or guarantee its final performance. Thus, a wealth of information has been gathered about the structure-property relationship of polyester film. Furthermore, process and material constitutive models, which are increasingly powerful, are also providing routes to predict specific properties. For example, understanding, predicting and controlling the dimensional stability of commercial polyester film has always been of key importance [51–54] and this property is most strongly influenced by conditions in the final heat set stage of manufacturing. Similarly, several other properties of the finished film which depend heavily on crystallinity or amorphous orientation will be highly correlated to the heat setting conditions of the process [55–59].

However, current and future applications are demanding film with increasingly stringent specifications. Consequently, it is becoming more important to account for the cumulative influence of all stages of the production process on final structure and properties of film, in order to manufacture to the required standard. Clearly, the strategy to consider the process as a whole is approached with the aid of statistical and phenomenological models. However, as this capability evolves and emphasises how many features of polyester film are primarily correlated to its chemical composition, it also becomes necessary to enlarge the process-property model and include information about the nature of the polyester. This is covered more fully in Section 5.3.

5.2.5 Slitting and winding

It can be argued that the production of high quality, consistent and stable rolls of film depends on the successful management of the film web from an early stage in the film line. Throughout the process, it is necessary to maintain tension in the film to stop it from sagging, wrinkling or ultimately wrapping around rollers. Good control of the speed of drives through the unit is therefore essential. Overall speed is referenced to that of the casting drum, so that adjustments are automatically made to all drives when the casting drum speed is changed. In practice, each drive is referenced to the preceding drive, so that speed adjustment at any point in the line results in adjustment of all downstream drives. At critical points in the line, notably between casting and forward draw, forward draw and tenter and tenter and wind up, the film tension is measured and the speed of the downstream drives adjusted—usually automatically—to maintain a preset tension level. When the film leaves the tenter, good control of tension is necessary to maintain good film flatness. The film is pulled from the tenter by the 'haul-off nip' (see Figure 5.12). Just before this nip, the film passes over a non-driven roller mounted on load cells in order to measure the tension in the film passing over the roller. Alternatively, a dancer roller is used, which relies on displacement from a balance point to determine the deviation from a set-point tension. During initial thread-up of the film, the haul-off nip is run in tension control mode. While the film is slack around the 'tension' roller, the haul-off nip will speed up to take up the slack in the film and restore the film tension. During normal running, the tension roller is used to indicate web tension and may be used to control the haul-off tension.

The haul-off nip consists of two rollers: a driven chrome-plated steel roller and a rubber-covered steel nip roll located in a pneumatically operated level arm. The nip force between the two rollers is used to prevent slip of the film over the roller surfaces, and must be uniformly loaded over the width to prevent film distortion and the potential for creasing.

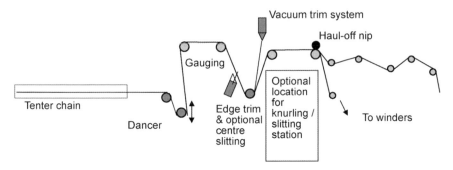

Figure 5.12 Take up (web transport) section.

5.2.5.1 In-line slitting and knurling

Some lines have a single winder, with the full width of film from the tenter wound up as a mill roll on a steel core for later batch slitting. Most lines manufacturing film above 23 μm, or wide thinner film lines above 3 m in width, carry out slitting in-line and subsequent winding of two or more reels at separate winding stations. Generally (but not always), thicker films are knurled at the edges before winding. Knurling thickens an area at the edge of the film so that it supports the wound in tension when the film is reeled. It also allows a layer of air to lie between adjacent layers of film, to prevent any surface damage by one layer moving relative to the next. Knurling heads are mounted on a cross-beam and can be positioned across the film web, where it will subsequently be slit (see Figure 5.13). They

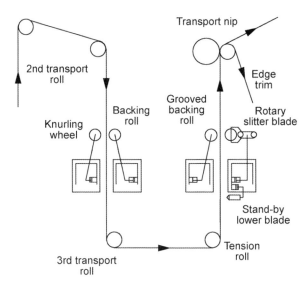

Figure 5.13 Knurling and slitting station.

consist of a knurling wheel, backing roller and a method of removing any debris generated. The knurling wheel is usually a narrow steel wheel with a patterned projection. This is forced against the film, which is supported on the other side by the wider backing roller. The pointed surface distorts the film by the amount of penetration of the pattern and, therefore, raises the previously flat surface. Both knurling and backing wheels are held in position by pneumatic cylinders. The knurling pressure is controlled pneumatically according to the height of film distortion required. If an interruption to the process, such as a split in the film occurs, the knurling and backing wheels are deactivated and move apart to prevent mechanical damage and subsequent re-threading to occur.

The slitting station is positioned after knurling and is designed to slit the main web into the width of film required for each winder. Slitting heads and sometimes backing blades or rollers are supported on cross-beams similar to the knurling heads. They are driven to precise location across the web. The knurling, slitting and backing systems must exactly match positions across the web and are usually geared together.

Film is slit by various methods, commonly employing razor blades for thinner films and rotary shear for thicker films. It is important that the slit edge of the film is flat to prevent a 'high edge' build-up and that generation of slitting debris is kept to a minimum. The razor blade is the simplest design and is sometimes oscillated so that the cut position moves along the blade edge. Rotary shear is more complex to set up. It consists of two circular blades, which are pushed together pneumatically. The small overlap and angular displacements between the blades needs to be precise; however, the quality and cleanliness of the slit edge for thicker film is far superior.

A final consideration of the slitting stage is the reclaim and recovery cycle. The film in and close to the clips in the stenter oven is very thick and cannot be wound in. It is slit off at the wind-up and sent to a recovery process. Here, it is combined with scrap film and is either cut up into flake and compacted into particulate form or is re-extruded and formed into pellets. This reclaimed polymer is either fed back with the virgin polymer at the start of the film process or is fed into the CP process.

5.2.5.2 *Winding conditions*

At the end of the tenter, the film may either be wound into reels on cardboard cores to be sent directly to the customer as 'direct hand-over' (DHO), or into mill rolls, which are subsequently slit out of line on a separate 'slitter-rewinder'. Much DHO film is knurled before winding, such that the winding conditions are then less demanding. Some DHO film and almost all film slit on slitter-rewinders are wound without knurl. Under these conditions, the winding conditions and surface properties of the film have a major effect on the reel quality. Winding conditions need to be carefully controlled to produce a reel free from faults and stable enough to be handled through packing, distribution and the customers' processes. Some tension is required to wind film onto a core and prevent looseness and flapping of the film through the machine. Layers of some smooth films stick (or 'block') to each other causing the film to 'buckle' on the reel and giving distortions in the film when wound. For this reason, knurling of the film is employed.

When 'filled' film is wound into a reel, air is drawn in between the layers as the surface roughness keeps layers a small distance apart. This provides a 'cushion', enabling some stress relaxation with a reduced risk of buckling, and also reduces the effect of variations in film thickness, which can locally distort the reel of film. If the winding speed is increased, more air is carried along with the film and entrained between the layers. When too much air is wound-in, the layers of film will move sideways relative to each other causing the reel to 'telescope' or 'dish'. To avoid this, the film is wound with a 'lay-on roller', which the film passes around before contacting the winding reel (see Figure 5.14). The pressure of the lay-on roller on the surface of the winding reel can be varied to alter the amount of air entrained. Normally, this pressure is kept constant through the winding of a reel but it will be increased if, for example, the winding speed is raised.

The tension applied to the film web during winding is also critical to good slit reel formation. Normally, a form of 'taper' tension is used, where the tension is reduced in a controlled manner as the diameter increases. In practice, although called taper tension, open loop torque control is used, where the power

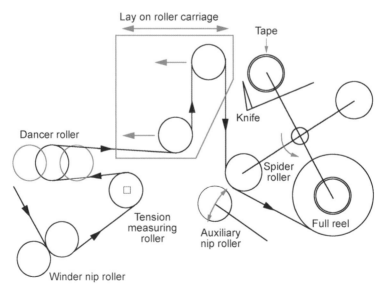

Figure 5.14 Typical winder layout.

to the winding motor is increased as the diameter increases but not to the extent required for constant tension winding. It is also normal to make some compensation for inertia and speed effects when accelerating and decelerating at the start and end of reels.

5.2.6   Reclaim and recovery

The film in and close to the clips is very thick and cannot be wound into film. It is slit off at the wind-up and sent to a recovery process. Here it is combined with scrap film and is either cut up into flake and compacted into particulate form or is re-extruded and formed into pellets. This reclaimed polymer is either fed back in with the virgin polymer at the start of the film process or is fed into the CP process.

5.3   Polymer, process and properties (3Ps)

The quality and consistency of the feedstock can have a significant effect on the efficiency of the film process, namely on the complicated interrelationship between polymer, process and properties [60, 61]. Subtle changes to the polymer feedstock can render the polymer unfilmable unless the film process parameters are altered to accommodate this change, and these issues are magnified if changes are made to the polymer formulation for the different grades of film produced. Because of the interdisciplinary skills required to understand and manage these changes through the film process, the '3Ps' interrelationship is often overlooked in discussions of this type. For this reason, it is being treated separately rather than included in the description of the process.

5.3.1   Polymer

The polymer architecture is controlled in the first place by the formulation, which will dictate the basic class of polymer and the chemistry and physical properties associated with it. Superimposed on this, however, is a series of subtle effects that can have a significant influence on the properties of the polymer. The molecular weight of the polymer, the molecular weight distribution, whether the polymer is linear or branched, filled or unfilled, the type and level of filler, and the type of end groups all contribute to the characteristics of the polymer.

The polymer architecture and recipe in turn controls whether the polymer is crystalline or amorphous and, if crystalline, the rate of crystallisation and the level of crystallinity. The polymer architecture also dictates the conformational changes that the polymer can undergo, and thus influences the Tg and Tm. These influences on the molecular morphology contribute to how the polymer behaves in the melt and to the polymer's rheology.

5.3.2 Process

As mentioned above, the architecture and rheology of the polymer are intimately related. Simplistically, these factors dictate the processing conditions. Superimposed on this, as described in Section 5.2, the process itself will further influence the polymer morphology and dictate the level of orientation and crystallinity in the finished article.

5.3.3 Properties

Polymer and process, by virtue of the interrelationship described above, essentially determine the properties of the finished article, namely the mechanical properties, the clarity, the dimensional stability, the enviromental resistance, the usage temperature, the barrier properties and so on.

The list of effects described under the three headings above is by no means complete but is meant to illustrate how a subtle change to the polymer can have knock-on effects to process and properties. In extreme cases, these subtle effects can render the polymer unprocessable. The interrelationship can be illustrated in more detail by revisiting the film process outlined above.

Extrusion. The purpose of the extrusion stage is to melt the polymer and provide a constant flow of molten polymer to the die. Any variation in the extruder conditions will cause fluctuations in the output and give variations in the cast film. These variations in thickness will last through the process and show up in the finished film—this is obviously undesirable. In addition to the process affecting thickness variations, polymer parameters that influence the rheology will also affect thickness for the same reason. Therefore, variations in molecular weight and molecular weight distribution and subtle alterations to the PET architecture that influence Tg and Tm, such as copolymer level or diethylene glycol content, can potentially influence thickness. The degradation reactions that the PET can undergo are influenced both by the polymer recipe and the extrusion conditions—these can lead to discolouration and surface defects in the film. Changes in molecular weight and branching due to degradation, and build-up of degraded polymer in the extruder will also influence how the polymer behaves in the melt and thus the stability of the extrusion process.

Casting. As discussed in Section 5.2, polymer factors such as copolymer level, filler level and molecular weight which influence crystallisation rates will also have a significant influence on the casting process.

Drawing. The drawing process is carried out above the Tg of the PET, and thus the polymer factors outlined above, which influence Tg, will have a subtle effect on the drawing process and therefore the mechanical properties of the film.

Heat setting. Polymer factors that influence the crystallisation characteristics will also have a significant effect on the heat setting process and therefore the final film properties, such as low shrinkage and dimensional stability, chemical resistance, flatness and delamination.

Wind up. The finished film is wound up after the heat set process. It is essential to obtain uniform tightly wound reels. This stage of the process is dominated by process conditions but even here the windability of the film will be affected if the inherent stiffness of the polymer has been altered, for example by the presence of comonomers.

In some cases the polymer factors dominate and in other cases the process factors dominate the interrelationship of the 3Ps. Furthermore, specific process or polymer properties can have a positive effect on some final properties and a negative effect on others, for example, increasing the draw ratio increases the ultimate tensile strength but has a detrimental effect on shrinkage, and increasing the crystallinity of the polymer increases the modulus of the film but can lead to web breaks during the filming process. These issues have to be understood and managed to successfully introduce new polymer variants to the film process or when altering process conditions to achieve new effects.

5.4 Surface and bulk properties

5.4.1 Film properties

The film process described produces rolls of PET film that have the properties required for a standard PET film, namely: high mechanical strength, good flexibility, excellent visual properties, flatness and dimensional stability and a thicknesses of 0.6–500 μm. The thermal characteristics of PET film (see Table 5.3) enable it to retain physical, chemical and electrical properties over a wide temperature range, with good resistance to the effects of heat ageing up to 150°C. (NB The typical properties listed in Tables 5.3–5.7 apply to 23 μm general purpose Mylar® film. They are for illustrative purposes only and are not intended to be used as design data.)

The mechanical properties of PET film are presented in Table 5.4. PET retains its physical properties, including tensile strength, folding endurance and tear strength over a wide temperature range (−70°C to 150°C). The toughness of the film is demonstrated by the very high tear-initiation strength. PET film does not contain plasticisers, a key advantage in the area of packaging consumable items, and does not become brittle with age under most conditions.

The chemical properties of PET film (see Tables 5.5 and 5.6) include excellent resistance to most chemicals. PET film is virtually impermeable to most chemical reagents, solvents, impregnants and varnishes. PET film is dissolved by hexafluoro-2-propanol, *m*-cresol and *o*-chlorophenol, and is attacked by

Table 5.3 Thermal properties of 23 μm Mylar® film

Property	Typical values	Test method
Melt point	253–255°C	ASTM D3418-82
Shrinkage (150°C, 30 min)		
MD	1.5%	ASTM D1204-78
TD	1.2%	
Coefficient of thermal expansion at 30–50°C	1.7×10^{-5} Cm/cm/°C	ASTM D696-44
Heat sealability	None unless coated or treated	
Specific heat (25°C)	1.32 J/g/°C	
Thermal conductivity	0.15 W/m°K	

Abbreviations: MD, machine direction; TD, transverse direction.

Table 5.4 Mechanical properties of 23 μm Mylar® film

Property	Typical values	Test method
Tensile strength (MD) at 25°C	29 000 psi 200 MPa	ASTM D882-80
Tensile modulus (MD) at 25°C	565 500 psi 3 900 MPa	ASTM D882-80
Elongation (MD) at 25°C	130%	ASTM D882-80
Stress to produce 5% elongation (MD) at 25°C	15 200 psi 105 MPa	ASTM D882-80
Density at 25°C	1 395 g/cm³	ASTM D1505-66
Folding endurance 1 kg loading, at 25°C	100 000 cycles	ASTM D2176-63
Tear strength		
Initial (Graves) at 25°C	294 N/mm	ASTM D1004-66
Propagating (Elmendorf) at 25°C	7.4 N/mm	ASTM D1922-67
Coefficient of friction (kinetic) at 25°C	0.33	ASTM D1003-61
Refractive index (AB 8E at 25°C)	1.64 ND25	ASTM D-542-50
Coefficient of hygroscopic expansion	1.0×10^{-5} mm/mm % RH	

Abbreviations: MD, machine direction; RH, relative humidity.

35% nitric acid, 10% ammonium hydroxide and *n*-propylamine. PET film also has very low moisture permeability and overall resistance to staining by various chemicals and food products. Gas and vapour barrier properties can be significantly improved by coating with a barrier coating, such as polyvinylidene chloride, or by vacuum metallisation.

Table 5.5 Chemical properties of 23 μm Mylar® film

Property	Typical values	Test method
Moisture absorption (immersion for 24 h at 23°C)	Less than 0.8%	ASTM D570-63
Permeability to gases		
Carbon dioxide at 25°C	16 cc/(100 in^2) (24 h) (atm)/(mil) 6 cc/(m^2) (24 h) (atm)/(mm)	ASTM D1434-72
Hydrogen at 25°C	100 cc/(100 in^2) (24 h) (atm)/(mil) 39 cc/(m^2) (24 h) (atm)/(mm)	ASTM D1434-72
Nitrogen at 25°C	1 cc/(100 in^2) (24 h) (atm)/(mil) 0.4 cc/(m^2) (24 h) (atm)/(mm)	ASTM D1434-72
Oxygen at 25°C	6 cc/(100 in^2) (24 h) (atm)/(mil) 2.3 cc/(m^2) (24 h) (atm)/(mm)	ASTM D1434-72
Permeability to vapours*		
Acetone at 40°C	2.22 g/(100 in^2) (24 h)/ (mil) 0.87 g/(m^2) (24 h)/(mm)	Mod. ASTM E96-80
Benzene at 25°C	0.36 g/(100 in^2) (24 h)/(mil) 0.14 g/(m^2) (24 h)/(mm)	Mod. ASTM E96-80
Carbon tetrachloride at 40°C	0.08 g/(100 in^2) (24 h)/(mil) 0.03 g/(m^2) (24 h)/(mm)	Mod. ASTM E96-80
Ethyl acetate at 40°C	0.08 g/(100 in^2) (24 h)/(mil) 0.03 g/(m^2) (24 h)/(mm)	Mod. ASTM E96-80
Hexane at 40°C	0.12 g/(100 in^2) (24 h)/(mil) 0.05 g/(m^2) (24 h)/(mm)	Mod. ASTM E96-80
Water at 37.8°C	1.80 g/(100 in^2) (24 h)/(mil) 0.70 g/(m^2) (24 h)/(mm)	Mod. ASTM E96-80

*Vapour permeabilities at the partial pressure of the vapour at the temperature of the test.

Table 5.6 Chemical resistance properties of 23 μm Mylar® film

Property	Tensile strength % retained	Elongation % retained	Tear strength % retained
Chemical resistance to:			
Acetic acid, glacial*	100	100	100
Hydrochloric acid (10%)*	100	100	100
Sodium hydroxide (2%)*	100	100	70
Ammonium hydroxide (10%)*	0	0	0
Trichloroethylene*	100	100	100
Hydrocarbon oil (immersion for 500 h at 100°C)	92	88	87
Ethanol*	100	100	100

*Immersion for 31 days at 23°C.

Table 5.7 Electrical properties of 23 μm Mylar® film

Property	Typical values	Test method
Dielectric strength 25°C, 50 Hz, 50 mm electrode	6400 volts	ASTM D149-64
Dielectric constant 25°C, 1 kHz	3.2	ASTM D150-81
Dissipation factor 25°C, 1 kHz	0.005	ASTM D150-65
Volume resistivity 25°C	10^{18} ohm cm	ASTM D257-78
Surface resistivity 25°C, 30% relative humidity	10^{16} ohm square	ASTM D257-78
Corona resistance 1000 volts	5 h	ASTM D2275-80

PET film has excellent electrical insulating properties, as shown in Table 5.7.

Because of its excellent thermal, insulating and moisture resistant properties, PET film is used in a wide variety of electrical applications. However, for many specialty applications further modification of either the surface or bulk properties are required, as illustrated in the sections below.

5.4.2   Coating

PET is a fairly inert polymer and for many applications the surface of the film is altered by coating or adhesive lamination to other materials, for example, film for packaging will be lacquered to accept inks or adhesives, and film for photographic applications will be primed to accept photosensitive overcoats. An untreated film would require treatment with aggressive chemicals to modify the surface sufficiently to allow such coatings to adhere to it. Coatings are also applied to achieve other surface effects, such as antistatic properties, barrier (to water, oxygen, carbon dioxide, flavour), release and frictional characteristics. The chemistry and colloid science of these formulations is vast and complex and outside the scope of the present article. However, it can be said that coatings often contain a 'primary effect' material, key to obtaining a given effect (more than one effect may be required). For example, in the case of an adhesion-promoting coating, this material is often a copolymer, such as an acrylic, vinyl, polyester or polyurethane. In the simplest case, all the other components of this coating are essentially there to facilitate the uniform application (and subsequent adherence) of the effect polymer to the film. Consequently, the wetting, flow, drying, film formation and cure kinetics of the coating during the surface treatment process are all important factors to be considered when formulating the coating material. A typical aqueous formulation may contain a solution or dispersion of an effect polymer, together with a wetting agent (surfactant or co-solvent), a cross-linking agent and catalyst, pH regulator, a film former (coalescent, plasticiser), defoamer, antiblock, etc. Clearly, a given ingredient may have an influence in more than one of these areas. Wetting of the PET surface ($\sim$43 dynes/cm) may be facilitated by flame or corona discharge treatment prior to coating. The cross-linker is a species that is simultaneously reactive towards the PET surface (hydroxyl and carboxyl end groups) and the effect polymer; multifunctional epoxides, aziridines, isocyanates and melamine formaldehyde resins are commonly used.

PET film is commonly coated either 'in-line' or 'off-line'. In-line coating usually involves aqueous-based materials and is carried out between the forward draw and before entering the tenter, while off-line coating involves unwinding and priming the surface of preformed film reels with aqueous or solvent-based coatings. In both cases, the coating may be applied to one or both sides, and some products have different coatings applied to either side of the film. Coatings are mostly applied by offset gravure or by direct gravure coating. Coat continuity and uniformity are important to ensure that the required properties are obtained. Uniform dry coating thicknesses as low as $\sim$10 nm can be routinely achieved in-line, while off-line coatings can be 10–100 times thicker. Too much coating on the film is to be avoided in the in-line process, since the water has to be dried off in the tenter preheat section and the film will only heat up to the required temperature for the SWD when it is completely dry. Excess wet coat thickness may cause the film to be at too low a temperature when drawing starts, resulting in poor profile and possibly breakage of the film web. Applying excessive amounts of coating can also result in unacceptable film optical properties, such as coat non-uniformity, haze and yellowness. Furthermore, because the tenter process often involves recycling some film, and some of this may be coated, thick coatings may adversely effect the colour, clarity and extrusion characteristics of the recycled polymer melt. Off-line coating does not have this restriction on the chemistry and thickness of the coating; the off-line process is better suited to coating high Tg coating materials (as the coating is not drawn), and those requiring longer cure schedules.

5.4.3 Co-extrusion

Co-extrusion is used to produce a film with two or more different polymer layers, so that one or both surfaces of the film have different properties to the 'core' polymer. In addition, co-extrusion allows manufacture of products with layers thinner than can be made and handled individually. Co-extrusion provides a means of flexibly configuring a wide range of film laminate structures which cost-effectively meet product requirements. An example of a co-extruded film is Mylar® 850, in the production of which a lower melting copolymer containing isophthalic acid is extruded onto a standard base film to give a film that can heat seal to itself and heat seal to thermoformed APET/CPET trays and APET-coated board. It will also heat seal to various other substrates, including polyvinylidine chloride (PVDC), PVC, paper and aluminium foil, but it does not heat seal to polyolefins. The co-extruded layer is about 15% of the total structure and by tailoring the copolymer chemistry a wide heat seal range, from 140–220°C, with outstanding hot tack properties can be obtained. The copolymer, if extruded as a film itself, would not have sufficient strength and the base PET homopolymer provides the mechanical properties required. The sealable surface of the film also acts as an excellent prime for water-based lattices. This film can be exploited in various ways, for example the film can be PVDC coated from an aqueous dispersion by convertors to produce a high barrier laminating film. Further examples of co-extrusion application include the use of reclaim polymer in the centre of films, so that more expensive virgin polymer can be used sparingly to provide surface properties.

The basic co-extrusion process consists of the generation of two or more melt streams and their confluence in the melt phase. The number of separate extrusion systems is determined by the number of polymer types, typically two but occasionally three and exceptionally up to ten. Each polymer type to be incorporated in the structure is separately melted, pressurised and (optionally) filtered in parallel extrusion systems before flowing into the co-extrusion hardware. The optimum method of bringing the separate melts together depends primarily on their respective flow behaviours. The melt layers must remain distinct but well bonded in the process, from the point(s) of confluence through to solidification. There are basically two hardware configurations in use for common polymers—the multimanifold die and the injector block. (Combinations of the two are also possible for complex structures.)

The multimanifold die is used when the viscosities of the melts are dissimilar, or low, or where there are many streams to combine. The streams are brought together at the die, just prior to solidification, thus allowing virtually no time for the weaker melts to be distorted in the short region of laminar flow. Co-extrusion profile (the relative thickness of layers across the web) is a key process requirement in many products. However, multimanifold dies are generally expensive, complex and can be difficult to incorporate in thermoviscous profile control systems. Polypropylene film processes typically utilise multimanifold dies because of the tendency of that polymer to encapsulate (that is distort the shape of) the secondary polymer in the laminar flow regions.

The second option, used extensively with PET, is injector block technology. This is restricted to two or three layer applications. Injector block systems consist, in basic form, of a main melt channel with one or two side entries through which secondary melts flow. The injected melt(s) then flow side-by-side with the main polymer, in the same pipe, towards the die. The mechanism is most effective when the viscosities of the melts are similar. The greater the viscosity difference the more pronounced will be the distortion of relative profile. For many packaging applications, however, profile distortions up to 10% are acceptable, and therefore the process has a degree of robustness towards viscosity variations. The basic injector technology can be enhanced by the addition of profiling cams and fins at the point of sidestream entry, which contour the flows in a way that leads to a more uniform profile. Adjustable cams can be included, which provide for *in situ* optimisation. Cam design is particularly important for uniform profile. Typically, cams take the form of a rectangular slot. A particular requirement for tenter film processes is often to avoid having sticky film under tenter clips, and hence to provide web edges clear of secondary polymer. Cams can be designed which restrict the application of secondary polymer to a centre width of the film web. However, the maintenance of uniform profile in such designs is very difficult, and companies that have developed good designs are often reluctant to share them. It is generally true to say that the wider the film die, the more difficult it is to achieve good profile and clear edges. A further enhancement to injector block technology is to include melt routing flexibility through reconfigurable transition blocks/flanges. This gives the option to redirect polymer streams to any entry point of a multilayer block through a simple operator selection procedure.

Co-extruded products that have low conversion efficiencies and have to include recovered scrap may present particular problems. Many packaging film applications can tolerate inclusion of reclaim in the base polymer layer, but those which cannot necessitate either scrap dumping or expensive chemical separation.

5.4.4 Fillers

Fillers are added to PET for two main reasons. First, to modify surface properties: particulate fillers, such as clays and silica, typically a few microns in diameter, are added to create surface roughness during the film-drawing process. A primary function of the surface roughness is to reduce the blocking or sticking propensity of the otherwise very smooth film surfaces during winding and reel formation. The roughness also enhances dynamic handling behaviour, particularly for high speed transport and winding of very thin films. Surface optical properties can also be regulated via filler-induced surface roughness, for example to control gloss or eliminate Newton's Ring fringes between adjacent film layers. Finite element methods have been applied to enable predictive capability in the design of roughened film surfaces [62].

Second, to modify bulk properties: although mechanical properties, such as softness, stiffness and toughness, can be addressed, it is most common for optical properties to be modified via particles. Opacity and whiteness are generated by two discrete mechanisms. Simple pigmentation (light scattering from the particle-polymer interface) can be achieved using similar titanium dioxide technology to that employed in the fibres and coatings industries. It is, however, more common for the anatase crystal form to be employed, since this is a less abrasive pigment than the more strongly scattering rutile. The second mechanism involves using the additive to generate micro-voiding during the film draw. The additive can be inorganic, for example barium sulphate or calcium carbonate, or polymeric, for example, polypropylene. In this mechanism, the opacity is derived from scattering between the polymer and the void. The use of micro-voiding confers the potential advantage of a softer film of reduced density.

5.4.5   Shrinkage

Standard PET film from the tenter process will shrink by 1–3% after 30 min at 150°C. For the high level of accuracy required by Membrane Touch Switch printers, 0.1–0.2% shrinkage is required. In order to meet this, the film is unwound and passed through a carefully temperature controlled oven, with almost no tension in the film. The amount of residual shrinkage in the film after this process is typically 0.1% in the MD and 0.06% in the TD.

5.4.6   Combination of effects

Typically, to achieve the wide range of properties required of polyester films the effects above can be combined, as illustrated in Figure 5.15. The structure of film can be further enhanced by coating on one or both sides. This is further illustrated in Table 5.8, which outlines the key properties required for packaging and other film applications. It can be clearly seen how the combination of base film properties, fillers, coatings and co-extrusion technologies have to be used to achieve the desired properties. In the packaging market, the film is commonly laminated to such substrates as polyethylene and polypropylene. This is illustrated in Figure 5.16, showing Mylar® 813, a clear film with a pretreat coating for enhanced print adhesion, which is typically used as a part of a laminate in flexible packaging. It is usually on the outside of the structure, where the excellent print quality achieved makes products stand out on retail display. Other key properties are: aroma barrier, rigidity and robust laminate performance. Other typical applications for this product, when laminated to polyethylene or polypropylene, include:

- stand-up pouches for such diverse products as soups and washing liquid, which is commonly a reverse-printed duplex structure (see Figure 5.16A)
- beverages and pet food pouches, where a high barrier is required for shelf-life (see Figure 5.16B) and where the film is reverse-printed and adhesively laminated to aluminium foil in a triplex structure
- coffee and other dried product packs, where the film is a metallised duplex (see Figure 5.16C) and can be surface printed

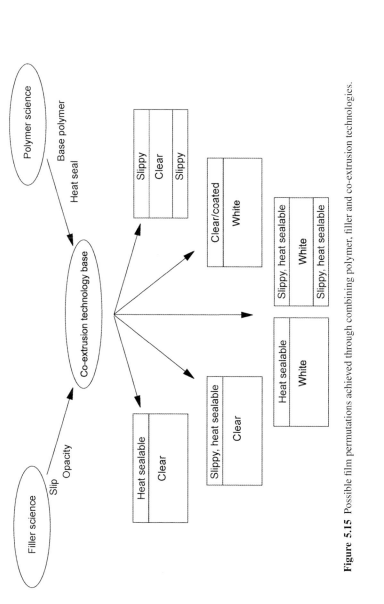

Figure 5.15 Possible film permutations achieved through combining polymer, filler and co-extrusion technologies.

Table 5.8 Examples of typical film applications and the key properties required

Application	Film properties required
Packaging	Seal (peelable through to permanent) Wide heat seal range Barrier (oxygen and aroma) White, clear Handleability Pretreats to allow metal, ink, sealant adhesion Stiffness 12–30 μm
Labels	Clear, white, matt, black Durability Adhesion with inks Silicone adhesion 23–175 μm
Imaging –montage –microfilm –digital technology media	High opacity, white through to high clarity, flatness Antistat Handleability Dimensional stability Adhesion with inks 50–175 μm
Casting and release	Wider range of surface textures from high gloss to very matte Release coats 12–50 μm
Capacitors	Thin film, low shrinkage, electrical – thermomechanical properties 0.9–23 μm
Electronics –flexible printed circuits, flat flexible cable –membrane touch switch –loudspeakers	Brilliant clarity-white film Electrical properties Handleability Low shrinkage Durability Adhesion with inks 12–175 μm
Coil coating/fibre reinforced plastics	Heat bondable UV stable Handleability 12–125 μm
Electrical insulation –motors –cable	High dielectric strength Thermal endurance at elevated temperatures Dimensional stability Durability Chemical resistance 12–350 μm

Table 5.8 (continued)

Application	Film properties required
Magnetic media	Tensilised film
	Surface quality
	Dimensional stability
	Sublayer for magnetic coating
Medical test strip	Dimensionally stable
	High stiffness
	Inert and plasticiser free
	Adhesion with ink
	Hydrophilic surface
	Clear (matt or glossy), white (high opacity or translucent)
Cards	Durability
	Adhesion with inks
	Heat bondable
	Temperature resistance
	100–350 μm

Further examples of PET film use in packaging are: films for convenience, such as lidding, where easy peel through to permanent seal, to a variety of container types, is a key feature in ready meal applications; and films for freshness, such as in meat and cheese packaging, where barrier and clarity are the key features required for modified atmosphere packaging to achieve excellent product shelf-life.

5.5   PET sheet

5.5.1   Extrusion of PET sheet

PET can be converted readily into containers suitable for a wide range of packaging applications by thermoforming of extruded sheet. The technology can be considered as comprising two process stages [63].

In the extrusion stage, the PET pellets are melted in a continuous screw extruder and the melt passes to a sheet die—ideally using a gear pump to control flow—and then on to a chilled three-roll stack, where the sheet is quenched rapidly into an amorphous cast sheet. The next stage reheats the sheet to the required temperature using an in-line, fully integrated system or a separate machine, and the sheet is thermoformed into the finished container.

Standard process conditions are used for extrusion of PET into sheet. The polymer pellets must be fully dried prior to processing, with air having a dewpoint below $-30°$C. This is recommended to ensure that the polymer pellets have a moisture content of less than about 10 ppm, to minimise degradation in

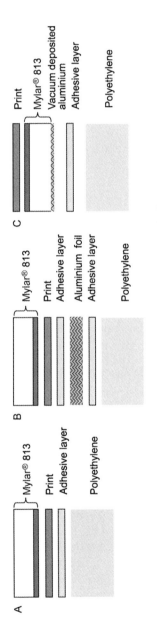

Figure 5.16 Illustration of laminates in flexible packaging.

the melt due to hydrolysis. The molecular weight of the polymer should be high enough to ensure reprocessed edge and web or skeletal regrind from the sheet. All regrind should be dried before re-extrusion. Levels of regrind up to about 60% can be used and should be maintained at a consistent level for the process. PET will show a small decrease in viscosity during extrusion and a high addition of regrind will give a sheet with low viscosity. Inclusion of regrind at variable levels may influence the resin melt-out characteristics. PET pellets are plasticised using a continuous screw extruder of a suitable screw design. Melt temperatures should be 270–285°C, and should be kept as low as possible to avoid degradation and generation of acetaldehyde (AA). It may be necessary to use a metering pump to ensure a uniform delivery of polymer melt and to minimise dimensional irregularities in the sheet. Static mixers can sometimes be used to reduce variability in intrinsic viscosity resulting from inadequate mixing of virgin and regrind material. Dies for sheet are flat, using either fixed or flex-lip openings. Flex-lip gives a better range of lip openings to produce sheet of required thickness, while fixed opening dies are more limited but give better gauge and temperature control.

PET melt should be cast from the die onto a three-roll stack with chilled feedwater to the rolls (see Figure 5.17). The calendering system is run with a melt bank in the nip area to produce a uniform polished sheet. Rapid quenching of the melt is essential to ensure that amorphous sheet is produced with high gloss, transparency and strength. Roll temperatures will vary between 10°C and 65°C, depending on the throughput and sheet thickness. The thickness of sheet cast will be influenced by: extruder output, die lip opening and casting roll speed. From the roll stack, the sheet can be transferred to a turret winder via edge trim slitters. Sheet can be transferred directly from the roll stack to a reheat oven and thermoformed in-line, or it can be wound and stored for forming at a later date. In direct processing, edge and web scrap is usually reground in-line and transferred back for redrying.

5.5.2 Thermoforming of PET sheet

PET can be thermoformed on most conventional equipment. The forming conditions depend on the specific container design and the degree of orientation. Low pressure containers need a shallow-to-medium draw process, while for pressure containers suitable for carbonated drinks and when good barrier properties are required, it is recommended that a deep draw process is used. The deep draw process induces a high degree of orientation into the container wall, which is essential for more demanding applications. Shallow-to-medium depth drawn containers, with stretch ratios up to about 2.5:1 in the axial direction, can be formed using vacuum (10 mbar) or with a combination of vacuum and compressed air (10 bar). Deep drawn containers require plug assist technique. The type of plug and conditions will vary depending on the sheet thickness and

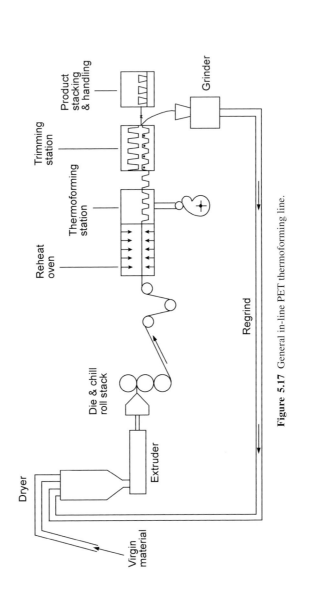

Figure 5.17 General in-line PET thermoforming line.

the stretch ratio. Stretch ratios up to 3:1 and sheet thicknesses up to 1.5 mm can use a cold nylon plug or polymer-coated metal plug. Deep drawn containers require oil-heated brass plugs with a temperature between 80°C and 110°C. Sheet produced in high IV PET will require the higher plug temperature. The sheet should be heated on both sides using top and bottom heaters. For shallow transparent containers, heating from the top is sufficient. The temperature of the sheet should be reheated between 90°C and 110°C; the deeper the draw and higher the PET IV, the higher the reheat temperature required.

5.5.3 Thermoforming of CPET sheet

PET can be modified with nucleating additives, including polyethylene and polypropylene polymers, such as low density polyethylene (LDPE) and linear low density polyethylene (LLDPE), which are incompatible with PET but which at low levels of addition (usually about 5%) can induce rapid crystallisation of PET during reheating and the forming process. Fast crystallising PET grades, called CPET, are specially formulated for the production of opaque containers, like ovenproof and microwaveable trays. CPET grades can be extruded on conventional sheet lines but are thermoformed into a heated mould, which heat sets the CPET container. Mould temperatures are about 160°C to optimise the crystallisation of CPET, with about 2 s contact with the hot mould, giving an overall cycle time of about 5 s. The clear sheet becomes milky white and opaque when reheated at 120–145°C, which indicates that crystallisation has occurred and it is ready to be formed. Uniform heating is critical to ensure that the finished container is uniformly crystallised. Containers that are not uniformly crystallised will tend to warp and distort when heated in an oven. Overheated sheet will produce too high a level of crystallinity and poor forming, while insufficient crystallinity will cause sticking in the mould. The ideal crystallinity range in the final container is 25–35%. Forming requires vacuum, pressure and plug assist, as well as a combination of these techniques. The containers can be cut from the sheet when it is below 65°C using heavy duty presses, matched dies with close tolerance fit, and sharp cut edges.

5.5.4 Materials

Dual-ovenable plastic packaging for cooking in microwave and conventional ovens was first commercial in the mid-1980s. CPET grades are available from Eastman Chemicals, as well as a more heat-resistant PCTA copolyester, 'Thermx'. The other major polymer producer is M&G (ex Shell), and DuPont and Kosa produce CPET grades. The PET includes a nucleating agent, which is incorporated to promote the very fast crystallisation needed for manufacture of high temperature resistant containers. High crystallinity ensures good dimensional stability of formed trays at high temperatures when cooked in an oven. Control of the level of crystallinity enables processors to produce trays which withstand temperature extremes. High crystallinity induces poor impact properties; processors compensate for this with a co-extruded sheet, nip polished during extrusion to improve impact properties. Impact modifiers and nucleating agents can be added to the PET to improve low temperature impact properties. CPET trays are designed to be used over a wide temperature band, from freezer storage (down to $-40°C$) to oven cooking (at $240°C$). The addition of impact modifier polymers can improve the impact strength at least twofold at ambient temperatures, and up to fivefold at freezer storage temperatures, compared to standard CPET.

5.5.5 New developments

A variation on CPET for packaging is expanded PET (EPET) for food trays, which can be deep drawn and are rigid enough for most foods, including frozen foods. An ethylene vinyl alcohol copolymer (EVOH) core layer co-extruded in the EPET structure provides barrier protection, and an adhesive layer bonds the complete structure. The expanded material achieved using a proprietary technique requires modification to extrusion equipment to ensure uniform distribution of the foamed material. Foaming can be achieved with an inert gas blowing agent. The EPET material can save up to 50% on weight compared to conventional CPET. Other techniques for extending shelf-life include the incorporation of oxygen scavenging additives. CIBA have now aquired the 'Amosorb 2000' range of inorganic absorbers developed by BP Amoco, and market the range under the trade name 'Shelfplus'.

The demand for dual-ovenable packaging is growing at a fast rate and the demand for microwaveable packaging is even higher. Because the temperature requirements for microwaveable packs is not as great, the choice of suitable materials can be extended beyond PET-based structures and competition is more intense.

Technological developments over recent years have included the 'Melt-to-Mold' system offered by Lawson Mardon Therma-Plate Corp. This process has an in-line rotary drum thermoforming process, where CPET sheet is extruded in the melt phase and fed directly onto the surface of a revolving drum, then into the thermoformer. The continuous process saves energy, the containers have better definition, and better process consistency is achieved. The technology offered includes modified polymers as part of the CPET material. The improved morphology results in smaller crystals, which can improve the gas barrier properties of CPET structures compared to conventional materials. Data on CPET trays produced by the 'Melt-to-Mold' technology claim that oxygen permeation is about five times better than that of standard co-extrusion CPET/APET structures and there is a 20% reduction in thickness. Impact resistance is also improved up to fourfold. Other advantages claimed are lower container shrinkage, improved dimensional stability, improved heat sealing properties and double seaming integrity [64].

Co-extruded CPET/APET with the amorphous PET as the inner layer hermetically seals to the PET film lidding material before retorting. Brown Machine offer various advanced systems designed for: quick mould change, advanced oven control, process flexibility for quick material changes, cooling programmes for faster cycle times, high speed vertical trim press for extended die life and quick and easy tool changes.

5.6 Conclusions—film

The process of making biaxially oriented polyester film is complex and requires a detailed understanding of the interrelationship between polymer properties and process to give the desired final properties in the film. The unique blend of properties of PET film makes it an extremely versatile product and the growth in the films market is predicted to be above 5% per annum. However, this masks the fact that some areas, such as packaging, industrial and electrical applications, are growing at a greater rate, whereas the growth in the more traditional markets, such as magnetic media or graphic arts materials, is more modest. Advances are continually being made in uprating the film process but, in addition, new applications for PET film in both packaging and elsewhere are continually being developed. There will be an increasing trend towards differentiation through the application of new process technologies and advances in the control of the process coupled with the combinations of base polymer, filler, coating and co-extrusion technologies.

5.7 Acknowledgements

The authors would like to thank their colleagues M. Ellam, J. Robinson, D. Wager, P. Mills, K. Looney, J. Lloyd and M. Jeffels of DuPont Teijin Films™ and P. Willcocks, S. Norval and A. Broadhurst of ICI for their contribution and comments.

References

1. *Encyclopedia of Polymer Science and Engineering* (ed. H.F. Mark), 2nd Edition, John Wiley and Sons, New York, Vol. 12, Polyester Films, p. 193 and references therein.
2. Ajji, A., Guevremont, J., Cole, K.C. and Dumoulin, M.M. (1996) *Polymer*, **37**, 3707.
3. Keller, A., Lester, G.R. and Morgan, L.B. (1954) *Phil. Trans. Roy. Soc. (Lond.)*, **A247**, 1.
4. Hartley, F.D., Lord, F.W. and Morgan, L.B. (1954) *Phil. Trans. Roy. Soc. (Lond.)*, **A247**, 23.
5. Tobin, M.C. (1974) *J. Polym. Sci. Polym. Phys. Ed.*, **12**, 399.
6. Tobin, M.C. (1976) *J. Polym. Sci. Polym. Phys. Ed.*, **14**, 2253.

7. Jog, J.P. (1995) *J. Macromol. Sci. Rev. Macromol. Chem. Phys.*, **C35**, 531.
8. Sheldon, R.P. (1963) *Polymer*, **4**, 213.
9. Zachmann, H.G. and Geumther, B. (1982) *Rheol. Acta*, **21**, 427.
10. Pilati, F., Toselli, M., Messori, M., Manzoni, C., Turturro, A. and Gattiglia, E.G. (1997) *Polymer*, **38**, 4469.
11. Guemther, B. and Zachmann, H.G. (1983) *Polymer*, **24**, 1008.
12. Gerlach, C., Buckley, C.P. and Jones, D.P. (1998) *Trans. IChemE.*, **76** Part A, 38.
13. Adam, A.M., Buckley, C.P. and Jones, D.P. (2000) *Polymer*, **41**, 771.
14. Boyce, M.C., Socrate, S. and Llana, P.G. (2000) *Polymer*, **41**, 2183.
15. Lorentz, G. and Tassin, J.F. (1994) *Polymer*, **35**, 3200.
16. LeBourvellec, G. and Beautemps, J. (1990) *J. Appl. Polym. Sci.*, **39**, 329.
17. Lapersonne, P., Bower, D.I. and Ward, I.M. (1992) *Polymer*, **33**, 1266.
18. Salem, D.R. (1998) *Polymer*, **39**, 7067.
19. Blundell, D.J., *et al.* (1996) *Polymer*, **37**, 3303.
20. Mahendrasingam, A., *et al.* (2000) *Polymer*, **41**, 7803.
21. Blundell, D.J., *et al.* (2000) *Polymer*, **41**, 7793.
22. Blundell, D.J., *et al.* (1999) *Polym. Bull.*, **42**, 357.
23. Ashford, E., Bachmann, M.A., Jones, D.P. and MacKerron, D.H. (2000) *Trans. IChemE.*, **78** Part A, 33.
24. Rule, R.J., MacKerron, D.H., Mahendrasingam, A., Martin, C. and Nye, T.M.W. (1995) *Macromolecules*, **28**, 8517.
25. Biangardi, H.J. and Zachmann, H.G. (1977) *Prog. Colloid Polym. Sci.*, **62**, 71.
26. Faisant de Champchesnel, J.B., Tassin, J.F., Bower, D.I., Ward, I.M. and Lorentz, G. (1994) *Polymer*, **35**, 4092.
27. Gehrke, R., Riekel, C. and Zachmann, H.G. (1989) *Polymer*, **30**, 1582.
28. Moscato, M.J. and Seyler, R.J. (1994) *ASTM Spec. Tech. Publ.*, **STP 1249**, 239.
29. Iwakura, K., Wang, Y.D. and Cakmak, M. (1992) *Intern. Polym. Processing VII*, **4**, 327.
30. Khan, M.B. and Keener, C. (1996) *Polym. Eng. Sci.*, **36**, 1290.
31. Yamada, T. and Nonomura, C. (1994) *J. Appl. Polym. Sci.*, **52**, 1393.
32. Chu, W.H. and Smith, T.L. (1973) *Polym. Sci. Tech.*, **1**, 67.
33. Kim, G.H., Kang, C.-K., Chang, C.G. and Ihm, D.W. (1997) *Eur. Polym. J.*, **33**, 1633.
34. Jungnickel, B.-J. (1984) *Makromol. Chem.*, **125**, 121.
35. Heffelfinger, C.J. and Burton, R.L. (1960) *J. Polym. Sci.*, **47**, 289.
36. Blundell, D.J. and Pendlebury, R. (1991) 'The Status of X-Ray Characterisation of Crystal Orientation in Polyester Film', ICI Company Research Report, IC 12763.
37. Chang, H., Schultz, J.M. and Gohil, R.M. (1993) *J. Macromol. Sci. Phys.*, **B32**, 99.
38. Pendlebury, R. (1977) PhD Thesis, University of Teesside.
39. Faisant de Champchesnel, J.B., Bower, D.I., Ward, I.M. and Tassin, J.F. (1993) *Polymer*, **34**, 3763.
40. Faisant de Champchesnel, J.B., Tassin, J.F., Monnerie, L., Sergot, P. and Lorentz, G. (1997) *Polymer*, **38**, 4165.
41. Tassin, J.F., Vigny, M. and Veyrat, D. (1999) *Macromolek. Chem. Macromol. Symp.*, **147**, 209.
42. Kugler, J., Gilmer, J.W., Wiswe, D., Zachmann, H.G., Hahn, K. and Fischer, E. (1987) *Macromolecules*, **20**, 1116.
43. Greener, J., Tsou, A.H. and Blanton, T.N. (1999) *Polym. Eng. Sci.*, **39**, 2403.
44. Bower, D.I., Jarvis, D.A. and Ward, I.M. (1986) *J. Rheol.*, **30**, 1459.
45. Bower, D.I., Jarvis, D.A., Lewis, E.L.V. and Ward, I.M. (1986) *J. Polym. Sci. Polym. Phys.*, **24**, 1481.
46. Gohil, R.M. and Salem, D.R. (1993) *J. Appl. Polym. Sci.*, **47**, 1989.
47. Gohil, R.M. (1994) *J. Appl. Polym. Sci.*, **52**, 925.
48. Okazaki, I. and Wunderlich, B. (1996) *J. Polym. Sci. Polym. Phys.*, **34**, 2941.
49. Wunderlich, B. (1976) *Macromolecular Physics*, Vol. 2, Academic Press, New York.

50. Greener, J., O'Reilly, J.M. and Contestable, B.A. (1997) *Polym. Mat. Sci. Eng.*, **76**, 266.
51. Wilson, M.P.W. (1974) *Polymer*, **15**, 277.
52. Shih, W.K. (1994) *Polym. Eng. Sci.*, **34**, 1121.
53. Watenabe, H., Asai, T. and Ouchi, I. (1980) *Proc. Jpn. Congr. Mater. Res.*, **23**, 282.
54. Haworth, B., Dong, Z.H. and Davidson, P. (1993) *Polym. Int.*, **32**, 325.
55. Gohil, R.M. (1993) *J. Appl. Polym. Sci.*, **48**, 1635.
56. Gohil, R.M. (1993) *J. Appl. Polym. Sci.*, **48**, 1649.
57. Varma, P., Lofgren, E.A. and Jabarin, S.A. (1998) *Polym. Eng. Sci.*, **38**, 245.
58. Tsou, A.H. and Tirrell, M. *ANTEC '93*, p. 1730.
59. Morel, J.F., Phung, N.D. and Joly, J.C. (1980) *IEEE Trans. Electr. Insul.*, **15**, 335.
60. MacDonald, W.A. (1995) *Trends Polym. Sci.*, **3**, 212.
61. Mills, P.J. (1997) *Structure and Properties of Oriented Polymers* (ed. I. M. Ward), 2nd Edition, Chapman & Hall, London, chapter 9.
62. Gerlach, C.G.F., Dunne, F.P.E., Jones, D.P., Mills, P.D.A. and Zahlan, N. (1996) *J. Strain Anal.*, **31/1**, 65.
63. 'Melinar' PET 14: Sheet Extrusion and Thermoforming. ICI.
64. Freundlich, R. and Johnson, D.C. (1992) *New CPET containers at Therma-Plate*. Therma-Plate Corp. FoodPlas, Orlando, Florida, USA.

6   Injection and co-injection preform technologies

Paul Swenson

6.1   Multilayer characteristics

Production of multilayer co-injection preforms is the most economical way to achieve enhanced properties in a rigid PET container. In the second half of the year 2000, the commercial introduction of several new oxygen barrier-scavenger materials improved multilayer O_2 performance to 62 times that of a monolayer PET bottle. This performance is a full order of magnitude better than was possible with barrier-only materials during the 1990s. Barrier-only materials, such as ethylene vinyl alcohol (EVOH) and polyamide (PA)-MXD6, in a multilayer PET container had provided only 2–4 times the O_2 performance of monolayer PET and, at that level of barrier improvement, large scale conversions of juices, beer and other oxygen-sensitive products did not make economic sense. Now, however, with the shelf-life extension provided by the 62 times enhancement conversion makes sense from the point of view of package cost, downstream supply line savings and consumer preference.

Two of the new barrier-scavengers were developed by material suppliers for general sale to all converters and three were developed by large converters for proprietary in-house use. A joint venture between Darex and Evalca (Kuraray) introduced an EVOH-based barrier-scavenger. Honeywell, formerly Allied Signal, developed a barrier-scavenger family of materials based on PA-6. The three converters with proprietary materials, Crown Cork & Seal, Schmalbach-Lubeca and Continental PET Technologies, based their new barrier-scavenger on PA-MXD6.

This exemplary shift in materials performance has almost coincided with the announcement of major developments in co-injection machinery and technology. Kortec, Inc., the supplier of the only co-injection machinery that operates at monolayer cycle times, has licensed certain multilayer technology from Pechiney. These two important advances—in materials and in converting machinery—will lead the next wave of conversions from glass to PET.

Multilayer co-injection is the process by which one or more interior layers of material are totally encapsulated by outer virgin PET layers. It is the only technology able to provide any combination of clarity, gas-barrier, gas-scavenger and recycled PET (RPET) in a single process carried out in the same cycle time as monolayer PET molding (see Figure 6.1 for materials and layer

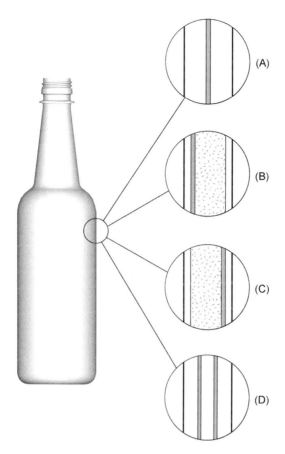

Figure 6.1 Wall sections of a multilayer container: A) three-layer wall with centered core; B) four-layer wall of barrier and recycled polyethylene terephthalate (RPET) with barrier near contents; C) four-layer wall with barrier towards outside; D) five-layer wall.

construction). Other technologies require either a second process on a separate machine (internal or external coatings, as shown in Figure 6.2) or a second injection cycle on the same machine (inject-over-inject, as shown in Figure 6.3). The net result is that a multilayer container is at least 20% less expensive than a coated or inject-over-inject container.

Multilayer co-injection not only produces the least expensive high-barrier container it also produces the best-performing container. Unlike multilayer containers, coated containers do not have gas-scavenging properties and, compared to multilayer containers, coated containers have either lower gas-barrier properties or have an unattractive tint or texture imparted to the container by the coating.

Multilayer co-injection technology, being simply a variant of injection molding technology, is also easier to introduce into an existing molding plant.

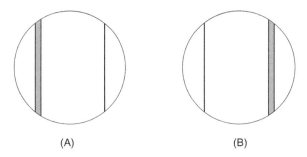

Figure 6.2 Wall sections of a coated container: A) internal coating; B) external coating.

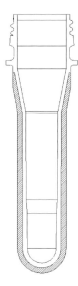

Figure 6.3 Preform cross-section made by inject-over-inject process.

Currently employed personnel already have the basic knowledge to enable them to learn multilayer co-injection, and co-injection machinery fits directly into existing preform handling and subsequent blow molding processes.

Multilayer co-injection has the advantage that the container properties can be changed by selection of the type and volume of each material injected into the preform. One of several commercially available materials having the clarity of PET could be used to provide both gas-barrier and gas-scavenger properties. One of a larger number of clear materials could be selected to provide a barrier-only property to the container. A single molding plant, or even a single machine, can easily produce containers for a variety of different applications requiring widely different properties by a simple change of material selection and volume.

6.2 Applications

Multilayer co-injection applications are characterized by the primary enhancement relative to monolayer PET needed for the use of the container. Performance-driven applications are those that require an increase in the gas-barrier or gas-scavenging properties relative to PET. Containers using RPET are considered to be economics-driven or legislative-driven applications.

6.2.1 Performance-driven applications

The most basic performance enhancement is to increase the gas-barrier property relative to PET. Small CSD (carbonated soft drinks) containers typically require a shelf-life increase by a factor of two to four times that of a similarly sized monolayer PET container. Barrier materials, EVOH or PA-MXD6, are traditionally used to block the CO_2 loss through the sidewall of the container. The thickness of the barrier layer is controlled to produce the desired shelf-life. Another gas barrier application is food containers for tomato products or green teas, which require a twofold or threefold improvement of the O_2 barrier of PET to prevent oxygen ingress into the container contents. The enhanced barrier performance of the multilayer container will reduce the O_2 ingress or CO_2 egress to extend the shelf-life of the packaged product. The shelf-life will expire when oxidation-induced flavor changes in the product exceed an acceptable threshold or when the CO_2 level in the product drops below its acceptable level.

A higher level of performance enhancement is achieved by co-injecting a material that has both gas-barrier and oxygen-scavenging properties. The barrier property of the material will reduce the permeation of both CO_2 and O_2, while the scavenging ability of the material will maintain the O_2 in the container at nearly zero during the life of the scavenger. Beer and small fruit juice are products that have stimulated the development of these combination barrier/scavenger materials, which have 62 times the O_2 performance of monolayer PET.

6.2.2 Economics- or legislative-driven applications

The use of RPET in a container is driven by the economic goal of using a material that is less costly than virgin PET or, perhaps conversely, is required by governmental mandate regardless of the cost. Multilayer co-injection can be used to incorporate up to 40% of the preform weight as an interior layer of RPET that is totally encapsulated by the virgin PET outer layer. Since the RPET does not contact either the contents or the consumer, a non-food-contact grade material can be used. The biggest impediment to the wide scale use of bottle-to-bottle RPET is the inability to secure consistent sources of RPET except within the few countries mandating such use. RPET in the form of flake, amorphous or crystallized pellets can be used in a multilayer container.

Flake RPET requires special drying equipment and in-line melt filtration in the co-injection machine to remove contaminents. Amorphous pelletized RPET also uses special drying equipment. Crystallized RPET pellets are dried in the same manner as virgin PET. If either of the pelletized forms of RPET is melt filtered during pelletization, added melt filtration is not required during the co-injection process.

6.2.3 Combination applications

Combination applications are those that combine the performance capabilities of a barrier or barrier-scavenger with the economic benefit of RPET or preform lightweighting. In an RPET application, virgin PET totally encapsulates a layer of RPET and a layer of barrier or barrier-scavenger. The lower cost of the non-contact RPET will usually offset the cost of the barrier enhancing material in order to satisfy either a cost- or legislation-mandated goal.

Small CSD containers, below 500 ml, may weigh the same as a 500 ml container because the shelf-life requirements of these containers can only be met by a thick sidewall to prevent CO_2 loss. A multilayer 23 g container with a CO_2 barrier layer will have more than twice the shelf-life of a 29 g monolayer PET container. The 20% reduction in container weight provides a multilayer container with a 20 week shelf-life at the same cost as a monolayer container with a 10 week shelf-life. The economic advantages of the barrier-enhanced lightweight container lie in improved product quality and extended distribution flexibility.

6.3   Closure *vs* bottle permeation

To create a barrier-enhanced container it is necessary to rethink the relative permeation through the closure as compared to that through the bottle itself. Table 6.1 lists O_2 and CO_2 permeation rates for typical 28 mm closures with standard liners and with O_2 scavenger liners. Also listed are typical permeation

Table 6.1 Oxygen and carbon dioxide permeation of different 500 ml containers and closures

	O_2 ingress (ml/day)	CO_2 egress (ml/day)
Monolayer bottle	0.025	1.60
Bottle with barrier EVOH-F101	0.006	0.50
Bottle with barrier-scavenger	0.0004	0.50
Closure with standard liner	0.0025	0.20
Closure with O_2 scavenger-liner	0.0002	0.20

Abbreviation: EVOH, ethylene vinyl alcohol polymer.

rates for monolayer PET, barrier-enhanced multilayer and barrier-scavenger-enhanced multilayer 500 ml containers. Similarly sized bottles and closures can be designed and manufactured to have permeation rates different from the listed values of these hypothetical ones but the table shows significant performance differences between monolayer and multilayer packages. A typical monolayer PET bottle has 4 times and 62 times the O_2 permeation rate of the barrier-enhanced and barrier-scavenger-enhanced multilayer bottle, respectively. The enhanced-bottle CO_2 permeation rate is less than 25% the rate of the monolayer bottle. Note that the presence of the O_2 scavenger has little effect on the CO_2 performance of the base barrier material. These significant reductions in the permeation rate of the enhanced bottle make the closure permeation rate a significant factor in overall package permeation.

A closure with standard liner may allow 9% and 17% of the respective O_2 and CO_2 permeation of the hypothetical monolayer package, but because the permeation of the enhanced bottle is so low, the closure will be responsible for nearly 30% and 45% of the respective O_2 and CO_2 permeation of the hypothetical barrier-enhanced multilayer package. The standard closure allows nearly 86% of the O_2 permeation of the barrier-scavenger-enhanced package. These large 'leaks' indicate the need to use O_2 scavenger liners to maintain the desired package performance of an enhanced bottle.

Table 6.2 lists the O_2 and CO_2 permeation rates of four different 500 ml bottle and closure combinations. It is clearly shown that the use of an O_2 scavenger in both the bottle and the closure will provide the extremely low permeation rate required by highly oxygen-sensitive foods or beverages. At the listed permeation rate of 0.0006 ml O_2/day, the total oxygen ingress over 300 days—less than 0.2 parts per million (ppm) of the container contents—makes the scavenger-enhanced package suitable for beer applications. This package provides 45 times the performance of a standard closure and monolayer PET bottle.

For allowable oxygen thresholds near 5 ppm, the barrier bottle and scavenger closure provides 37% longer shelf-life than the barrier bottle with a standard closure, and one of these combinations would normally be suitable for juice and other food products.

Table 6.2 Total oxygen and carbon dioxide permeation of different 500 ml container and closure combinations

	O_2 ingress (ml/day)	CO_2 egress (ml/day)
Monolayer bottle + standard closure	0.0275	1.90
Barrier bottle + standard closure	0.0085	0.70
Barrier bottle + scavenger closure	0.0062	0.70
Scavenger bottle + scavenger closure	0.0006	0.70

6.4   Container performance

As discussed in the previous section, the overall package performance is a combination of the closure and the container and the only way to achieve the most economical overall performance is by engineering both the closure and container as an overall system.

The ability to engineer a wide range of properties is one of the key advantages of multilayer technology. The primary variables are the properties of the materials (gas permeation rates for O_2 and CO_2, and O_2 scavenger capacity), and the amount and placement of these properties in the preform and container.

6.4.1   Barrier properties

The barrier properties of a barrier material (or the barrier component of a barrier scavenger) are dependent on the relative humidity (RH) of the material experienced during its shelf-life. Figure 6.4 shows the humidity relationship of O_2 permeability of EVOH-F101 and PA-MXD6. The effect of RH on CO_2 permeability is similar to the oxygen relationship. In a beverage container, the

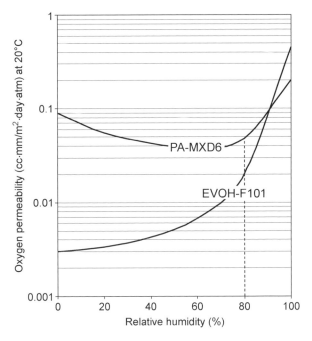

Figure 6.4 Coefficient of permeability of oxygen *vs* relative humidity for PA-MXD6 and EVOH-F101. Abbreviations: PA, polyamide; EVOH, ethylene vinyl alcohol.

Table 6.3 Enhancing-layer relative humidity and resulting coefficient of permeability in different wall structures

Wall structure (see Fig. 6.1)	Layer relative humidity	Coefficient of permeability ($cm^3 \cdot mm/M^2 \cdot day \cdot atm$)			
		EVOH-F101		PA-MXD6	
		O_2	CO_2	O_2	CO_2
1A	70%	0.010	0.08	0.040	0.310
1B	80%	0.018	0.13	0.046	0.360
1C	90%	0.070	0.24	0.070	0.550
1D	1/2 at 90%				
	1/2 at 70%	0.18	0.12	0.052	0.410

Abbreviations: EVOH, ethylene vinyl alcohol polymer; PA, polyamide.

RH across the container wall changes linearly from 100% RH at the inside surface of the wall to ambient humidity at the outside surface of the wall—assumed to be 60% RH for the purposes of engineering most containers. The radial placement of the barrier layer in the preform wall dictates the average RH of the barrier layer and, therefore, the effective permeation coefficient.

Table 6.3 shows how the position of the barrier in the cross-section of the container wall affects the average RH of the barrier layer and the permeability for EVOH and PA-MXD6. Note that the O_2 permeability of EVOH is less than 40% that of MXD6 at 80% RH but is equal to that of MXD6 at 90% RH. For the container performance shown in the remaining tables and figures, the position of the barrier layer is assumed to be as shown in Figure 6.1A. Interestingly, for MXD6, the performance of the five-layer wall shown in Figure 6.1D provides less barrier effect than the three-layer wall shown in Figure 6.1A. This is because the effective permeability of the sum of the two barrier layers (RH 90% and RH 70%) is about 13% higher than the permeability of a central layer at RH 80%.

6.4.2   Oxygen barrier

The oxygen ingress of typical 500 ml containers is shown in Figure 6.5. The PET-only container has a relatively high permeation rate, reaching 1 ppm within about 16 days. The MXD6 barrier at 5% of container weight allows O_2 permeation at a rate which reaches 1 ppm in about 30 days, nearly 2 times the PET-only time. The container with 5% EVOH-F101 barrier reaches 1 ppm in about 62 days, more than 4 times the PET-only time. A greater thickness of EVOH can be used to further extend the time to reach 1 ppm, but applications needing more enhancement than 5% EVOH-F101 typically use O_2 barrier-scavenger rather than thicker barrier-only to extend the product shelf-life.

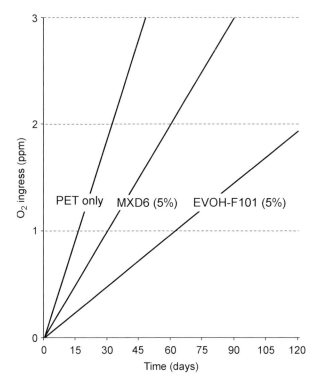

Figure 6.5 Oxygen ingress *vs* time for 500 ml containers enhanced with PA-MXD6 and EVOH-F101. Abbreviations: PA, polyamide; EVOH, ethylene vinyl alcohol.

Different foods and beverages have different sensitivity to oxygen ingress, and this sensitivity must be matched by the container and closure permeation rate to keep the total below the threshold level for the desired shelf-life of the package. Barrier-only containers are ideal for applications that allow oxygen ingress of 5–10 ppm. If the allowable ingress is less than 5 ppm, an oxygen barrier-scavenger is the preferred material (see Section 6.4.4).

6.4.3   Carbon dioxide barrier

The carbon dioxide egress from several different 500 ml containers is presented in Figure 6.6. Acceptable shelf-life is often defined as the time at which the volume of CO_2 remaining reaches some percentage of the volume when the package is initially filled.

The egress rate is affected by the type of barrier material, the thickness of barrier layer in the container sidewall and the distribution of the layer in the container. The difference in CO_2 permeability between PA-MXD6 and

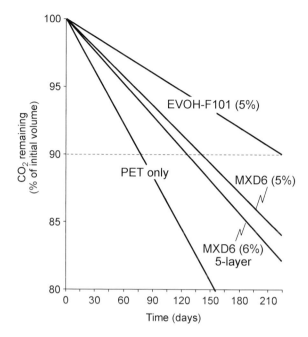

Figure 6.6 Carbon dioxide egress *vs* time for 500 ml containers enhanced with PA-MXD6 and EVOH-F101. Abbreviations: PA, polyamide; EVOH, ethylene vinyl alcohol.

EVOH-F101 is clearly shown at the barrier layer RH = 80% used in this example. The relative differences decrease when the barrier RH approaches 90% (see Figure 6.4 showing O_2 relationships).

The five-layer container with 6% volume of MXD6 will have an average barrier thickness of 4% in the container sidewall. Therefore, it has a shorter CO_2 shelf-life than the 5% volume of the three-layer container having an average of 7% thickness in the sidewall and nearly zero in the thick finish and base.

6.4.4   Scavenger property

The addition of an O_2 scavenger property to a barrier material completely changes the total package O_2, as shown in Figure 6.7. By chemically binding O_2, the scavenger maintains the O_2 level close to the amount that was initially entrained in the contents during the filling process.

The addition of a scavenger extends the glass-like O_2 level to a period exceeding 300 days for the 5% barrier-scavenger volumes shown in Figure 6.7. A thicker or thinner barrier-scavenger layer will, respectively, increase or decrease the glass-like life. When the scavenger is completely bound with O_2, the permeation rate is similar to the 'barrier' rate of the base material.

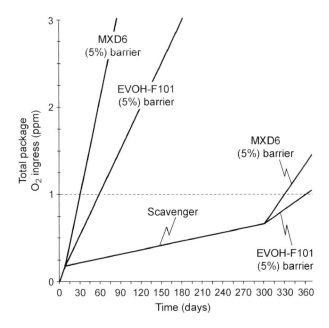

Figure 6.7 Total package oxygen ingress *vs* time for 500 ml containers and closures enhanced with barrier or barrier-scavenger materials.

6.5   Wall structure

The scavenging property of commercial scavengers is fairly independent of RH in the 70–90% range that typically occurs in container walls. This fact, coupled with the humidity dependence of the barrier property, leads to the concept that the optimal placement of the layer within the cross-section of the container wall is different for different classes of application.

Figure 6.8 shows various container wall sections that maximize the container performance by the optimal placement of the barrier or barrier-scavenger layer. Figure 6.8A is recommended for most applications requiring a combination of O_2 scavenger and CO_2 barrier. The central placement provides lower CO_2 permeability than placement closer to the inside wall and the scavenger layer is protected by the thickness of the outer PET layer equal to nearly half the container wall. In the wall shown in Figure 6.8B, the barrier-scavenger layer is biased toward the inside of the container, and while the thicker outside PET layer provides maximum protection of the scavenger property, the higher RH reduces the CO_2 barrier effect of the layer.

Figure 6.8C shows placement of a barrier-only material layer closer to the outside wall. The CO_2 and O_2 permeability are both lower because the RH at that position is lower than in the central position. While the lower RH enhances

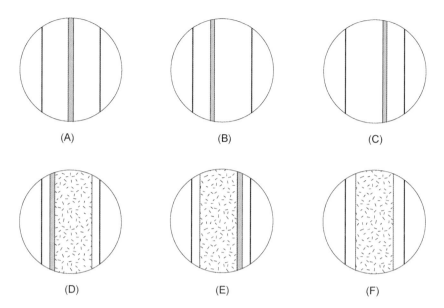

Figure 6.8 Wall sections of multilayer containers: A) with centered layer; B) with bias-inside layer; C) bias-outside layer; D) recycled polyethylene terephthalate (RPET) and biased-inside layer; E) RPET and bias-outside layer; F) centered RPET layer.

the barrier property, this bias to the outside is not recommended for an O_2 scavenging layer because the thinner outer PET wall leads to quicker saturation of the scavenger capacity.

Figure 6.8D and 6.8E show four-layer structures of RPET along with either barrier or barrier-scavenger layer. For a barrier-scavenger layer, the structure shown in Figure 6.8D is recommended if the decrease in CO_2 barrier caused by the higher RH location does not make the CO_2 shelf-life too short. The thickness of the RPET and outer PET layer protects the scavenging capacity of the barrier-scavenger layer, and the closer proximity to the container slightly reduces the amount of O_2 desorbed into the contents from the inner PET wall. Compared to the structure shown in Figure 6.8A, the reduction in desorbed oxygen will be approximately 0.08 ppm for a typical 500 ml container. Figure 6.8E may be used to maximize the CO_2 and O_2 barrier effect by placing the layer in a lower RH position. Figure 6.8F shows the wall section of a PET/RPET container.

6.6   Preform and bottle design

Apart from the position and thickness of the core layer, the design of a multilayer preform and container is no different from that of a monolayer PET container. The mechanical properties, such as top load and vacuum test, are not affected by the presence of the small thickness of a barrier layer or the layer thickness of a RPET layer with intrinsic viscosity (IV) similar to the virgin PET skin. Impact failure in the gate area will be unchanged from monolayer impact tests if the barrier or RPET layer does not extend completely into the preform base (see Figure 6.9A).

The design of the external dimensions of the multilayer preform is the same as that of monolayer PET. The relative thickness of the core layer in the

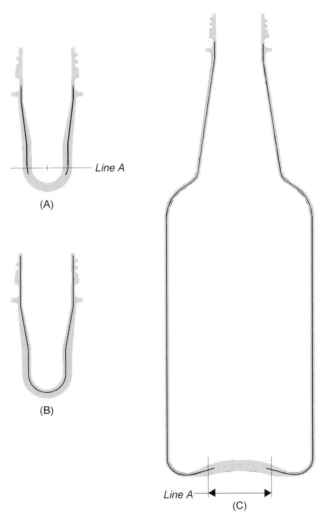

Figure 6.9 Layer distribution in preforms and blown container: A) preform with 'controlled fill' layer distribution; B) preform with complete fill distribution; C) container blown from preform (A).

container follows the relative thickness as molded in the preform. For barrier and barrier-scavenger applications, the design of the core layer thickness and position should be such as to maximize the core thickness in that portion of the preform wall which forms the sidewall of the container.

6.6.1 Permeation through finish, sidewall and base

Figure 6.9A and 6.9C show the ideal distribution of a barrier or barrier-scavenger layer in a preform and subsequent container. The tip of the preform below line 'A' corresponds to the thick amorphous portion of the base in the container. Although the preform tip and finish are a significant part of the preform weight, they form a very small portion of the surface area of the container. The small surface area in combination with the 5–10 times greater thickness than the container sidewall means that no barrier enhancement is needed in those areas of the container. In fact, any barrier material molded in those portions of the preform only increases the cost of the container without improving its performance. Table 6.4 presents data for the same CO_2 permeation rate from 500 ml containers with different MXD6 barrier layer distributions. To obtain the same permeation rate in the blown container, the total volume of MXD6 barrier material must be greater in the preform shown in Figure 6.9B than in the controlled distribution of the preform shown in Figure 6.9A, because in both cases virtually all of the permeation occurs though the thin sidewall rather than through the thick finish and base. The data show that in the structure of either bottle, the proportion of permeation through the finish and base is less than their corresponding proportion of the total surface area. Thus, there is no benefit to enhancing the barrier of the finish and base unless the permeation of the sidewall is much less than the permeation of the finish and base on a per unit area basis. It is evident that the barrier molded into the finish and base shown in Figure 6.9B does not effectively reduce the permeation rate through those portions but simply

Table 6.4 Percentage of permeation, surface area and container weight for the finish, sidewall and base portions of 500 ml containers of different structures

	Permeation (%)		Surface area (%)	Weight(%)
	Fig. 6.9A	Fig. 6.9B		
Finish	2.4	1.3	4.8	16.2
Sidewall	96.4	98.2	93.4	74.0
Base < line A	1.2	0.5	1.8	9.8
Total barrier volume	**5.0%**	**6.0%**		
Weeks to 10% CO_2 loss (including closure loss)	20.9	20.9		
Weeks to 10% CO_2 loss (excluding closure loss)	29.6	29.6		

increases the MXD6 volume in the container by 20% (from 5.0% to 6.0% of the preform weight).

6.6.2   Controlled fill

The concept of 'controlled fill' has been developed because the overall performance of the container is improved more by enhancing the thin sidewall (92–95% of container surface area) than by enhancing the thick finish and base (both of which total about 5–8% of the container surface area). This general rule holds true until the container barrier performance exceeds more than 15 times PET.

The optimal 'controlled fill' location of the leading edge of the core layer is just above the neck support ring, as shown in Figure 6.9A, and the optimal location of the trailing edge of the core layer is just below the intersection of the preform body and spherical tip, shown as line 'A' in Figure 6.9A. These limits ensure that the enhancing layer provides coverage of the container sidewall and thin portion of the base (shown outside of line 'A' in Figure 6.9C) and extends partially into the thick portion of the finish and amorphous portion of the base.

The core layer position in the preform can be described in terms of coverage along the axial length of the preform, 'controlled fill' as shown in Figure 6.9A and 'complete fill' as shown in Figure 6.9B, and separately in terms of the placement between the inside and outside surfaces of the preform wall, 'centered', 'bias-inside', 'bias-outside' and 'five-layer', as shown in Figures 6.8A, B and C and 6.1D respectively. Multilayer co-injection technology capable of 'controlled fill' typically places the core layer 'centered' within the preform wall. However, one company, Kortec, has also developed the ability to place it 'biased-inside' or 'biased-outside' as well as the standard 'centered'. The multilayer technology that can produce a 'complete fill' preform places the layer 'biased-inside' or 'five-layer'.

Table 6.5 lists different preform structures and the corresponding core layer volumes needed to produce the same container performance as a baseline structure 'A' of 'controlled fill' with the core layer 'centered' in the wall. The data were generated using MXD6 as a barrier layer and as the base component for the barrier-scavenger layer. Similar results occur for an EVOH or PA-6 barrier and barrier-scavenger material. Structure 'A', a bottle made from a 'controlled fill' and 'centered' preform, has the optimal overall O_2 and CO_2 enhancement because the enhancing layer is concentrated in the thin sidewall of the container. For the bias-inside structure of container 'B', a scavenger layer requires less volume but a barrier layer requires more volume to equal the performance of container 'A'. Conversely, for the bias-outside structure of container 'C', a scavenger layer requires more volume and a barrier layer requires less volume to duplicate the performance of the 'A' container.

Table 6.5 The enhancing-layer volumes used to maintain the same permeation rate for different layer distributions in 500 ml containers

	Fill profile	Structure		O_2 Scavenger (%)	O_2 Barrier (%)	O_2 Desorption (ppm)	CO_2 Barrier (%)	Comments
		Wall						
A	Controlled fill Figure 6.9A & 6.9C	Centered Figure 6.8A RH = 80%		5	5	0.18	5	Optimize overall enhancement by maximizing protection in thin sidewall
B	Controlled fill Figure 6.9A & 6.9C	Bias-inside Figure 6.8B & 6.8D RH = 90%		3.9	7.5	0.17	7.5	Higher RH decreases barrier properties
C	Controlled fill Figure 6.9A & 6.9C	Bias-outside Figure 6.8C & 6.8E RH = 70%		6.5	4.2	0.24	4.2	Lower RH improves barrier properties
D	Complete fill Figure 6.9B	Centered Figure 6.8A RH = 80%		6	6	0.14	6	20% of core layer is wasted in thick finish and base
E	Complete fill Figure 6.9B	Bias-inside Figure 6.8B & 6.8D RH = 80%		5	9	0.10	9	Higher RH decreases barrier properties, and 20% of core layer is wasted in finish and base
F	Complete fill Figure 6.9B	Bias-outside Figure 6.8C & 6.8E		7.5	5.2	0.20	5.2	Benefit of lower RH is wasted in thick finish and base
G	Complete fill Figure 6.9B	5-layer Figure 6.1D ½ at RH = 90% ½ at RH = 70%		6.0	6.8	0.10	7.5	Split RH decreases barrier property, and 20% of core layer is wasted in finish

Abbreviations: RH, relative humidity; ppm, parts per million.

All of the structures using 'complete fill' require a greater volume of enhancing layer than the corresponding 'controlled fill' structures. For example, while a 'bias-inside' scavenger layer of container 'E' requires less volume than the 'centered' layer of container 'D', it requires 20% more volume (6% *vs* 5%) than the corresponding 'controlled fill' container 'B'.

6.7 Headspace oxygen absorption

Oxygen enters a container by entrainment during the filling process or by permeation through the container or closure. Any of this O_2 in excess of the amount absorbed by the contents remains in the container headspace.

While the amount of headspace O_2 in a multilayer package with a properly applied closure is extremely small, for highly oxygen-sensitive foods or beverages, this O_2 could be enough in some instances to produce oxidized flavors. The question then becomes, can this O_2 be absorbed by a scavenger layer in the container or by a scavenger liner of the closure before it reacts with the package contents. To be absorbed by the scavenging components of the closure or container, the O_2 must permeate through the contacting surface of the closure or container; the contacting surface of the container being the inside wall of the finish and headspace portion of the neck. The thickness of the inner PET layer in the finish is at least five times greater than in the neck, so that scavenging of headspace O_2, if any, will occur primarily through the thinner neck sidewall. However, the partial pressure of O_2, even if equivalent to as much as 1 part per million (ppm) of the container contents, is so low that the driving force to cause permeation through the PET layer into the scavenging layer of the container is only 1/50 of that driving O_2 into the container from the atmosphere. The active components in the closure scavenger liner are also below the contact surface of the liner, and the lack of driving force makes the permeation rate from the headspace to the liner effectively zero. The net result is that the reaction rate of O_2 with the contents will most likely absorb the headspace O_2 far faster than it could possibly be absorbed by either the closure or container scavenger components.

6.8 Oxygen desorption from PET

A portion of the small volume of oxygen that is absorbed into the PET wall of the container prior to filling will desorb into the contents after filling. Under normal atmospheric conditions, each gram of oxygen-saturated PET will contain 0.008 ml of oxygen if no oxygen scavenger is present in the container structure.

Figure 6.10A and B show a uniform concentration of oxygen across the wall section of a monolayer wall and barrier layer wall, respectively, before filling the container. After filling with a low O_2 product, such as beer or fruit juices,

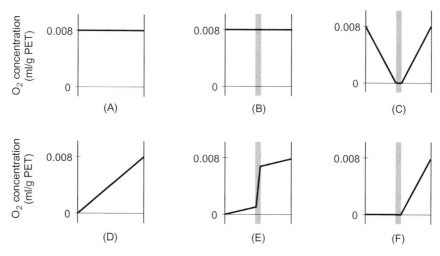

Figure 6.10 Oxygen concentration in container wall before and after filling: A) PET-only wall before filling; B) barrier-enhanced wall before filling; C) barrier-scavenger enhanced wall before filling; D) PET-only wall after filling; E) barrier enhanced wall after filling; F) barrier-scavenger enhanced wall after filling.

the soluble oxygen in the wall will desorb from the PET into the contents. For a monolayer wall, the concentration gradient will be as shown in Figure 6.10D, reflecting the fact that half the oxygen originally in the wall has desorbed out of the wall. The concentration of oxygen in the inner wall of Figure 6.10E is nearly zero because it is easier for O_2 to permeate through the inner PET layer than it is to permeate through the barrier layer. Approximately 50% of the original oxygen will have desorbed into the contents, and although the amount of desorption is similar to the monolayer wall, the concentration gradient after filling is different.

The placement of a scavenger layer in the wall produces an O_2 concentration gradient across the wall even before the container is filled. As shown in Figure 6.10C, the concentration is zero at the scavenger/PET interface. Figure 6.10F depicts the concentration gradient after the oxygen has desorbed from the PET. In a scavenger structure, approximately 25% of the oxygen in the inside PET wall at the time of filling migrates into the scavenger layer. The remaining 75% desorbs into the contents.

Table 6.6 lists the amount of oxygen desorbed from the walls of different container structures. Scavenger-enhanced containers desorb less oxygen than containers enhanced with only a barrier layer. As expressed in terms of ppm of the contents, the amount of desorbed O_2 is only significant if the contents have a shelf-life O_2 threshold of less than 5 ppm. If the threshold is less than 3 ppm, the amount desorbed by a barrier-only container (0.37 ppm) will begin

Table 6.6 Oxygen desorbed from PET into contents for different 500 ml container structures

Structure	Figure	Oxygen desorbed (O_2 ppm in contents)
Center barrier	6.8A	0.37
Center scavenger	6.8A	0.18
Offset scavenger-inside	6.8B	0.10

to have a significant effect on shelf-life. The difference between the center and the bias-inside scavenger layer is a relatively small portion (0.08 ppm) of the threshold of even the most sensitive food or beverage product.

6.9   Beer containers

Beer is probably the most oxygen-sensitive beverage that is packaged in plastic. The total package O_2 threshold is typically in the 1–2 ppm range, and some beers are even below 1 ppm. Thus, a beer package requires the use of scavenger components in both the closure and the container. Table 6.2 shows a total package O_2 permeation rate of 0.0006 ml/day. This rate will allow 0.09 ppm O_2 ingress over a 150 day life. If this permeated oxygen is added to the oxygen desorbed from the PET in a container with a central barrier-scavenger, the total O_2 ingress will be less than 0.3 ppm (see Figure 6.7).

Oxygen ingress is only part of the permeation design for a beer container. Carbon dioxide egress defines the requirements of the barrier component of the barrier-scavenger. When using any of the commercially available materials, the thickness of the barrier-scavenger layer is dictated by the requirement to slow CO_2 egress rather than by the capacity of the O_2 scavenging component. For example, the thickness that would provide 150 days of CO_2 life has enough oxygen scavenging capacity to prevent O_2 ingress for more than 300 days.

The placement of the barrier-scavenger layer in the wall must balance several factors: oxygen desorption, CO_2 ingress *vs* relative humidity of the barrier layer and cost. Scavenger layers are effective when located in the center of the wall or when biased toward the inside wall surface of the container. However, the higher RH of a layer closer to the inside of the container increases the CO_2 permeability of the barrier layer by 50%. For the preform shown in Figure 6.9A, Table 6.7 shows the changes in O_2 desorption and CO_2 shelf-life caused by the barrier position in the container wall. This position change will decrease the CO_2 shelf-life by 14%, or conversely, require a 50% increase (from 5% to 7.5%) in barrier-scavenger volume to maintain the same CO_2 life as does a centrally placed layer.

Table 6.7 Oxygen desorption, oxygen permeation and carbon dioxide shelf-life for 500 ml containers with different wall structures and layer volumes

Barrier scavenger position	Layer volume (%)	Desorbed O_2 (ppm)	O_2 permeation (ml/day)	CO_2 life (weeks)
Center of wall	5.0	0.18	0.0006	20.9
Bias-inside	5.0	0.10	0.0006	18.0
Bias-inside	7.5	0.10	0.0006	20.9

The bias-inside scavenger layer halves the already low O_2 desorption, from 0.18 ppm to 0.10 ppm, but requires a thicker layer to meet the CO_2 shelf-life requirements. The cost of the layer thickness increase to achieve 0.08 ppm desorption decrease adds about 3% to the price of a barrier-scavenger-enhanced bottle. Thus, the ideal structure to optimize the O_2 and CO_2 performance for a product such as beer is the central barrier-scavenger layer shown in Figure 6.8A. For a typical scavenger-lined closure and 500 ml beer bottle with 5% volume of barrier-scavenger centrally placed in the wall, the total package oxygen is plotted in Figure 6.11. The initial oxygen level is the sum of the O_2 entrained in

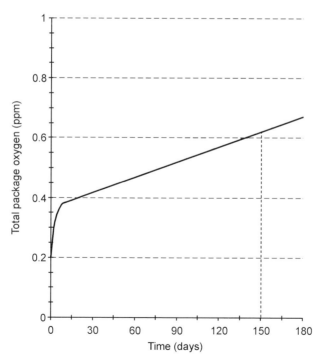

Figure 6.11 Total package oxygen *vs* time for 500 ml barrier-scavenger enhanced container.

the beer at the time of filling (typically 0.20 ppm) and the O_2 desorbed from the PET (0.18 ppm). The graph shows the desorption taking place in the first week after filling. For purposes of comparison, the CO_2 shelf-life expiration time is shown at 150 days.

If a 'complete fill' with 'bias-inside' layer is used to distribute the barrier-scavenger in the preform, the desorbed O_2 is reduced to 0.10 ppm, but the core volume must be increased to 9.0% of the preform to achieve the same CO_2 shelf-life. A thicker barrier layer is required by the humidity affect of the 'bias-inside' position, and the extra core volume in the finish and base does not extend the CO_2 shelf-life. The additional core volume increases the price of the bottle by about 6%, and while it does decrease the O_2 desorption by 0.08 ppm (from 0.18 ppm to 0.10 ppm), it does not improve either the CO_2 shelf-life or the O_2 ingress rate after desorption.

6.10   Small juice containers

Fruit juices typically have oxygen thresholds in the range of 2–10 ppm. To achieve a desired shelf-life for juices at the lower end of the range (< 5 ppm), an oxygen barrier-scavenger core layer is usually used in bottle sizes of 500 ml or smaller. At the higher end of the allowable oxygen threshold, a barrier-only core layer usually provides sufficient shelf-life. For example, the 500 ml container shown in Figure 6.5 using 5% volume of EVOH-F101 has less than 3 ppm O_2 ingress at 180 days. That same container with a barrier-scavenger layer is suitable for the most sensitive juice product, and will have less than 1 ppm ingress over the same time period, as illustrated in Figure 6.7.

Because juices require only oxygen protection, it is worthwhile to consider whether 'bias-inside' or 'bias-outside' placement of the enhancing layer will improve the economics of package performance. The most economical combination of layer material and preform wall structure is 3–4% volume of a 'centered' barrier-scavenger. This combination will be less expensive, yet give a longer shelf-life, than as much as 5% EVOH barrier 'biased-outside' to maximize its barrier property. Since 3% is the lower limit of core layer volume in a multilayer preform, there is usually no advantage to a 'bias-inside' placement of the scavenger except for shelf-life requirements that approach one year duration.

6.11   Small CSD containers

Small containers for carbonated soft drinks have not been widely used because their CO_2 shelf-life has been too short for the capabilities of even the best soft drink distribution systems. Monolayer PET packages as small as 400 ml have been made with the same preforms normally used on 600 ml containers in order

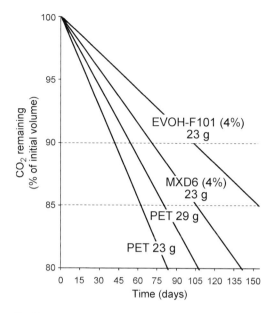

Figure 6.12 Carbon dioxide egress *vs* time for 400 ml carbonated soft drink (CSD) containers weighing 23 g and 29 g. Abbreviations: PET, polyethylene terephthalate; EVOH, ethylene vinyl alcohol.

to produce a thicker PET sidewall that compensates for the adversely increasing surface area-to-volume ratio of the smaller package. The heavier walled container typically has a marginally acceptable shelf-life and is relatively expensive because of the extra PET weight and longer cycle times for injection molding.

Multilayer preforms with a CO_2 barrier layer can provide in excess of 20 weeks shelf-life while reducing preform weights by 6 g. Figure 6.12 shows the CO_2 barrier performance of a 400 ml container made from a standard heavy preform, 29 g, and the corresponding performance of a 23 g preform with and without barrier enhancement. Enhancement with MXD6 extends the shelf-life to 15 weeks from the 10 week shelf-life of the heavyweight PET monolayer container. The higher barrier properties of an EVOH layer extend the shelf-life to more than 20 weeks. When using multilayer technology that can run at monolayer cycle times, the cost of the enhanced container can actually be less than the cost of the heavyweight monolayer container, while doubling its shelf-life. The resulting economic benefits of longer shelf-life on bottling distribution efficiencies flow to the bottom line of the bottlers.

6.12   Core layer volumes

The performance of every multilayer container can be optimized to provide the most economical package to meet the performance demanded by its specific

Table 6.8 Typical volumes of enhancing layers for different applications

Type of layer	Performance	Enhancing-layer volume (% of preform)			
		Beer	Juice	CSD	Foods
O_2 barrier-scavenger	$60 \times PET$	4–6	2–4	–	3–4
O_2 or CO_2 barrier	2–$4 \times PET$	–	4–6	4–6	4–6
RPET	$\approx PET$	≤ 40	≤ 40	≤ 40	≤ 40

Abbreviations: CSD, carbonated soft drink; PET, polyethylene terephthalate; RPET, recycled polyethylene terephthalate.

food or beverage application. The primary multilayer design variables are:

- type of enhancing layer
- percentage volume of enhancing layer in the container
- distribution of the enhancing layer in the sidewall of the container

Notwithstanding the optimization carried out for each specific container, some generalizations are listed in Table 6.8 as guidelines in the design of containers for the typical range of multilayer applications.

6.13 Recycling

Numerous studies have demonstrated the recyclability of multilayer PET containers. The enhancing layers that are being used on a commercial basis do not adhere to PET after the grinding of bottles into flake. The typical recycling processes used in North America and Europe will remove more than 80% of the enhancing layer volume present in multilayer containers, and the remaining 20% is diluted by the greater number of monolayer containers in the recycle stream to a level less than 100 ppm. Even when post-consumer bales have been loaded with five times the expected level of multilayer bottles, the standard recycling processes have produced reclaimed PET flake of good quality in terms of color, haze, intrinsic viscosity or fiber tenacity, tensile strain or modulus. The reclaimed PET is of similarly good quality for the barrier and barrier-scavenger versions of PA-MXD6, EVOH or PA-6 with nanocomposites. The suppliers of these materials have performed extensive testing in actual post-consumer recycling at commercial recycling plants.

6.14 Comparison of co-injection technologies

Four different multilayer co-injection technologies are in use in commercial production. There are differences in the container structures produced by each of these technologies, and there are differences in the businesses of each of the

Table 6.9 Comparison of the four different multilayer technologies

Company	Preform fill	Wall structure	Cycle time	Business
Kortec, Inc.	Figure 6.13A	Table 6.5A, B or C	= monolayer PET	Complete turnkey system
Pechiney (ANC)	Figure 6.13A	Table 6.5A	= monolayer +15%	Merchant bottle producer
Owens-Illinois (CPT)	Figure 6.13C	Table 6.5G	= monolayer +15%	Merchant bottle producer
Otto Hofstetter Gmbh	Figure 6.13B	Table 6.5E	= monolayer +15%	Hot runner & mold making

four companies. Table 6.9 lists the companies and the important differences between their technologies.

Although Kortec and Hofstetter sell multilayer equipment to PET converters, they each do so in different ways. Kortec sells completely integrated turnkey molding cells (co-injection machine, molds, parts-handling robot, dryers) to its customers. The scope of the Kortec turnkey system includes materials consultation, co-injection training, in-plant start-up and ultrasonic quality assurance instrumentation to verify layer thickness and position. Co-injection turnkey systems are Kortec's primary business. Kortec has developed its own patent portfolio and is also a licensee of certain Pechiney (ANC) multilayer patents. A key benefit of the Kortec approach is that its machines operate at the same cycle times as monolayer PET preform molding machines. The fundamental principle of the Kortec technology assures that as monolayer molding evolves in terms of cycle times and output, its co-injection equipment will do the same. Figure 6.13A shows a typical layer distribution of a Kortec preform. In addition

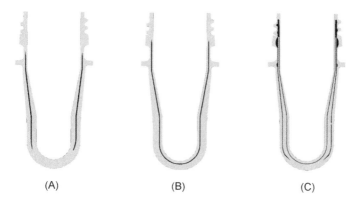

(A)   (B)   (C)

Figure 6.13 Preform cross-sections: A) 'controlled fill' with centered layer; B) complete fill with bias-inside layer; C) complete fill with five-layer.

to the centered layer shown, Kortec technology can produce bias-inside or bias-outside structures.

In contrast, Hofstetter is primarily a mold maker, and has developed a hot runner system that it sells in conjunction with its molds. The purchase and specification of the co-injection machine and its integration with the mold, robot and other auxiliary equipment is the responsibility of their customer. The Hofstetter equipment produces either a bias-inside or five-layer wall structure, as shown in Figure 6.13B and 6.13C, respectively. The resulting cycle time is about 15% slower than the equivalent monolayer PET cycle time.

Pechiney, formerly American National Can, is an early developer of multi-layer technology with a large multilayer patent portfolio, and it uses its pro-prietary technology as a merchant seller of multilayer containers made from preforms, as shown in Figure 6.13A. It has not licensed its multilayer patents to any company except Kortec, which became a licensee in exchange for royalties as part of the out-of-court settlement of litigation between them.

In the early 1990s, Continental PET Technologies, now owned by Owens-Illinois, was the first company to successfully commercialize co-injected mul-tilayer containers. Its proprietary process and equipment produces five-layer containers for merchant sale to its customers. These containers are made from preforms, as shown in Figure 6.13C. Most of the multilayer containers it sells use its proprietary version of a PA-MXD6 oxygen scavenger.

All of the technologies can produce two-material preforms, and only Kortec and Owens-Illinois (CPT) have developed three-material capability, Kortec in a four-layer preform and CPT in a five-layer preform.

6.15   Co-injection molding equipment

All current multilayer co-injection is being carried out on two-stage equipment. The differences between each of the four co-injection technologies lies in each company's intellectual property relating to nozzle design and the injection process controls that follow therefrom. The co-injection nozzle at each cavity is where the two (or three) materials are joined to create the combined flowstream that forms the fill distribution and wall structures of the layers in the preform (see Table 6.5 and 6.9 and Figure 6.13). The underlying fundamental principles of each co-injection nozzle design dictates whether the respective co-injection molding process operates at monolayer cycle time or at 10–15% slower cycles.

The basic equipment for multilayer co-injection is the same as for monolayer PET molding: machine, mold, robotic takeout equipment. Although multilayer preforms have been molded in the past on a number of different machine platforms, currently all of the technologies use Husky multimaterial machines as the base on which to build their production co-injection processes. The Husky equipment produces the fastest cycle times for monolayer preforms, and the same cooled multi-position takeout plates, Cool-jet technology or Dual- or Quad-faced Index turrets are used to speed the molding cycle for multimaterial co-injection. Each technology uses different machine control software and different hardware to adapt the basic Husky multimaterial machine to its processes and molds.

Although all multilayer co-injection is currently being performed on two-stage preform equipment, it is expected that single-stage machinery for multilayer bottles will be developed in the near future.

6.16 The future

In the five years leading up to 2001, three significant developments have placed multilayer co-injected containers as the high performance PET package leading the conversion of sensitive products from glass to plastic. One development is the emergence of several different clear barrier-scavenger materials that enhance package O_2 barrier performance to more than 60 times monolayer PET.

The second significant development is that multilayer co-injection is now being performed by leading-edge converters around the world. Whereas previously co-injection was carried out by only two converters, American National Can (now Pechiney) and Continental PET Technologies (now Owens-Illinois), the emergence of Kortec and Hofstetter as commercial suppliers of co-injection equipment has provided the technology to a global market.

The third development is that Kortec is the first to develop multilayer co-injection systems that operate at monolayer cycle times, and that will continue to be so even as monolayer operates at faster cycles in the future.

These three developments have completely changed what is possible: a PET container with a clear, barrier-scavenger, at monolayer cycle times, at 60 times PET performance, by leading converters anywhere in the world. It is expected that by 2005, the annual production volumes of these containers will be several billion units, that new material developments will continue to extend the performance envelope for both oxygen and carbon dioxide, and that further development in monolayer PET molding productivity will be translated into multimaterial co-injection systems up to 96-cavity.

7   One-stage injection stretch blow moulding

Bob Blakeborough

7.1   Introduction

Glass containers have been manufactured for many centuries by a blow moulding process. Blow moulding may be defined as 'a process in which a hollow article is manufactured by enclosing a preform in a mould and expanding it by internal fluid pressure to the shape of the mould cavity'. The particular material property which favours this process is an absence of a sharp transition from the solid to liquid form, with instead a wide 'plastic' range in which the viscosity of the molten glass is determined by its temperature.

Thermoplastic materials have a similar characteristic to glass and can be successfully blow moulded in a like fashion. As the term implies, in the extrusion blow moulding process the plastic is presented to the mould as a freely extruded, open-ended tube. Where the blow moulds close and pinch the end of the tube, the material is hot enough to 'weld' together and create a preform with a closed end, which can then be expanded by blowing with pressurised air to the shape of the mould cavity. With a properly shaped cavity, it is even possible to blow containers with an integral handle. This process is very widely used for making high-density polyethylene (HDPE), polypropylene (PP) and polyvinyl chloride (PVC) bottles.

As described in Chapter 1 of this volume, the development of the PET bottle was driven by the need of the carbonated soft drink industry to find a material suitable for making larger family-size bottles. Up to the 1970s, almost all carbonated soft drinks (CSDs) had been packaged in metal cans or small returnable glass bottles. Attempts to introduce family-size glass bottles were short-lived due to their heavy weight and the dangers of breakage. What was needed was a plastic that could be used to make cost-effective, lightweight, family-size bottles for CSDs. The bottle-making materials, such as HDPE, that were available at that time were unsuitable for two reasons: low tensile strength and a gas barrier insufficient to retain the carbon dioxide 'fizz' for a viable shelf-life. High tensile strength is needed because the carbon dioxide gas that is dissolved under pressure in a CSD pushes outwards against the walls of the container, and the more carbon dioxide dissolved (or the higher the ambient temperature) the greater the pressure that it exerts. Glass bottles, being rigid, withstand these expansion forces without visible deformation but an HDPE bottle will 'creep' and blow up like a balloon. Any plastic used to make CSD bottles must be strong enough to resist these expansion forces.

The reason that PET succeeded where other plastic materials did not, is that PET (like nylon) has a very special characteristic: when heated to a temperature where its chain-like molecules are sufficiently mobile to unfold instead of breaking when stretched, PET can be 'oriented'. Stretching applied from two directions at right angles gives 'biaxial orientation'. Oriented material contains closely packed chains aligned in the directions of stretch. It is stronger because the molecules work together to support a load that would break individual molecular chains. Not only is the tensile strength of oriented PET several times that of the basic material, the impact strength, gas barrier and chemical resistance are also improved, which means that bottles made from biaxially oriented PET can be lighter without sacrificing performance.

Conventional extrusion blow moulding techniques were found to be unsuitable for producing a highly oriented PET container. This was partly due to the low cohesion of PET in the melt state, but more crucially, biaxial orientation can only occur in a narrow temperature band well below the temperature necessary for an extruded PET tube to form a 'weld'. The two-step injection stretch blow moulding process was therefore developed to overcome these problems. The first step requires the injection moulding of a test-tube-like 'preform'. In a second step, the warm preform is inflated inside a larger blow mould to form a bottle. The two steps may be completed on two separate machines, referred to as the two-stage process, or may be combined in one machine, known as a one-stage machine. A comparison of the main features of the one- and two-stage processes is given in Table 7.1. The two-stage process is described in Chapter 8 of the present volume.

The injection stretch blow moulding process has a number of important advantages:

- biaxial orientation provides optimum material properties, so that bottles can be lighter without compromising performance
- the initial injection moulding step provides a high-precision neck finish that requires no additional finishing operations and is ready for immediate filling
- the process does not generate scrap with each container made

Due to the lower temperature at which blowing must be performed in order to realise the benefits of orientation, it is not possible to produce PET bottles with blown handles (although various means of providing handles have been devised, as shown in Figure 7.1).

Mr Katashi Aoki, Chairman of Nissei Plastic Industrial (the Japanese manufacturer of injection moulding machines) unveiled the first commercial machine for making PET bottles, the one-stage ASB-150 injection stretch blow moulding machine, in December 1975. All injection stretch blow moulding machines deriving from this original 'Aoki Stretch Blow' design are hereafter referred to as 'classic' one-stage machines. The main distinguishing features of the

Table 7.1 Comparison of classic one-stage and two-stage systems

Classic one-stage	Two-stage
Lower initial investment *for a complete system.*	Higher initial investment *for a complete system,* although choosing not to make preforms but to buy them can minimise initial costs.
On classic one-stage machines, the number of injection cavities and blow moulds is the same.	The cycle-times of the injection and blow moulding machines are usually different, so their outputs are matched by increasing the number of injection cavities relative to the blow moulds. Typical ratios are 3 or 4 injection cavities to one blow mould.
Traditionally considered to be more suitable for low and medium volumes and for production of more technically demanding containers.	Considered to be more suitable for high-volume production of less technically demanding containers, such as carbonated soft drink bottles.
Classic one-stage machines are more versatile, being capable of making both bottles and wide-mouth jars in a very wide variety of sizes and styles.	Frequently two-stage machines are only capable of making bottles.
Since container dimensions generally determine preform spacing, differences in production costs between bottles and jars are minimal.	The production costs of wide-mouth preforms for two-stage production are high because of the very large moulds and injection moulding machines that are needed.
The thermal efficiency of the one-stage process is better because the residual heat in the preforms from the injection moulding process is used for blowing.	The thermal efficiency of the two-stage process is lower since the preforms are first cooled to room temperature for storage purposes and later reheated to the correct blowing temperature.
Preforms are blown immediately they are moulded, so there is no opportunity for preforms to 'age' or absorb moisture from the atmosphere, which would result in an alteration in stretch-blowing characteristics.	The relatively compact preforms can be stockpiled to cater for fluctuations in demand, but care must be taken in controlling storage conditions to ensure consistent aging and moisture absorption of different preform batches. – Centralised preform production facility can be created to supply several blow moulders. – Bottlemakers can make or buy preforms.
Preforms are not re-gripped before blowing so the bottles are free from scuff marks.	Preforms are normally stored before blowing and may thus bear minor surface blemishes.
Since preforms are not re-gripped, a neck support ring is not essential.	A neck support ring is required to facilitate preform handling.
Conventional one-stage machines can make containers of highly oval or rectangular cross-section.	Degree of ovality restricted owing to the difficulty of temperature profiling the preforms.
Preform designs are different: preform wall is *thicker* to retain heat where stretching is wanted.	Preform wall is *thinner* so that it heats up faster where stretching is wanted.

Figure 7.1 PET bottles can be made to virtually any capacity and any size of neck. PET bottles can even be produced with integral carrying handles that are moulded in PET together with the preform.

Figure 7.2 Three lip cavity halves manufactured by R&D Tool and Engineering for a mould to be installed on a Nissei ASB machine. Each pair of lip cavity halves forms the outside of one preform neck. An eight-cavity toolset for a classic four-station machine will have $4 \times 8 = 32$ pairs of lip cavities. The internal and external tapers that are required to ensure that the lip cavities, injection core pins and injection cavities remain concentric are clearly visible.

classic one-stage machine design are the vertical press to clamp the injection mould (which may have from 1 to 24 preform cavities) and the rotary table, beneath which are installed a complete set of lip cavities (see Figure 7.2) for each 'station.'

The basic steps in the classic one-stage injection stretch blow process are:

Drying the PET. PET absorbs moisture from the atmosphere. This must be reduced to levels of less than 40 parts per million (ppm) before processing. At temperatures above the melting point of PET, any water that is present in the material rapidly hydrolyses the polymer, thus reducing its molecular weight and decreasing its physical properties.

Plasticising the PET. The dried PET pellets are fed into the rotating screw of an injection moulding machine, where they are compressed and melted. Typically the temperature of the melt is about 290°C.

Injection moulding the preform. The molten PET is injected via a hot runner system into the mould cavities, where it is cooled as fast as possible. PET forms white crystals very quickly when hot and so the injection mould is chilled to around 10–15°C to promote rapid cooling in order to form clear, amorphous preforms. However, the preform is not allowed to remain inside the cavity until it reaches this temperature. Instead, once the preform surface has hardened sufficiently, the injection cavity is opened and, gripped by the neck, the preforms are swung round to the conditioning station while they are still hot and malleable.

Optimising the preform temperature. The preform temperature is adjusted to the correct level for blowing (95–120°C) at the conditioning station. The temperature of the preform must be exactly right to achieve maximum strength enhancement without compromising the clarity of the bottle through stress whitening (too cold) or crystalline haze (too hot).

Stretch blow moulding the container. At the blowing station, the hot preform is enclosed in a bottle-shaped mould. A stretch rod inserted through the neck of the preform rapidly stretches it to the bottom of the mould, while simultaneously high-pressure air inflates the preform like a balloon. As the expanding PET bubble touches the cooled metal surface of the blow mould, it is transformed into a rigid bottle with an injection-moulded neck-finish. The simultaneous stretching and blowing biaxially orientates the PET and enables the peak physical properties of the PET material to be achieved. An important by-product of the strength-enhancing stretching process is the self-levelling effect that assures an even wall thickness in a fully oriented bottle.

Container ejection. After the blowing air has been vented, the blow mould opens to reveal a sparkling transparent bottle and the tough, lightweight, fully formed container is ejected from the machine.

7.2   One-stage machines

7.2.1   One-stage machine construction

The classic one-stage design is extremely versatile in that the same basic machine can be used to make a wide variety of bottles and jars in all shapes and sizes. The machine consists of the following basic components (Figure 7.3A presents

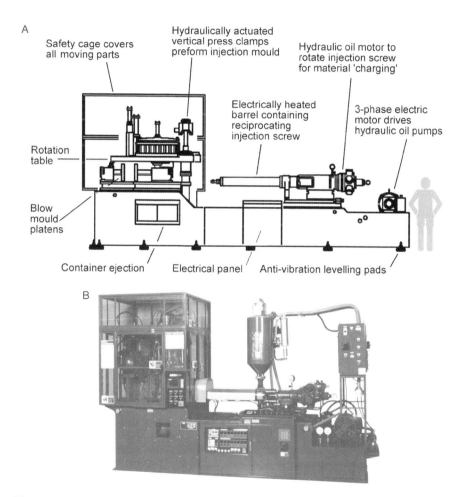

Figure 7.3 A) Schematic diagram of a one-stage, four-station machine. B) A classic ASB-50 one-stage machine produced by Nissei ASB of Japan showing the PET dehumidifying dryer mounted above the main electric motor at the back of the machine, and the drying hopper mounted above the injection barrel. This is the smallest model produced by Nissei ASB and was envisaged as a laboratory machine for test moulding, but is now widely used for the production of smaller bottles, such as the miniature spirit bottles used by airlines.

a schematic diagram and Figure 7.3B a photograph of a four-station, one-stage machine):

1. The prime mover, typically an electric motor driving one or more hydraulic pumps.
2. An injection unit, consisting of a reciprocating injection screw inside an electrically heated cylinder or 'barrel'. The screw is rotated by an oil motor to 'charge' the screw with molten PET prior to the injection stroke, when a hydraulic ram forces the screw forward in order to inject the molten PET via the hot runner system into the preform mould cavities.
3. A vertical, hydraulically actuated press which opens, closes and applies clamp forces to the injection mould (see Figure 7.4A).
4. An electrically driven or hydraulically actuated rotary table mechanism to carry the preforms from one station to the next. Classic machines may have three or four stations depending upon the design. Every design will have injection, blowing and take-out stations but some also have a conditioning station before the blowing station.

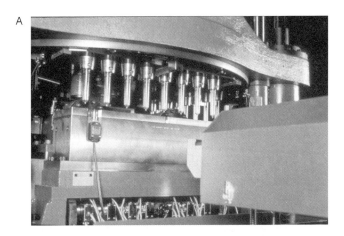

Figure 7.4 A) Hot preforms are visible immediately after the eight-cavity injection mould has opened on a classic one-stage injection stretch blow moulding machine produced by Nissei ASB. B) The set of preforms is swung round to the conditioning station by the movement of the rotary table. Electrically heated 'pots' rise up around the preforms, while heated conditioning cores descend inside for precise temperature adjustment before blowing. C) As the blow moulds begin to close, preforms—now at the correct temperature for blowing—are just visible. They are still being carried in the injection mould parts known as the 'lip cavities', which form the screw thread on the preform neck. D) Inside the blow moulds, the preforms are blown with 35 bar air, so that they expand like a balloon, but once they touch the cold metal surface of the mould, they rapidly harden to form the bottles that are visible here at the take-out station.

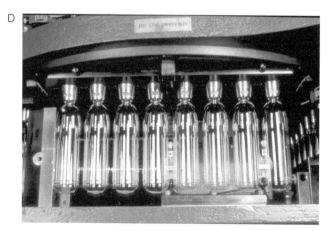

Figure 7.4 (continued)

5. The conditioning station is used to adjust the temperature of the preforms produced on a multicavity mould before they are blown into bottles. Typically, the conditioning station consists of heated 'pots' that rise up and surround the preforms, together with heated 'cores' that are inserted into the preforms from above. Individual preform temperature adjustment is possible at the conditioning station (see Figure 7.4B).

6. Two-piece or three-piece (depending upon the design of the base of the bottle) water-cooled aluminium blow-moulds are installed at the blow station. A tightly fitting blow core is pressed into the preform neck. The steel stretch rod pierces the centre of the blow core, leaving an annular opening for the blow air to enter the preform (see Figure 7.4C and D).

7. The blown bottles may simply be dropped into a waiting box at the take-out station, or more commonly will be transferred to a bottle conveying system.

A wide variety of classic one-stage machines is available from single-cavity up to 32-cavity models. Tables showing the number of containers of a given height, diameter and neck size that can be made on each model are given in manufacturers' catalogues. Names and addresses of one-stage machine manufacturers are given at the end of this chapter.

7.3   Process stations on a one-stage machine

7.3.1   Injection mould and hot runner

The injection mould is an assembly of parts containing within it an 'impression' into which the PET material is injected. The impression gives the preform its shape and consists of several mould parts:

- The 'cavity', which is the main female portion of the mould, gives the preform body its external form (see Figure 7.5)
- The 'core', which is the male portion of the mould, forms the internal shape of the preform
- The 'lip cavities', which are female portions of the mould, form the outside of the neck of the preform (see Figure 7.2)

Each injection core and cavity is chilled (by circulating water at a temperature of 10–15°C through passages cut in the surrounding steel), so that the molten PET (at 290°C when it is injected) is cooled below its peak crystallisation temperature of 160°C as rapidly as possible. During the injection process, the PET material is delivered to the cavities through a 'hot runner system' (see Figure 7.6). The hot runner, which is thermally insulated from the cavity, is maintained at a closely controlled elevated temperature similar to that of the injection barrel, in order to keep the PET material inside it permanently molten.

Figure 7.5 A single injection cavity insert manufactured by R&D Tool and Engineering for a mould to be installed on a Nissei ASB machine. An eight-cavity mould requires eight inserts, which are fitted into a 'cavity block.' The passageways that circulate chilled water around the insert are clearly visible.

Figure 7.6 A 12-cavity Kona hot runner system manufactured by Synventive Molding Solutions. Temperature uniformity is critical in one-stage stretch-blow moulding where the heat content of the preform largely determines the distribution of the material in the blown container. Kona hot runner manifolds and nozzles contain heat pipes which effectively make them into super thermal-conductors to evenly distribute the heat regardless of non-uniform heat losses from the manifold.

The hot runner is possibly the most critical, yet sometimes most neglected, part of a PET preform moulding system. The one-stage process in particular places very high demands upon the hot runner system because the physical properties of a given container depend upon two factors: the degree of biaxial orientation of the PET and the placement of the PET material in the container wall—both of which are a function of the preform heat content, which in turn is largely controlled by the hot runner system. It determines the mould start-up time, the cavity-to-cavity consistency of the preforms and the appearance of the gate vestige or sprue, and affects the moulding cycle time, the level of acetaldehyde (AA) in the preform, etc. (For example, a uniform flow of PET into each cavity is essential since the PET starts to cool as soon as it enters the mould; thus, a preform from a cavity that had filled faster would be colder than one from a cavity that had filled more slowly.)

The effect of the high-pressure melt entering the cavities is to exert an opening force on the injection mould. The magnitude of this opening force is proportional to the total projected area of the preforms at the parting surface (i.e. the larger the number of cavities—or the wider the neck diameter—the greater is the opening force that must be resisted by the machine clamping force). Therefore, machines designed for making preforms for jars have greater mould clamping capability than those intended only for making narrow-neck bottle preforms.

7.3.1.1 *Process conditions affecting preform quality*

The injection moulding process can be divided into the following stages:

Injection. Pressure is applied behind the injection ram to force the screw for-ward. This, in turn, forces the molten PET (via the hot runner) through the opening in the bottom of the cavity known as the 'gate'. The injection stroke ends when the cavity is almost full of molten material.

Holding. At the end of the injection stroke, the molten PET in the cavity cools and starts to solidify and the relatively high injection pressure is reduced to a lower holding pressure. The holding pressure has two functions: to add additional material to replace the volume lost due to shrinkage of the cooling PET in the cavity, and to partially counteract the volumetric expansion that takes place when PET is heated by compressing the material. If nothing were done to rectify this, the final preforms would show sink marks and could be underweight and undersized. The holding pressure is maintained until the gate freezes off. Time is an important factor here as the pressure will not be effective after the gate has frozen, while taking it off before the gate freezes will result in variable preforms.

Cooling time. After the gate has frozen, the mould remains closed while further cooling of the PET inside the cavity takes place. Note that the preform wall thickness determines the overall duration of the injection moulding cycle, since the preform cannot be withdrawn from the cavity until the interior of the preform has been cooled sufficiently not to crystallise and the surface has cooled and hardened enough that it will not to be damaged during the removal process.

Typically, a one-stage preform will have a temperature of 120°C or more when it is removed from the cavity.

Key injection moulding settings affecting the preform temperature include the following:

- Melt temperature. Low temperature melt will freeze quickly in the mould (it may even result in 'short shots' if the gate freezes) and higher pressures will be required to fill the cavity in a given time. High temperature melt will flow more easily but may cause 'flashing' or 'sink' marks (the hotter the material the more it will shrink on cooling). Sink marks are small depressions in the preform surface caused by inadequate filling of the cavity. Flash is the result of too much material being packed into the cavity, so that PET flows into the vents that are intended to allow the air displaced by the incoming PET to escape. The aim is to operate with the lowest possible melt temperature that is consistent with fully acceptable preform and bottle quality.

- Screw speed (rpm). The screw speed should be kept as low as possible to avoid excessive shearing of the material, which would cause overheating and lead to degradation of the PET and generation of excessive levels of acetaldehyde.

- Injection speed (time). The proper injection speed varies from bottle to bottle and depends on preform length and wall thickness. It is normally best to achieve a fast but constant rate of fill, so that the preform has an even temperature gradient. If the injection speed is low, the preform temperature will vary along its length, with the last part to fill (immediately adjacent to the gate) being hotter than the neck part.

7.3.2 Conditioning station

The classic one-stage machines produced by Nissei ASB of Japan incorporate a conditioning station (see Figure 7.4B) that is used for fine control of the preform temperature. If the preform temperature is too high, the preform will blow unevenly because the designed level of orientation will not occur and the bottle may even be hazy in appearance due to crystal growth. If the preform temperature is too low, the internal layers in the PET material will separate, resulting in a 'pearlescent' appearance (or the expanding preform may even burst inside the blow cavity). Furthermore, the correct temperature 'profile' is required along the length of the preform in order to provide the desired material distribution in the container (since the hotter parts of a preform stretch more easily than the cooler areas).

Since different bottle designs and performance specifications require different material distributions in the finished container, Nissei ASB have developed a range of conditioning systems and methods to provide the appropriate degree

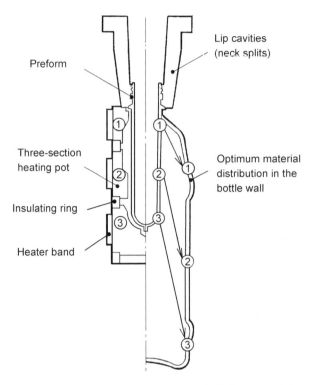

Figure 7.7 Schematic representation of a three-section electrically heated 'pot' that may be used to make fine adjustments to the preform temperature at the conditioning station of a Nissei ASB machine. The three sections are individually contoured, so that the heat is directed precisely where it is needed in order to achieve the optimum material distribution in the bottle.

of temperature adjustment for each application. These include electrically or thermo-oil heated (or cooled) 'core pins' to control the inside surface temperature, and 'pots' to control the outside surface temperature of the preform (see Figure 7.7).

7.3.3 Blowing station

At the blowing station, the hot preforms are stretched and blown into cooled, polished, aluminium blow moulds (see Figures 7.4C and 7.8). (Note that, in comparison to the extrusion blow moulding process, much less cooling is required in the blow mould because of the lower temperature of PET when it is stretch blown—typically only 95–110°C.) The stretch rod not only keeps the preform gate centred in the base of the container but, together with the primary and secondary blow air, places the PET material in the correct position in the blow mould. The precise time in the stretching and blowing sequence when

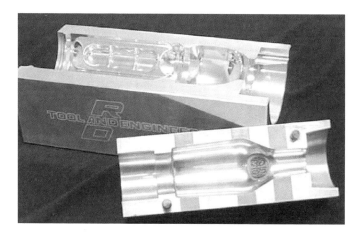

Figure 7.8 Two blow mould halves manufactured by R&D Tool and Engineering for making a 1500 ml hot-fill bottle and a 500 ml spirit flask on Nissei ASB machines. The shallow vents that allow the air to escape from the mould as the preform expands are visible as light and dark bands on the spirit flask mould.

the primary and secondary blow are initiated is critical, since this, together with the temperature of the preform, controls the distribution of the PET in the container. For example, the amount of material in the lower half of a container can be increased at the expense of that in the upper half, by delaying the start of the primary blow and allowing the stretch rods more time to drag material towards the base of the bottle before the preform begins to expand sideways and touches the blow mould surface (once the expanding preform has touched the blow mould, the PET hardens and no more stretching is possible).

7.4 'Integrated two-stage' machines

A classic one-stage machine will have one blow cavity for each injection cavity. In consequence, the blow moulds are under-utilised since the stretch blow process is normally completed three or even four times faster than the preform injection moulding step. For high output volumes, it may be more efficient to separate the two steps of the PET bottle production process and run each one on a dedicated machine: an injection moulding machine with a large number of cavities, to make the preforms, and a 'reheat' stretch blow moulding machine having rather fewer moulds, to blow the bottles (see Chapter 8, Two-stage injection stretch blow moulding).

However, in recent years a new type of one-stage process, sometimes referred to as 'integrated two-stage', has been developed, which combines the advantages of the one- and two-stage processes. Unlike classic one-stage machines, these machines have different numbers of injection cavities and blow moulds in the ratio of 3, 4 or even 5 preform cavities to one blow mould. After moulding, the preforms on integrated machines are first cooled slightly below typical blowing temperatures and then reheated before blowing using infrared quartz lamps. It is claimed that, as a result, every preform achieves exactly the same optimal temperature before blowing and so the bottles are consistent, despite the disparity in the time elapsed between moulding and blowing of consecutive preforms.

The integrated approach has several advantages, including:

- a reduction in the number of mould parts, particularly blow moulds and lip cavities.
- smaller injection moulds and hot runner systems. (The spacing between preform cavities in a classic one-stage injection mould is determined by the size of the bottle, as can be seen by comparing Figure 7.4A and D. No such constraint applies to an 'integrated two-stage' machine.) As a result, the actuation systems and the machine itself are more compact.
- preform temperature adjustment using infrared quartz lamps is quick and precise.

One possible disadvantage of the integrated approach is that current machines of this type are designed specifically for bottle production and are less versatile than classic one-stage machines, which can be used to produce both bottles and wide-mouth jars.

7.5 Drying system

PET in pellet form is hygroscopic, which means it absorbs water from the atmosphere. This will eventually diffuse throughout the pellet, reaching an equilibrium level with the surroundings. The water in the pellet will do no harm but it is essential to reduce this moisture level to less than 40 ppm by weight and ideally lower before melting the PET. This is because any moisture that is present in the molten PET rapidly hydrolyses (breaks down) the polymer, thereby reducing its intrinsic viscosity (IV) and the associated physical properties. Experiment has shown that when 0.76 IV PET melts, an IV loss of 0.01 occurs for every 16 ppm of moisture retained in the material; thus, if 40 ppm by weight of water is present in the PET material, the IV will be reduced by about 0.03.

The drying of PET is the reverse of the moisture absorption process. The absorbed moisture will have diffused towards the centre of the pellet. To reverse this effect, the moist pellets are placed in extremely dry conditions and heated to accelerate the outward diffusion. However, there is a limit to the maximum drying temperature and the time for which the PET can be dried. Firstly, hydrolytic degradation is temperature dependent: it begins to occur in solid PET at temperatures as low as 150°C (although only very slowly) and the rate increases with temperature. Secondly, excessive drying times can lead to oxidative degradation (especially at higher temperatures), which can give a yellow tint to the preform.

Typically, PET is dried in a specially designed dehumidifying dryer operating at a temperature of 160°C for a period of at least 4 h. This temperature is considerably above the glass transition temperature (T_g), of 70°C (above this temperature amorphous PET begins to soften and becomes sticky). Therefore, it is essential that the PET pellets placed inside the dehumidifying hopper-dryer are crystalline because material in the amorphous state would stick together, preventing the free flow of the pellets.

7.5.1   Requirements for a reliable drying system

Correct dehumidified air temperature.   This should not normally exceed 150°C, measured at the air inlet to the dryer, but the material supplier and the manufacturer of the dehumidifying hopper-dryer system should recommend the correct PET drying temperature for a particular application.

Note that it is essential to turn down the dryer set temperature to 120°C should moulding be interrupted, otherwise the recommended pellet residence time for the selected drying temperature could be exceeded and the PET would be degraded.

Adequate flow of dehumidified air through the pellets.   Airflow of 1 ft^3/min per lb/h material throughput (0.06 m^3/min per kg/h) is a minimum requirement. If the airflow rate through the material is less than this, the material at the top of the hopper may not reach the required temperature (because of heat losses from the system), which would mean that the material would not be adequately dried. A 25% reduction in temperature can result in material taking three times longer to dry than at the correct temperature.

Pellet residence time (drying time).   Most PET producers recommend a residence time of not less than 4 h. The residence time is calculated by dividing dryer capacity (in kg of PET) by polymer throughput (in kg/h).

The correct dehumidified air dewpoint (according to dryer type).   The dryness of the air is defined in terms of the dewpoint temperature. This is the temperature at which water vapour in the air will form droplets of water on a cold surface— the lower the temperature, the dryer the air. Most new dryers will deliver air with a dewpoint of lower than −40°C when new but eventually drying efficiency will be reduced and the drying medium (desiccant material, molecular sieve, etc.) must be replaced.

7.5.2 Drying process monitoring

The following areas should be closely monitored:

Air filter. Routine filter cleaning is essential. The filter prevents the desiccant material inside the dehumidifying dryer from becoming clogged with PET dust that might impair its efficiency.

Heat exchanger. The moisture removing properties of the desiccant material will be lowered if the moist return air is too hot, so a heat exchanger is normally fitted to PET dryers to cool the return air before it passes through the desiccant.

Air inlet temperature. Where the dehumidified air temperature is measured at a distance from the hopper inlet, the temperature setting may need to be raised to compensate for heat losses from the system and so achieve the required 'material' temperature.

Ingress of ambient air. Dried PET reabsorbs moisture very rapidly when it is exposed to ambient air (which is very wet compared to the drying air), so that it is important to prevent any ingress that will reduce drying efficiency. If the PET is dried away from the machine, it is essential that the dried material be conveyed to the machine in dry air with a temperature and dewpoint similar to that of the drying air.

Drying process control. Most dryers monitor the drying air temperature; some also monitor the pressure drop across the filter (sounding an alarm if the filter becomes clogged) and the dewpoint of the drying air. Another useful check is to monitor the temperature of the material entering the feed throat of the machine. A change in temperature provides an early indication of dryer malfunction.

7.6   Preform design

To achieve the maximum strength improvement through orientation, PET needs to be stretched at the correct temperature by an amount known as the 'natural stretch ratio'. It is therefore essential to have the correct preform design. A selection of preforms is illustrated in Figure 7.9.

Figure 7.9 A selection of PET bottle preforms, ranging from 250 ml with a 28 mm neck and weighing 23 g (lying on its side) up to 2000 ml with a 32 mm neck and weighing 57 g (standing upright at the centre).

The preform design is constrained by several factors, listed below in order of importance.

7.6.1 Neck finish

The neck finish determines the maximum internal diameter of the preform, since clearly the internal diameter of the preform cannot be larger than the neck opening.

The minimum internal preform diameter is limited by the need to remove heat from the preform. Injection core pins are hollow and have a tube inside them that carries the cooling water directly to the tip of the preform. The water then flows back out between the tube and the inner surface of the injection core pin. The injection core pin must therefore be large enough to contain the tube and permit adequate chilled water flow, and be strong enough not to be easily bent by heavy-handed cleaning efforts.

In preforms with very small necks, a special 'stepped-out' preform design may be necessary to achieve the desired weight, rather than the conventional 'stepped-in' design (see Figure 7.10).

One of the most difficult parts of a preform design is the interface between the neck and the preform body. There are few, if any, written rules for deciding how thick or thin this area should be, the usual approach being to base the design on previous successful preform designs. Ideally, the preform will have a uniform wall thickness, with any transitions that are necessary between thinner and thicker sections being very 'smooth'. Abrupt changes in preform wall thickness must be avoided because they will retain heat unevenly and cause ridges and other witness marks in the bottle.

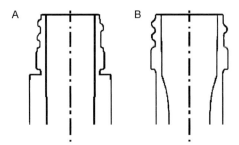

Figure 7.10 Types of preform. A) 'stepped-out' preform (non-standard); B) 'stepped-in' preform (conventional).

Table 7.2 PET densities

Section of container	Density (g/cm^2)
Amorphous PET (preform neck)	1.334
Average figure for the body part of a container (oriented and semi-crystalline)	1.368

7.6.2 Preform weight

The preform weight determines the container wall thickness, and thus the strength and 'feel' of the bottle. The proper weight can often be estimated by comparison with existing containers or, for a new container design, may be found by calculating the surface area of the bottle using a computer-aided design (CAD) system and multiplying the area by the average wall thickness to give the volume of PET. The approximate weight can then be determined by multiplying the volume of PET by its density (see Table 7.2):

With PET more weight is not always better. If a bottle is too heavy, the stretch ratios will be lower—making it more difficult to produce a bottle with an even material distribution—while the cycle time will be longer.

7.6.3 Cycle time and preform wall thickness

Within broad limits, the time required to mould a certain preform or to stretch blow a particular bottle is determined by two variables: the preform wall thickness and the weight of PET (see Figure 7.11).

Preforms with thick walls require longer injection and curing times, so if a fast cycle time is desired, thin wall preforms are best. However, if the preform wall is too thin, the melt could freeze during mould filling so that it becomes impossible to make a preform. Furthermore, a very thin-walled preform would need a highly uniform wall thickness in order to avoid material distribution problems in the finished container, but some variation is always present due to

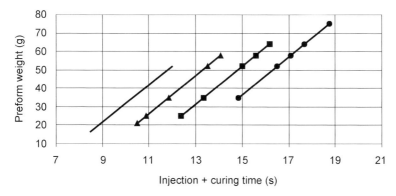

Figure 7.11 Chart showing preform weight and thickness *v* injection moulding time. To calculate an approximate cycle time for a given preform, find the weight on the vertical axis, move right across the horizontal line until the diagonal line approximating to the preform wall thickness is reached, then move down the vertical line and read the injection moulding time from the horizontal axis. To calculate the approximate cycle time, add this to the machine 'dry cycle' time provided by the machine manufacturer. Key to preform wall thickness: ———, 2.7 mm; —▲—, 3.1 mm; —■—, 3.5 mm; —●—, 3.9 mm.

mould manufacturing tolerances. Experience dictates that with standard grades of PET the minimum preform wall thickness is about 2.3 mm (possibly less for a very small preform), while the maximum is about 4 mm. Typically, the wall thickness will be within the range 2.7–3.3 mm. Preforms with a thickness up to 10 mm are possible with specially formulated grades of PET but they are only used in exceptional cases due to the lengthy cycle time.

7.6.4   Stretch ratios

The physical properties of a PET bottle, such as the strength and, to a lesser extent, the 'feel' of the container, depend upon achieving the correct level of orientation. Thus, the preform dimensions are closely related to the bottle dimensions—the larger the bottle, the bigger the preform. Three 'stretch ratios' are used to define the degree of orientation: axial, hoop and area (the product of hoop and axial).

The axial stretch ratio is calculated by dividing the height of the stretched part of the bottle by the length of the stretched part of the preform. With conical wide-mouth preforms, the contour height and contour length are used to give a more accurate figure.

Hoop stretch ratios are defined as the bottle diameter divided by the preform (mid-wall) diameter. Since some preforms have extreme tapers, judgement is required when deciding at which part of the preform the diameter should be measured. A similar problem arises with bottles having large changes in section, and also with square or oval containers. In such cases, it is better to determine the hoop stretch ratio by comparing the preform and bottle circumferences (see Figure 7.12).

Points worthy of note are listed below:

- if the axial stretch ratio is low then the hoop ratio should be higher
- smaller stretch ratios mean that the PET material can be blown into the corners of the container more easily—because the orientation is less—but also mean that the bottle wall thickness will be more difficult to control
- in most cases, the 'ideal' stretch ratios (see Table 7.3) cannot be used because an overriding requirement for a fast cycle demands a thin wall, which for a given weight generally means a longer preform

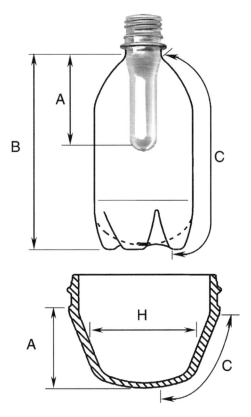

Figure 7.12 A, preform axial stretch length; B, container axial stretch length; C, contour axial stretch length; H, average hoop diameter of conical preform.

Table 7.3 Typical and ideal stretch ratios

	Typical	Ideal stretch ratios
Axial	2.0–2.5	2.2
Hoop (mid-wall)	3.0–4.0	3.6
Area	6–9	8.0

7.6.5   Injection mould design and manufacture

To facilitate removal of the preform from the mould, preforms are tapered internally and externally. Typically cavities have a minimum 0.5° taper, while the cores may be slightly more tapered at around 0.6°. This means that the preform wall is slightly thicker towards the gate, which compensates for the longer cooling time experienced by being the first part of the cavity to fill.

The PET melt must be cooled fast enough to prevent crystallisation, and so the chilled water channels must be as close as possible to the cavity. This is easier to achieve in straight-wall preforms. Wide-mouth preforms with large concave radii tend to suffer from uneven cooling, leading to thickness variation in the container, and are best avoided if possible. Therefore, preforms for jars are generally conical but the need for good cooling must be balanced against the need to maintain the compression strength of the cavity (see Figure 7.13).

The gate area of the preform must be strong enough not to rupture during stretching. Therefore, the thickness in the gate area is normally only 60–70% of the thickness of the preform wall in order that it will cool faster and not stretch so easily.

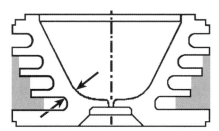

Figure 7.13 Sectional view of a jar preform injection cavity. The water channels must be close enough to the moulding surface to provide adequate cooling, especially in the gate area, but without weakening the cavity, which needs to be strong enough to support the injection mould clamp forces. The arrows indicate the site of minimum cavity wall thickness and thus the weakest point. If necessary, the outer circumference of the cavity may be fitted with a contoured 'sleeve' (shown here in grey) to reduce the cross-sectional area of the channel and maintain turbulent flow, concentrate the flow close to the gate and improve the load-carrying capacity.

7.6.6   Preform design for varying container sizes

Clearly, the neck design and bottle weight will be determined by the chosen preform; only the container size, shape and therefore wall thickness can be altered.

Preforms are designed to be stretched a specific amount. If this is not achieved, then the self-levelling effect that normally occurs when PET is biaxially stretched will be less effective, while if the preform is stretched too much, the result will be pearlescence. (When PET is biaxially oriented, the molecular chains are aligned in the directions of stretching to form very thin 'layers' of PET with low interfacial strength. If the preform is stretched too far—beyond the natural stretch ratio—slippage occurs between layers, which becomes visible as pearlescence.) If the preform is blown hotter, then the stretching that is possible before causing pearlescence will be increased. The limit will be reached when the area stretch ratio (axial × hoop ratio) reaches about 12. Extreme (mid-wall) stretch ratios:

- axial 2.75
- hoop 4.25
- area 12

7.6.7   Preform weight adjustment

The preform weight can only be changed by replacing or modifying parts of the injection mould. An injection core with a slightly smaller or larger body diameter can be used in order to change the preform body weight (not including neck) by up to $\pm10\%$. More than this is not usually recommended. Modification of the cavity is more difficult and generally not attempted. When changing the wall thickness, it is important to recognise the implications for injection moulding problems (preform wall too thin) or increased cycle time (preform wall too thick).

7.6.8   Differences between one- and two-stage preform designs

In a classic one-stage machine (hot preform method), the residual heat retained in the preform from the injection moulding step is sufficient for the preform to be blown. In contrast, the two-stage (cold preform) system relies on the preform being reheated from room temperature prior to blowing. This means that in a classic one-stage preform, thicker areas of the preform that retain heat better will stretch more easily to become thinner areas in the bottle, whereas in a two-stage preform the opposite is true—thin areas of the preform that absorb heat faster will become thinner in the blown bottle. This means that preform designs for one system cannot be used in the other.

7.7 Container design

PET bottle and preform designers must not only be familiar with the injection stretch blow moulding process, they also need to understand the physical properties of PET material and how these relate to the container performance. Some key points are defined in Table 7.4. For a more detailed treatment of PET material refer to Chapter 3 of this volume.

Table 7.4 Container design considerations

Overflow and fill point capacities/fill height
Dimensional constraints (filling/transit/retailing)
Shape constraints (pressurised/non-pressurised):

- base style (base cup, champagne, petaloid style, flat)
- shaping to enhance rigidity (vacuum/capping forces)
- neck finish (dimensions/manufacturer's identification)

Closure type:

- retention system/seal features
- tamper evident/child resistant/dispensing

Label/decoration requirements:

- label size/position (location indent in base)
- label protection features/manufacturer's logo/recycling mark on base

Performance specification:

- container weight
- thermal stability (creep resistance)
- top load resistance (filling/stacking)
- drop impact strength
- wall thickness requirements

Special product requirements:

- barrier requirement
- carbonation retention/oxygen ingress/moisture
- chemical compatibility
- stress crack resistance

Filling line considerations:

- fill temperature (% shrinkage)
- filling line stability/handling
- cap application method (shape constraints)
- label application/direct printing system

Consumer requirements:

- container stiffness—'feel'
- ease of handling—'pourability'

7.8   Hot-fill PET bottles

Some beverages, such as fruit juices, need to be sterilised by pasteurisation in order to destroy any microorganisms that could affect the stability and shelf-life of the product. The bottle and closure also need to be sterile. Sterilisation can be achieved by filling the container with the beverage while it is still hot from the pasteurisation process. However, sterilisation by hot filling poses special problems for conventional PET bottles, which shrink and distort at higher temperatures, and so special techniques have been developed to produce hot-fillable PET bottles.

The earliest techniques to raise the temperature at which PET bottles could be filled involved blowing the bottle and then holding it for a time inside heated blow moulds to raise the level of crystallinity of the PET and also relax some of the strains induced in the PET by the stretching process (strains that would otherwise cause the bottles to shrink when hot filled). The resulting increase in allowable fill temperature was a function of both the blow mould temperature and the time for which the bottle remained in contact with the hot surface. Classic one-stage machines were particularly suitable for this simple 'heat setting' process because it was possible to slightly lengthen the blow mould closed time without increasing the overall cycle time (which is controlled by the preform injection moulding step). For a detailed description of the various heat setting technologies see Chapter 10 of this volume.

7.9   Quality control procedures

Routine monitoring and assessment of container performance against an agreed specification is an essential and integral part of overall quality assurance. The aim of the quality checks at the manufacturing stage should be to give confidence that the containers will perform as required and to remove substandard containers. Furthermore, to provide sufficient information for determining any process changes that may be needed to prevent the production of containers which do not meet the required specification. When preparing a container specification, it is sometimes forgotten that the ultimate acceptance criterion for a PET container should be whether it will perform satisfactorily during its life. Any limits and tolerances defined in the specification, while reflecting the customer's requirements, must also take into account the process capability and the technical proficiency of those who are responsible for the machine and its operation.

Many methods have been devised for assessing container quality; those described below have been selected as being of general application, although not all of these tests will be applicable to every container type, while in some cases additional tests will be required. It may be found useful for the more subjective tests to keep representative sample containers and preforms of both acceptable and unacceptable quality for comparison with current production.

How often a particular test should be performed will depend on its complexity (how difficult, time-consuming and expensive), and how useful it is in providing information as to whether the process is under control. Normally, a full and comprehensive test programme will be carried out each time a mould is installed in a machine. During a production run, an abbreviated test programme will commonly be used, often divided between the quicker checks performed by the machine operator and the more involved tests undertaken by the Quality Assurance Inspector.

As both preforms and bottles will shrink a very small amount with time, to permit valid comparisons all preform measurements should be made 24 h after moulding, while measurements on bottles should be made after the bottles have been stored for 72 h in controlled conditions.

7.10   Preform examination

Checks on appearance and shape, weight, neck dimensions, wall thickness and polarised light inspection should be made on preforms taken from each set of lip cavities. When required, intrinsic viscosity needs to be measured in only one sample from each injection cavity (although two or even three replicate samples may be required for greater accuracy).

7.10.1   Appearance and shape

Minor imperfections are likely to be present in every preform, so the specification should quantify or set limits on what is acceptable (Table 7.5 may be used as a guide).

Table 7.5 Acceptable limits to minor imperfections in preforms

Description	Typical limits
Mismatch of mould parts	0.05 mm maximum 'step' height
Flash	0.05 mm maximum
Crystalline haze/streaks	Not generally acceptable, although a lace-like collar of crystallinity within a Ø12.5 mm 'disc' centred on the gate will not usually impair container performance
Short shots/sink marks	Normally, the small sink marks often found in the thicker parts of the neck, such as the pilfer proof or roll-out band are considered to be acceptable
Bubbles	Not acceptable
Grease/surface or embedded dirt	Not acceptable
Crystalline slugs	Not acceptable
Gate stringing	Not acceptable

7.10.2 Preform weight

Variation or fluctuation in preform weight will affect container wall thickness (and thus the top load and vacuum strength) and should be checked using an electronic balance to ensure that each cavity is filling correctly during injection:

- Is the weight of each preform within the specified range? Typically the tolerance is ±1% of the preform weight. Small variations in preform weight may be due to mould design, mould machining tolerances, flash, short shots, etc.
- Is the cavity-to-cavity weight variation (on a multicavity mould) within the specified range? Typically, the tolerance is ±0.5% of the preform weight. If the variation is more than this, check the preforms for sink marks, short shots or flashing; also check the gate balance and make sure the air vents are clean. Having confirmed these items, check the condition of the mould and finally the mould dimensions.
- Is the shot-to-shot weight fluctuation within the specified range? The fluctuation in weight of consecutive preforms from the same cavity (i.e. between one shot and the next) should be less than ±0.5% of the preform weight. If the fluctuation is more than this, check the preforms for sink marks, short shots or flashing and check that the moulding conditions are correct.

7.10.3 Neck dimensions

Preform neck dimensions should be within the tolerances specified on the drawing. If a dimension appears to be wrong, the following should be checked first:

- Was the measurement correct?
- Is the preform appearance abnormal?
- Is there evidence of flash or a short shot?
- In the case of diameter measurements, does the result differ according to the direction of measurement (i.e. parallel or perpendicular to the parting line)?
- Do the minimum and maximum preform dimensions vary according to the injection and blow cavities or according to the lip cavity?

Having confirmed the above items, the moulding conditions, then the condition of the mould and finally the mould dimensions should be checked.

7.10.4 Preform eccentricity

The difference between the minimum and maximum wall thickness measured near the preform gate should be no more than 0.1 mm for preforms of up to 120 mm in length (and *pro rata* for longer preforms).

Possible causes of uneven wall thickness include:

- mould misalignment
- injection core pin is bent
- injection core pin taper and lip cavity taper are not matched and not heavily contacting
- lip cavity taper and injection cavity taper are not matched and not heavily contacting
- injection core and lip cavity flat contact surfaces are not in heavy contact
- contamination on mould surfaces
- nozzle touch force is too strong
- lip cavity manufacturing process was faulty (centres not aligned)
- injection core pin manufacturing process was faulty (centres not aligned)

When determining the cause of uneven wall thickness, one should consider whether the minimum and maximum wall thickness and their relative positions remain the same from shot to shot (if they change it would indicate that the mould parts are able to move during injection).

7.10.5   Polarised light inspection

Examination of the preform under polarised light indicates the stresses set up in the preform during injection moulding and curing. A well-made preform will have a regular stress pattern, while a random stress pattern indicates a poorly made preform. It is far easier to get good consistent material distribution throughout the bottle walls when using a well-made preform. However, inter-pretation of stress patterns is difficult and this test is not commonly used except to reveal the 'moisture rings' that occur when moisture has condensed on the injection core pins during preform moulding.

7.10.6   Intrinsic viscosity (IV)

Measuring the IV of the material in the preform and comparing it with that of the virgin material can give an indication of dryer effectiveness. In the molten state, PET is very susceptible to hydrolytic degradation and must therefore be correctly dried before processing. Any moisture remaining in the polymer chip will lead to a drop in viscosity during processing, resulting in reduced physical properties, such as top load, and increased creep in the finished container. Experiment has shown that an IV drop of 0.01 will occur for every 16 ppm of moisture retained in the PET on melting. IV testing is normally available from the PET supplier.

IV is related to the average length of the molecular chains from which the polymer is formed. The viscosity of the molten polymer is a function of the IV, which also determines the toughness and drop impact strength of the blown container.

7.10.7 Acetaldehyde (AA)

Acetaldehyde is a substance produced in small amounts during the melt processing of PET. However, the ability of AA to diffuse out of a bottle wall and impart a flavour to certain products means that its generation during preform moulding must be carefully controlled (mineral waters and cola-type beverages are particularly sensitive to tainting). AA is created during the melt-phase polymerisation of PET and is trapped in the amorphous material by the cooling and pelletising operation. A proportion of the trapped AA is driven off in the subsequent solid phase polymerisation process, leaving a residual level of around 1.5 ppm in the virgin pellets. Some of this is driven off when the PET is dried prior to moulding but the injection moulding process then generates more AA.

The generation of AA is not associated with any significant IV loss but is a thermal decomposition reaction. This means that the amount generated can be minimised by adopting the mildest possible moulding conditions to give the minimum possible melt temperature. Therefore, it is vital to use:

- low barrel temperatures
- minimum screw speed (rpm), back pressure and injection rate—all of which contribute to shear heating
- minimum possible melt residence time (i.e. the shorter the cycle time, the better), since the length of time for which the PET is maintained at an elevated temperature also affects the AA generation

7.11   Container examination

Containers will shrink after moulding by an amount that will depend on the ambient conditions. Samples should be stored and measured under controlled conditions so that test results may be compared. Containers from each blow cavity and pair of lip cavities should be tested.

7.11.1 Shape and appearance

The visual examination of a container is an essential first part of any testing programme. Containers from each blow cavity and each pair of lip cavities should be examined for colour, clarity, process faults and shape definition (especially of corners or 'feet' and freedom from dents in flat faces). Common faults and their probable cause are listed in Table 7.6.

7.11.2 Dimensions

Container dimensions should be within the tolerances specified on the drawing (typically ±1 mm on diameter and ±1.5 mm on height for a 2 litre bottle).

Table 7.6 Common faults found on examining containers and their probable cause

Fault	Probable cause
Cloudy appearance	Blow mould requires polishing
Crystalline haze	Preform too hot
Pearlescence	Preform too cold, wrong material distribution
Scratches from the blow mould	Damaged blow mould, wrong blow timing, wrong blowing rate or stretch rod speed or inadequate exhaust
Yellow colouration	Over drying

Many factors can affect the container dimensions but, in general, if the preform is fully blown, the container dimensions should be correct. Diameters can be quickly measured with a vernier PI tape, while 'go/no go' gauges simplify height measurements. Container dimensions are affected by:

Height

- position of bottom mould
- material distribution
- blow mould temperature
- preform temperature
- short shot
- blowing pressure (petaloid style bottles)

Diameter

- blow mould opening during blowing
- material distribution
- blow mould temperature
- preform temperature

7.11.3 Capacity

The container capacity should be within the tolerance specified on the drawing. If the shape and dimensions are accurate but the bottle capacity is wrong, there may be a problem with the mould design or manufacture. The factors affecting container capacity are the same as those listed above for dimensions (Section 7.11.2), plus any effects due to the shape of the container (such as rectangular or oval containers with large panels which bulge outward due to the weight of the contents, or the reduction in capacity that occurs when a container is dented).

Container capacity is affected by storage time and environment. Shrinkage of a typical PET bottle is 0.5% on volume after 72 h storage, rising to a maximum

Table 7.7 Typical bottle capacity tolerances as derived by an international soft drink company

Bottle capacity (ml)	Individual tolerance as delivered (ml)	Individual tolerance as manufactured (ml)	Arithmetic average as manufactured (ml)
330–500	+8, −11	+8, −6	+2.5, −0
501–1000	+9.5, −14.5	+9.5, −4.5	+5, −0
1001–1250	+11, −19	+11, −6	+6, −0
1251–1500	+13, −22	+13, −7	+6, −0
1501–2000	+17, −28	+17, −9	+8, −0
2001–3000	+25, −28	+25, −13	+12, −0

of about 1.5% after 150 days at 22°C (higher temperatures will increase shrinkage rates but not the peak values). Table 7.7 presents typical bottle capacity tolerances.

7.11.4 Container wall thickness and material distribution

When a bottle is blown, it is crucial that the material is consistently and accurately placed in the correct position since a variable wall thickness will result in poorer top load, increased carbonation loss and, in extreme cases, distortion of the bottle caused by the carbonation pressure. For these reasons, the specifications for CSD bottles will often define the minimum wall thickness in each part of the container.

Very accurate (although time-consuming) thickness measurements can be performed on sections cut from a bottle wall using a micrometer. For large quantities of containers automatic measuring equipment is commercially available, which can measure the wall thickness without damaging the bottle. Normally, however, a check on the average material distribution is all that is required. One method is to cut the bottle into sections and compare the sectional weights with those of sample containers previously found to have acceptable material distribution. An electrically heated hot-wire cutter can provide the precision cutting required. Uneven material distribution may be due to the factors listed in Table 7.8.

Table 7.8 Factors causing uneven material distribution

Circumferential	Vertical
Container shape (oval designs)	Container shape (waisted designs)
Uneven wall thickness of preform	Uneven gate balance
Wrong temperature distribution in preform	Wrong conditioning temperatures
Misalignment of conditioning parts	Wrong stretch speed
Wrong stretch ratio or grade of PET resin	Wrong primary and secondary blow timing
	Wrong stretch ratio or grade of PET resin

7.11.5   Top load strength

PET containers must be capable of withstanding stacking loads when empty plus any load imposed by the filling or capping machine. Containers with shoulders of low angle of slope or horizontal ribs (especially those with sharp corners) tend to have poor top load resistance and should be avoided for maximum top load performance. The best shapes are those with large radii.

Top load strength is a function of a container's shape and wall thickness. Testing involves subjecting an empty bottle to an increasing load applied vertically down on the neck. The top load is defined as the load being applied at the point when the bottle starts to collapse. For a bottle to be considered acceptable this load should be greater than the maximum load applied during the life of the container.

During production, top load strength should be measured regularly, perhaps twice per shift, taking bottles from each blow cavity. The results should be recorded and analysed since a reduction in top load can provide an early warning of production problems, such as inadequate drying (which causes an increased viscosity drop in processing leading to variability in wall thickness). Typically, the top load strength is measured on an empty, uncapped bottle under standard conditions (see Table 7.9).

7.11.6   Impact resistance (drop) test

Ideally, filled and capped containers should not break when dropped accidentally. Drop testing for CSD bottles is typically carried out as follows:

- two sets of test bottles are filled with carbonated water and stored for 24 h at 24°C and 4°C, respectively
- a bottle, held at an angle of approximately 30° from the vertical and 2.0 m above a concrete surface, is released and allowed to fall freely
- the bottle is inspected—to pass the test there should be no leakage and the bottle should be able to stand upright, unaided

7.11.7   Leakage of liquid (seal integrity)

The seal integrity of CSD bottles is very important for the retention of the CO_2 pressure. Fortunately, because the neck finish of a stretch blow moulded container is injection moulded, leakage problems are very rare. If they do

Table 7.9 Standard conditions for measurement of top load strength

Test temperature	18–25°C
Top load application speed	50 mm/min
Measurement point	First peak value (load at initial deformation point)

arise, they are normally due to a poor cap sealing surface, the wrong cap application torque, a poor bottle sealing surface (short shot or weld line across sealing surface), a poor fit between the bottle and cap or incorrect cap material. Automatic, in-line, leak-testing machines are available to match any line speed.

7.11.8 Vacuum strength

Bottles intended for hot-filling or for holding products that will oxidise the headspace air must be able to withstand or accommodate the ensuing reduction in internal pressure as the product cools or the headspace oxygen is consumed, without collapse of the container sidewall. (Note that while stiffening ribs can be helpful, the improvement in stiffness is often at the cost of a reduction in top load resistance.)

The cooling of hot-filled products or the oxidation of certain products, such as oil, by the headspace oxygen leads to a reduction in the pressure inside the container that causes it to buckle or 'panel'. Bottle shape and wall thickness contribute to vacuum strength, which can be measured by lowering the pressure inside an empty container and recording the pressure at which collapse first occurs. The test temperature and the rate of pressure reduction both have an effect on the vacuum strength.

7.11.9 Acetaldehyde (AA)

Acetaldehyde (AA) is formed by the thermal degradation of PET during melt processing and so will be present in the preform and thus the bottle wall. AA in the bottle wall will diffuse into the contents and may cause a taint problem, particularly with very bland products, such as mineral water. Thus, it is very important that the injection cycle is set up with the optimum processing conditions to minimise the quantity of AA generated.

The most common method of comparing AA levels is by analysing the quantity that accumulates inside a sealed bottle over a set period. Immediately after moulding, the bottle is purged with nitrogen gas, capped and stored at a temperature of 22°C (71.6°F). After 24 h, the amount of AA that has diffused into the bottle is measured using gas chromatography. For ease of comparison between bottles of different capacity, the measurement is then quoted in micrograms per litre (µg/l). How much AA can be permitted in a particular bottle will depend upon the type of beverage and how sensitive it is to tainting. Typically, bottles that are intended for cola products may impart no more than 3 µg/l of AA to the contents, while for mineral water, levels as low as 1 µg/l are required.

It is possible to measure the amount of AA in the preform but, as this requires the preform to be ground up under cryogenic conditions to prevent the generation of additional AA, the test is not very common.

7.11.10   Oxygen permeation

Many food and some beverage products require the exclusion of oxygen to keep them in good condition. The amount of oxygen that permeates into a container depends upon its wall thickness and surface area to volume ratio. Machines are commercially available to measure the oxygen transmission rate through pieces cut from a bottle wall. The permeation into a container can be estimated from this, provided the material distribution is known. Normally, however, these machines are modified to allow them to directly measure the permeation into a complete package. The units of permeation are $cm^3 \cdot mm/m^2 \cdot 24\,h/atmosphere$. PET has reasonably good barrier properties but it is important to recognise when designing a container for a specific product that actual permeation levels (the shelf-life) depend upon the surface area to volume ratio. This means that smaller bottles, which by definition will have a less favourable ratio than a larger bottle, will have a shorter shelf-life.

7.11.11   Moisture vapour transmission rate

The moisture vapour transmission rate (MVTR) is a measure of how fast moisture will pass through a piece cut from a bottle wall and, like oxygen transmission, is a function of the thickness of the sample. Usually, the moisture permeability of the complete pack is more important. This can be measured by filling the container with dry silica gel and storing it in a temperature- and humidity-controlled environment. The increase in weight due to moisture ingress can then be monitored over a period of time; the units are $g \cdot mm/m^2 \cdot 24\,h$.

7.11.12   Product filling temperature

Conventional PET bottles cannot be filled at temperatures above 60°C without distorting, and so special processes and machines have been developed for producing bottles that may be filled at higher temperatures (see Chapter 10 of this volume).

7.11.13   Container weight

A common mistake is to make the bottle heavier than necessary in order to better emulate the rigidity of an original glass container. As a result, the preform dimensions are so large it is impossible to stretch it sufficiently to induce orientation and the concomitant self-levelling effect, and so under production conditions the wall thickness of consecutive bottles varies significantly and may even negate the original purpose of increasing the weight. A better solution would be to ensure the weight is such that the dimensions of the preform permit proper stretching and uniform wall thickness in the bottles.

7.12 Bottles for carbonated beverages

The design of bottles for carbonated beverages must take into account that PET is a viscoelastic material that will creep and distort when pressurised—the rate being a function of time, temperature and the level of applied stress. The recoverable, elastic, part of the deformation happens almost immediately the bottle is filled, but the remainder, which takes place over a period of time, is not reversible. Such deformation must be tightly controlled in order to avoid excessive drops in fill level, apparent loss of carbonation due to the increased container capacity, and unacceptable dimensional changes. Good material distribution, high levels of orientation and appropriate bottle designs all help to minimise such problems.

7.12.1 Burst pressure

This is the pressure required to burst or split a container. The test is carried out to demonstrate that the bottle is capable of withstanding a higher pressure than it could possibly be subjected to in use. The bottle is pressurised until it bursts in some form of test apparatus (preferably a proprietary test unit especially designed to withstand a bottle bursting). Safety procedures should be observed, especially protection of ears and eyes.

7.12.2 Thermal stability

Creep is the expansion of the bottle due to the carbonation pressure. A high level of creep would lead to an unacceptable fall in fill-height, while variable wall thickness could lead to non-uniform creep (i.e. distortion of the bottle shape). Well-oriented and uniformly distributed material in the bottle wall is the prime factor in minimising creep.

To test for creep, bottle height, diameter and fill point are measured before and after carbonation and storage under controlled conditions, and the percentage change determined. After the test, the bottles should still be able to stand upright unaided and should not have developed a poor appearance.

The creep test may be performed in conjunction with the carbonation loss test described below, but usually an elevated temperature accelerated test is performed in which the bottle is stored for 24 h at 38°C (100°F).

7.12.3 Carbonation retention

PET bottles were originally developed for the packaging of CSDs because they have the strength to withstand the carbonation pressure without undue distortion and because PET provides a reasonable barrier to carbon dioxide

(CO_2). Nevertheless, CO_2 slowly permeates through the bottle wall resulting in a loss of carbonation pressure. Note that there is also a reduction in pressure as a result of the volume increase that takes place as the bottle creeps (the apparent carbonation loss referred to in Section 7.12) but this is normally off-set by slightly over-pressurising the bottles on filling.

To test for carbonation loss, bottles are filled with carbonated water and stored at a closely controlled constant temperature. The pressure in the bottles is monitored regularly for a set period of time determined by the bottler.

7.13 Additional tests for hot-fill containers

Additional tests are required for hot-fill containers to determine whether the two major hurdles to hot filling (shrinkage and leakage at the neck) have been overcome. PET containers shrink when heated because the biaxially oriented molecular chains of which they are composed have, in effect, been frozen in an unstable state and the heating provides energy for the chains to begin reverting to their original randomly folded state. However, the very small crystallites present in the oriented structure anchor the molecules in their stretched positions, so the container cannot turn back into a preform (although distortion of the bottle and a significant capacity reduction will occur). Furthermore, the amorphous PET neck will soften at temperatures around the T_g and may allow the product to leak out or contamination to enter the bottle.

The various techniques for improving the high temperature strength of PET bottles involve increasing the number of very small crystallites present in the oriented material by holding it at elevated temperatures and sometimes crystallising the necks. A test has been developed to check the efficiency of this heat-setting process for producing hot-fillable bottles (see Table 7.10).

Table 7.10 Test for efficiency of heat setting process for producing hot-fillable bottles

Time (s)	Action
0	Fill container to the brim with hot water at product fill temperature and leave to stand
60	Secure closure on bottle with correct torque and lay the pack on its side to sterilize inside the neck and closure
120	Stand the container upright
240	Place the pack under a cold water spray until it has cooled to room temperature
-	Determine percentage change in overflow capacity, overall height and diameter and neck dimensions
-	Check for neck leakage

7.14 Additional tests for returnable/refillable PET bottles

Returnable/refillable (RR) PET bottles are containers that are designed to be re-used. Typically, they are washed in a hot caustic solution before rinsing and refilling and are therefore heat set to prevent them from shrinking.

A simple test has been developed to check that the degree of heat setting is sufficient to prevent unacceptable shrinkage during the hot wash. The bottle is submerged in hot water at $59 \pm 1°C$ for 5 h. To pass the test, the capacity reduction should be less than 1%.

RR bottles tend to be thick-wall PET containers with a low degree of orientation in the shoulder and base regions and are therefore susceptible to environmental stress crazing (ESC) in these areas. A test has been developed to check the resistance of a bottle to ESC (see Table 7.11).

Table 7.11 Test to confirm resistance to environmental stress crazing

Time (min)	Action
15	Submerge empty bottles in a 2.5% solution of sodium hydroxide in water at $59 \pm 1°C$
-	Remove from the solution and allow to cool naturally to room temperature $(21 \pm 5°C)$
15	Internally pressurise the bottles using compressed air at 0.68 MPa
-	Depressurise and examine visually for signs of crazing
Repeat 25 times	To pass the test there should be no leakage from the pressurised bottles after the test has been repeated 25 times

Appendix: one-stage and integrated PET machine manufacturers

Aoki Technical Laboratory Inc.
4963-3 Minamijo, Sakaki-machi
Hanishina-gun
Nagano-ken 389-0603
Japan
Tel. +81 268 82 0111
Fax +81 268 82 3699
www.aokitech.co.jp

Automa SpA
Via Chiesaccia, 38
Località Calcara
Ponte Samoggia, 40056 Crespellano
Bologna, Italy
Tel. +39 051 739 597
Fax +39 051 739 578
www.automa.net

Gerosa3 SpA
Via G. Pascoli
21 - 24030 Mapello
Bergamo, Italy
Tel. +39 035 499 4011
Fax +39 035 494 5894

Golfang Mfg & Development Co., Ltd.
51 Tay-Yi Road
Jente Hsiang 717
Tainan
Taiwan
Tel. +886 6 279 5551
Fax +886 6 279 5553
www.golfang.com

Husky Injection Molding Systems Ltd.
500 Queen Street South
Bolton
Ontario
Canada L7E 5S5
Tel. +1 905 951 5000
Fax +1 905 951 5384
www.husky.ca

Jomar Corporation
115E Parkway Drive
Offshore Commercial Park
Pleasantville
NJ 08232
US
Tel. +1 609 646 8000
Fax +1 609 645 9166
www.jomarcorp.com

Magic MP SpA
Via Medici, 40
20052 Monza
Italy
Tel. +39 039 230 1096
Fax +39 039 230 1017
www.magicmp.it

Multiplas Enginery Co., Ltd.
No. 10, Lane 51
Fu Yng Road
Hsinchuang
Taipei
Taiwan
Tel. +886 2 2904 3652
Fax +886 2 2904 3276

Nissei ASB Machine Co., Ltd.
4586-3 Koo, Komoro-shi
Nagano-ken 384-8585
Japan
Tel. +81 267 23 1565
Fax +81 267 23 1564
www.nisseiasb.co.jp

SIG PETtec GmbH & Co. KG
Brüsseler Straße 13
D-53842 Troisdorf
Germany
Tel. +49 2241 489 100
Fax +49 2241 489 255
www.sigplastics.com

SIPA SpA
Plastic Packaging Systems
Via Caduti del Lavoro, 3
31029 Vittorio Veneto
Treviso, Italy
Tel. +39 0438 911 511
Fax +39 0438 912 273
www.sipa.it

Uniloy Milacron, Inc.
10501 Highway M-52
Manchester
MI 48158
US
Tel. +1 734 428 8371
Fax +1 734 428 1165
www.uniloy.com

The illustrations in this chapter have been provided courtesy of the following companies:

Nissei ASB Machine Co., Ltd.
4586-3 Koo, Komoro-Shi
Nagano-Ken 384-8585
Japan
Tel. +81 267 23 1565
Fax +81 267 23 1564
www.nisseiasb.co.jp

R&D Tool and Engineering Group
1018 Browning Street
PO Box 247
Lee's Summit
MO 64063, US
Tel. +1 816 525 0353
Fax +1 816 246 1337
www.rdtool.com

Synventive Molding Solutions Inc.
10 Centennial Drive
Peabody
MA 01960, US
Tel. +1 978 750 8065
Fax +1 978 646 3600
www.synventive.com

8 Two-stage injection stretch blow moulding

Michael Koch

8.1 Introduction

8.1.1 The principles of the two-stage process

The two-stage stretch blow moulding process is targeted to produce a hollow, plastic container in two distinct, consecutive steps: the first stage is to injection mould a preform and the second is to reheat and then stretch blow mould a container. It is sometimes also referred to as the cold process, as opposed to the hot process which is a single-stage process carried out on a machine that integrates the preform and bottle making. In the two-stage process, the preform is cooled, after injection moulding, to ambient temperature. In a separate operation, it is subsequently reheated to its optimum stretch blow moulding temperature to be formed into a container. These two distinct steps are illustrated in Figure 8.1.

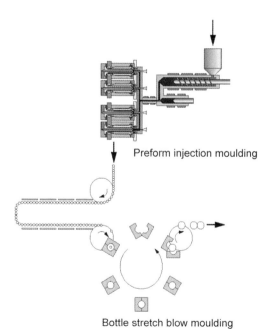

Preform injection moulding

Bottle stretch blow moulding

Figure 8.1 The two steps of two-stage injection stretch blow moulding: preform injection moulding and bottle stretch blowing.

Table 8.1 Advantages of PET injection stretch blow moulding

Consumer accepted packaging

   convenient
   low weight
   unbreakable
   clarity and gloss
   easy shape differentiation
   easy to recycle

Container properties

   full food compatibility
   strength
   good barrier properties
   thermal stability
   chemical resistance

Cost efficient

   low weight
   easy to produce and highly efficient to manufacture
   low investment hurdle

$\longrightarrow$ **Combination of all the above**

Two-stage injection stretch blow moulding technology for PET resin has so far been used mainly for higher output production, serving a broad variety of applications. This technology offers many advantages and has had tremendous market success, demonstrating its superior technological and commercial benefits for a broad variety of liquid food packaging applications. Some of these advantages are summarised in Table 8.1 for the two-stage process, as well as for PET rigid-container packaging of liquid foods. This only gives an indication of the reasons for its superiority, while many other niche or speciality applications are appropriate to the highly dedicated and efficient two-stage technology.

8.1.1.1 *Preform moulding*

During the preform moulding process, the PET material is suitably dried, plasticized to the melt temperature and injected into multicavity injection moulds. Over the years, improvements in cycle time have led to the introduction of post-mould cavity cooling, in separate product-handling systems, as a subsequent but integrated step. These four steps in converting resin are essential to the manufacture of a preform. Their individual nature is not as simple as might be expected because of the nature of PET and the fact that efficient productivity of such commodity components uses moulds with as many as 96 cavities.

Productivity demands low cycle times with equipment tailored to meet that need. Current industrial operations run fast and include the cooling of even

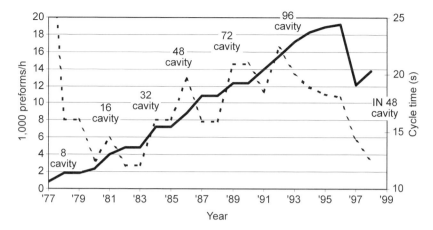

Figure 8.2 Productivity improvements over 25 years. Key: — productivity; - - - cycle time. (Source: Husky Injection Molding Systems.)

thick-walled preforms within the cavity. Post-mould preform cooling has developed to maintain the visual quality required in a highly visible consumer package. Cycle times are squeezed to an extent that drives the output per capital invested ratio (OCR) to a maximum.

The OCR is the key factor in running a profitable preform moulding operation. High quality is the limiting factor and is considered to be essential. The production capability of preform moulding systems has advanced considerably over the past 25 years (see Figure 8.2). Not only has the number of cavities increased but also the productivity of such systems has improved dramatically. Cycle times of 25–30 s were normal in the early days of 2 litre carbonated soft drink (CSD) preform production but today such preforms can be produced with a cycle time of less than 15 s. This is possible because the weight of the preforms has been reduced while the wall thickness limitations on the cooling time are still much the same as in the early days of PET production.

Preform injection moulding equipment is now highly specialized and consists of machine, mould, product handling equipment and auxiliary devices, which together supply the needs of the four process steps mentioned above.

8.1.1.2   Container stretch blow moulding

Prior to the blow moulding stage, the preform is reheated to its thermoelastic state. It is then conditioned and stretch blown into the final container. Typically, the stretching takes place prior to the blowing step, although the timing of both processes can overlap in their sequence. Stretching primarily predetermines the axial material distribution in the container; blowing mainly has an impact on the circumference during the time overlap of stretching and blowing. The combination of stretching (axially) and blowing (radial stretch) leads to the high level of orientation that is key to this process. With axial and radial orientation, the stretching process is known as biaxial orientation. Orientation enhances the material strength, thus producing lightweight containers with high strength and excellent barrier properties.

Again, cycle time is crucial in operating the stretch blow moulding system efficiently. The limits are set by the reheating capability of the machine on the one hand and the required mould-closed time on the other, which are the main determining factors for production speed. Typically, machines should be arranged so that the reheating system delivers as many conditioned preforms as the stretch-blow-cool process in the blow mould will allow through the system. The duration of the stretch-blow-cool process is determined primarily by the cooling time required to freeze the shape of the container to its final dimensions, thus being product dependent. The other variable elements of the mould-closed time are limited, while the handling times remain an essential area of optimization. The cycle time is usually determined by the number of bottles per hour per mould cavity (bphpm) that a machine is capable of handling. This number was 400–500 bphpm in the early days of PET processing and equipment but today can be as high as 1500 bphpm in some production lines. Competitive figures for a broad range of products up to one-way 2 litre containers are 1200–1300 bphpm on large scale production equipment. This improvement in productivity is the result of lighter weight containers and larger scale equipment that allows a reduction in utility and handling times in the machines. Even on smaller machines, these limiting factors have been addressed efficiently.

Heating is achieved by means of high-temperature, infrared oven systems, and the layout is crucial to effect optimal preform heating and the desired temperature distribution. The target temperature range to be achieved is 90–110°C. The key to high performance machines is not the time in the oven but the penetration of heat and the temperature profiling capability. Much of this is achieved by positioning of lamps, timing between heating, and equilibration of the heat.

Typically, the blow mould serves to cool the container to a stable temperature which allows it to retain the designated shape as defined by the blow mould. Application of high pressure during the cooling stage provides the final shaping, contour forging and freezing of the material in contact with the chilled surface of the mould. The natural tendency of the material would be to retract back towards its original preform shape as long as it is in a thermoelastic state above the glass transition temperature (T_g) of around 70°C.

Specialty processes may provide thermal treatment to the container while in the blow mould (i.e. thermal relaxation, heat setting, thermo-fixation, etc.); however, it is essential to cool the container to below the temperature at which stable shape retention occurs. Once cooled, or defined to the final shape, the container is ready to be filled.

8.1.1.3 Preform and container design

Process cycles in both steps, preform making and stretch blow moulding, rely on the design of preform and bottle. There are key features to address in the design, and the successful production of a PET container is the result of careful consideration in this field. While PET offers a broad variety of shape and design potential, there are a few features that are particular to the two-stage process, such as the neck support ring for fast and efficient handling. The key advantage of the two-stage process is the high orientation of the material, and thus the potential for lightweight containers. The interface of design and process in the two-stage approach is primarily through temperature processing, which stresses the importance of parameters that are often difficult to control and are sensitive to variation. The design of preform and bottle can help to limit the sensitivity in processing to temperature alone to such an extent that the superiority of the two-stage process exceeds the capability of other blow or stretch blow processes.

8.1.2 Technological basics of PET as a stretch blow moulding material

PET is a semi-crystalline thermoplastic polyester with alternating aromatic rings and ester bridges (Figure 8.3). It is polymerized either from dimethyl terephthalate (DMT) or terephthalic acid (TPA) plus ethylene glycol (EG). Industrially, the use of the TPA based material is more widespread. Polymerization reactions

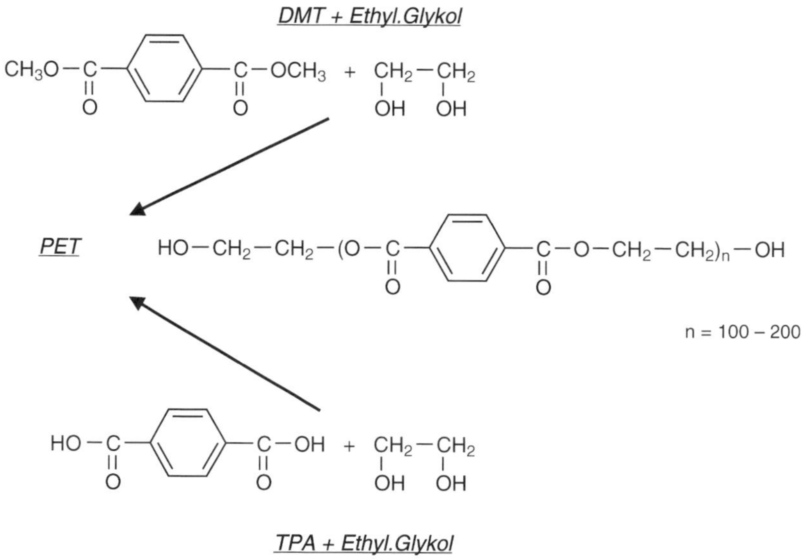

Figure 8.3 Chemical formula of PET.

Table 8.2 Intrinsic viscosity (IV) and applications

			IV (g/cm^3)		
	0.40–0.60	0.70–0.78	0.74–0.80	0.76–0.85	0.85–1.05
Grade	fibre	bottle	bottle	bottle	extrusion
Polymer type		homopolymer	light copolymer	copolymer	
Application		MW, EOS, other	MW, CSD, heat-set	hot-fill	
'Solid stating'	───→				

MW, mineral water; EOS, edible oils and sauces; CSD, carbonated soft drink.

are typically driven and controlled by catalysts such as antimony, germanium and titanium. Recent developments of improved grades have focussed on better heat absorption during the reheating of preforms in stretch blow moulding machines. After the basic poly-condensation process the material is 'solid stated' to increase the chain length, the material is fully crystallized and conditioned as bottle grade.

The length of a molecular chain determines the molecular weight of the material and determines the properties and usage. The intrinsic viscosity (IV) is a practical measure of the molecular weight of the PET macromolecules. It indicates potential applications based on required properties and processing conditions (Table 8.2). The IV does not, however, fully describe the chain structure, which can comprise homopolymers (typically linear chains) or copolymers with branches off the main linear chain. The length and complexity of these branches can vary and have an impact on the IV, so that a homopolymer with IV 0.78 will stretch significantly differently compared to a high copolymer grade with the same IV. Technological properties that differ are, for example, strength, stretching behaviour, melt strength and thermal behaviour.

Table 8.3 summarizes other PET properties with relevance to processing and eventual container properties. Some of the parameters vary subject to resin supplier and grades.

PET is a semi-crystalline polymer with crystallinity levels up to a maximum of 50%. Crystallinity is the arrangement of molecular chains with geometric regularity. The amorphous phase can be considered to be continuous, acting as a matrix in which the crystallites are enclosed. The density of PET at room temperature is a measure of the level of crystallinity (Figure 8.4).

A PET differential scanning calorimetry (DSC) curve is shown in Figure 8.5 and displays four distinct sections that are characterized by typical temperature ranges:

- glassy state ($< 70°C$)
- thermo-elastic/rubbery state ($85–120°C$)
- crystallization range ($120–200°C$)
- melt point ($>255°C$).

Table 8.3 Properties of PET material

	Units	Value
Resin properties		
Intrinsic viscosity	g/cm^3	0.800
Crystalline density	g/cm^3	1.400
Amorphous density	g/cm^3	1.335
Bulk density	g/cm^3	785
Molecular weight		
M_n		26 000
M_w		52 000
Crystallinity	%	50
Melting point	°C	245
Heat of fusion	kJ/kg	59
Thermal conductivity	W/mK	0.25
Specific heat		
ambient temperature	kJ/kgK	0.27
100°C	kJ/kgK	0.36
280°C	kJ/kgK	0.49
Material properties in a container		
Wall thickness	mm	0.30
Density	g/cm^3	1.363
Crystallinity	%	25
Tensile strength at yield		
hoop	MPa	172
axial	MPa	69
Tensile strength at break		
hoop	MPa	193
axial	MPa	117
Tensile modulus of elasticity		
hoop	MPa	4.275
axial	MPa	2.206
Water transmission rate	$g/m^2/24\,h$	2.3
Gas transmission rate		
O_2	$cm^3/m^2/24\,h$	31.0
CO_2	$cm^3/m^2/24\,h$	6.2

Source: Eastman Chemical.

It is essential to understand the effect of these temperature ranges on the performance of PET containers and their related processing conditions. The transitions between the sections are not very distinct, while glass transition temperature and melt temperature show the sharpest transition.

Normally, containers are usable without major deformation below the glass transition temperature at 65–70°C. Above the glass transition temperature PET can be deformed, unless there are forces that retain the desired shape, such as thermally-induced crystallinity.

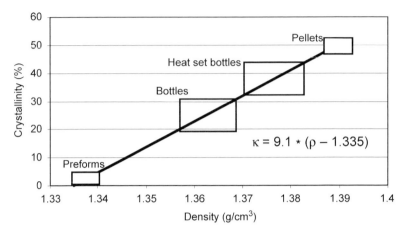

Figure 8.4 Density and crystallinity.

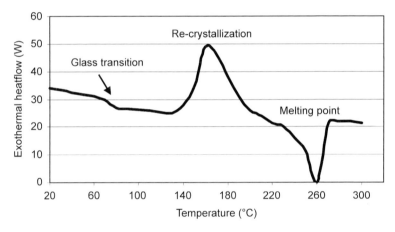

Figure 8.5 DSC heating curve of PET.

The thermo-elastic temperature range is important for the stretch blow mould-ing process, inducing stress in order to deform the material to a point of strain hardening, resulting in uniform material thickness. The stress in the material is caused by the deformation or stretch extension and is subject to temperature as well as deformation speed (Figure 8.6). The higher the deformation speed (strain rate) the higher the resultant stress, limiting the maximum achievable extension or 'stretch ratio'. The higher the deformation temperature the lower the induced stress; this gives a higher maximum stretch ratio.

The optimum stretching process has the right balance between the applied temperature, the strain rate and the allowable stretch ratio beyond the strain

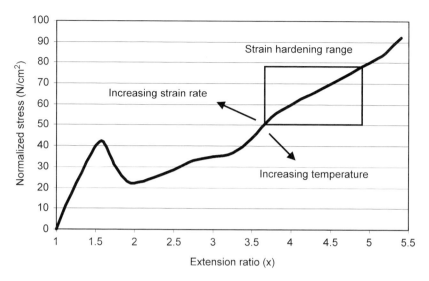

Figure 8.6 Stress strain curve of PET.

hardening point but before the point of rupturing the molecular chain. The stretching process causes strain hardening, this strengthening each portion of the material above its initial material strength. The portions of material extended beyond this point are no longer stretched while the neighbouring segments are still extended until they reach the strain hardening threshold and so on, leading to the uniform material distribution desired. The broader the temperature range, the easier and more consistent the process, determining the process window for the stretch blow moulding process and resulting in bi-axial orientation in the material. The bi-axial orientation is a result of the axial stretching process and a circumferential expansion caused by blown air.

Induced stress in the thermo-elastic state causes a structured alignment of the molecules in the PET material that is frozen during the cooling of the material below the glass transition temperature and gives PET bottles their unique properties. This alignment effect is called orientation. Orientation is partially reversible and promotes a density increase much like thermal crystallization. However, the effect of orientation is different—the molecular structure is changed providing a different set of properties in the container. Mechanically induced orientation is used to achieve crystallinity levels, and therefore levels of density that thermal crystallinity would induce. The processes effecting either mechanically induced orientation and thermal crystallinity are not only different but also the achieved property profile is considerably different.

Crystallization in PET is a result of exposure to high temperatures in the range that is above the thermo-elastic temperature for PET. Often, however, these temperature ranges can overlap depending on the grade of PET resin,

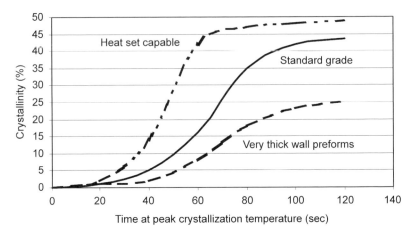

Figure 8.7 Different PET grades and their crystallization.

making the process delicate as the material can start to crystallize during the reheating process up to the thermo-elastic temperature range. Crystallization is an effect of both temperature exposure and time. Material grades can differ considerably (Figure 8.7). Maximum crystallization rate is reached typically around 160°C. It is an exothermal process as can be seen from DSC analysis (Figure 8.5). It is irreversible unless the material is transferred above the melt temperature. As a result, crystallinity can provide thermal stability to a container within given limits of the achievable thermal crystallinity (max. for PET around 50%). Copolymer grades typically crystallize differently from homopolymers, while not all copolymers have the same crystallization behaviour. The crystallization process initiates around particles in the material called nucleating points. Crystallization extends from these points and can grow either many smaller crystals or few big crystals. There are nucleating agents that can be added to the material to both advance and accelerate crystallization; inhibitors are available to cause the opposite. The purity of the PET has an impact in any event; there is also a resin cost consideration.

Above the melt temperature, PET has less melt strength—it easily flows. Degradation of chains starts around 300°C while at temperatures above 350°C the material is completely degraded. Best processing conditions are found to be above, and as close as possible to, the melt temperature, when good and uniform flowability is required. PET generates acetaldehyde (AA) in the melt stage as a result of exposure to heat (temperature and time) and shear. AA is a gas that is generated as part of a molecular chain degradation process. It is then captured within the material and migrates out over time. It may have a flavour impact on the product filled into a container which is made from a preform with a high AA content.

Moisture absorbed by PET causes degradation of the molecular chain. There-fore, storage of resin, preforms and bottles must be carefully controlled. Time and temperature may accelerate this effect, causing considerable change in the processing behaviour of the material and preform to the extent that it is not at all compatible with fresh products but also can be the basis of completely different container properties in a final bottle. Oriented structures and crystallinity are considerably affected. Recommended storage conditions are in a controlled environment around $20°C$ at 60% relative humidity (RH). Preforms that are older than three months can age and may have detrimental effects on the processing behaviour, while bottles may start to change their properties after few weeks, subject to the above parameters.

8.1.3   Production concepts and target markets

The PET stretch blow moulding industry was initially simple in its structure. There were raw material producers, resin suppliers, machine manufacturers and converters, who were selling finished packages, sometimes labelled and together with closures, to filling companies or even directly to food and drinks manufac-turers. The value chain was clearly defined and every participant had scope for a successful business and a defined area of responsibility for innovation.

Over the past 10–12 years, changes have occurred that have driven the competitiveness of PET container-making to the edge of profitability for many of the participants who started out and grew in the traditional structure of the industry. Consolidation has taken place in the converting industry and many medium-sized entrepreneurial converting companies have disappeared. Standard container manufacturing, its design and all the related requirements became a commodity easily accessible to any who dared to venture into the activity and saw value in handling in-house rather than outsourcing. Differentia-tion has become more difficult over time and today only leading edge technology is a guarantee of success in the industry. Eroding margins in an increasingly crowded and competitive, while still fast-growing, commodity business make it difficult for many to remain focused on the key aspects of differentiation and technological innovation. The structure of the industry today is scattered and it has become a global business over all continents, with the lead markets in the US, France, UK, Italy and Japan becoming less of a keydriver.

In addition to these effects, advanced and extended production concepts have contributed to a change in the value chain as many food and drinks companies started to integrate processes to improve profits. At first, companies bought containers from the converting industry, thus concentrating on the filling and packaging. Soon afterwards, they understood that a blow moulding machine could be run in-line with a filling machine (Figure 8.8) relatively easily with the support of the equipment suppliers who continued to provide ever more reliable machines. This pushed forward their business volumes, since utilization in food

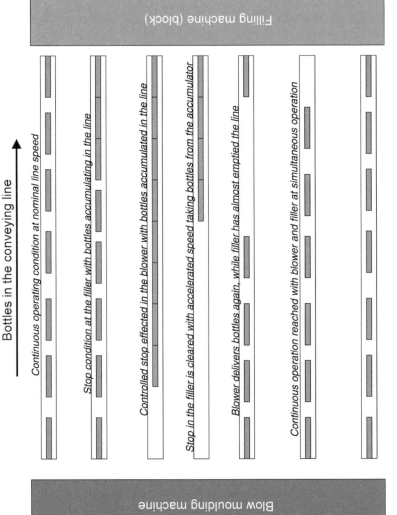

Figure 8.8 Accumulation characteristics of blow moulder/filler combination.

and drinks companies with integrated filling lines was naturally reduced by use of dedicated equipment but also brought economies of scale, not only to the size of machines but also to the value that the equipment supplier could create. Interestingly, the concept of in-line bottle manufacturing and filling is still not fully implemented in various parts of the world, where:

- either transportation costs are so low that dedicated lines with only 60–70% utilization cannot compete with bottle manufacturing sites running for more than 90% of the time throughout the year (typically highly populated areas, in particular North America)
- or technological and quality requirements make bottle production enough of a dedicated focus that integrated line operations will suffer efficiency reductions and lack of technological expertise due to the primary manufacturing needs.

In some instances, this trend has continued even further into manufacturing of preforms and bottles in mega-plants for the CSD or mineral water industry. Meanwhile economy of scale (more than 1 billion bottles per year) has been the main driver in the profitability of these operations. In other instances, food and drinks companies, particularly smaller businesses and local brands, still need to buy preforms, but this market has become primarily a commodity environment.

The usual concept of producing bottles in-line with filling machines is based on the assumption that the filling machine has an almost instant start/stop time behaviour. In contrast, the blow moulder needs more time to clear the preforms and bottles in the system or to be restarted (filling the machine with preforms) with its continuous heating process. Moreover, the filler often runs, or should run, at a higher speed than the blow moulding machine. The blow moulding machine, however, runs at about the same speed for different container sizes, while the filling machine is dependent on the liquid flow for its output capacity. As a result, the link between the two production modules is carried out with accumulation systems. These can take the form of a long conveyor or other means of controlled accumulation (e.g. parallel line accumulator or spiral configuration), or even uncontrolled accumulation in a silo combined with unscrambling and refeeding systems. Since about 90% of all filling machine stops can be cleared in less than 3 min, an accumulation system should cover the supply of at least this 3 min production time. The length (L) of a conveying line is, accordingly, to be calculated by:

$$L_{conveyor} = Output\ rate\ (\text{bph}) * 3/60 * D_{bottle} \qquad (1)$$

where, bph is bottles per hour and D is bottle diameter. In case of a filling machine stop, the blow moulder continues to produce into the conveying line or the accumulator, while after clearance of the filler stop, the filler has to produce at a slightly higher speed than the blow mould to empty the accumulator—typically, a 10% higher speed is recommended.

The next wave of innovation in these conceptual issues is the integration of blowing and filling into one machine. A close link between the process of bottle making and filling a container may have cost advantages, while it certainly adds to the complexity of handling the operation. Relative to cycle time and handling characteristics, this close linkage between two processes appears to make sense, unlike the link between preform moulding and bottle making, where cycle times are substantially different and capital utilization becomes a competitive hurdle. As a result, the single-stage, integrated process is becoming increasingly a niche application for specialty products and low volume production.

The direct machine linkage for the production of CSD bottles is more difficult to accomplish, since the bottle, in particular the thicker base section, may not be fully cooled. Upon application of the fill pressure and the pressure in the product the base may pop out. Appropriate support systems have been designed. It is considered that such integrated blow-and-fill machines make particular sense for still liquids, with the need to maintain a clean environment with no intermediate storage and no need to clean the bottle prior to filling. For this, appropriate over-pressurized linkage chambers have been developed.

Much more integration can be seen from the raw material to the solid stated resin and even to integrated preform manufacturing where economies of scale determine the logistic cost through energy and capital reductions.

PET packages have conquered a broad variety of rigid container packaging applications. The use of PET resin is shown in Figure 8.9, by application. Of the total various packaging solutions, PET takes a share of 22.5% (Figure 8.10) of all liquid packaging with over 110 billion containers produced in the year 2000. The container requirements for each of the applications are summarized in Table 8.4. It is interesting to see that PET is suitable for a broad variety

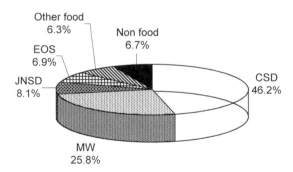

Figure 8.9 PET usage by application : 1999 – 133 billion containers. Abbreviations: CSD, carbonated soft drink; MW, mineral water; JNSD, juices, nectars and still drinks; EOS, edible oils and sauces.

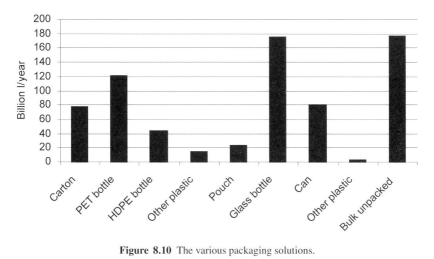

Figure 8.10 The various packaging solutions.

Table 8.4 Applications and requirements

Application	Requirement	PET property
CSD	Pressure resistence	Strength
	Carbonation retention	Barrier property
MW	Flavour neutrality	FDA approved
	Carbonation retention	Barrier property
	Low cost	Low weight
	Color	Transparent-blue
JNSD	Hot-filling	Thermal stability
	Oxygen sensitive	Barrier property
	Flavour neutrality	FDA approved
EOS	Oxygen sensitive	Barrier property
	Low cost	Low weight
Beer	Pressure resistence	Strength
	Flavor sensitivity	Barrier solutions
	Oxygen sensitivity	Barrier solutions
	Carbonation retention	Barrier property
Other	Various	Covering a broad range

of applications through the two stage stretch blow moulding process, which explains its success as a rigid container packaging material.

Consumption growth is predicted to be in the range of 10–15% in the coming five years. One of the restricting factors may be the availability of material, since resin suppliers have been more reluctant to invest in new capacities in times of lower demand. Figure 8.11 shows a prediction of the growth segments over a five year period and gives an indication that PET not only advances its position in the established segments, i.e. carbonated soft drinks (CSD), mineral water

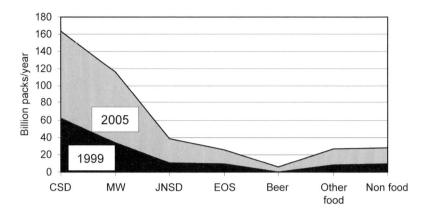

Figure 8.11 Prediction of growth by category (source: PCI reports, Tetra Pak *et al.*).

Table 8.5 PET market drivers

Single-serve packaging for all segments
Emerging market for CSD, MW and EOS
Ongoing package conversion for MW
Applications in hot-fill and aseptic packaging of JNSD
Beer and fresh milk are future candidates
Product differentiation in container shape and size
Multimaterial barrier applications
Changing industrial concepts
Population growth and changing consumer habits
Environmental concerns and easy PET recycling
High cost competitiveness

(MW) and edible oils and sauces (EOS), but also penetrates new applications such as juices, nectars and still drinks (JNSD), liquid dairy products (LDP) and beer packaging. PET will likely capture most of the growth pattern in the consumption of these categories but will also grow through replacement of existing competitive package materials like glass, cans and cartons. Table 8.5 summarizes the overall market drivers for PET application growth.

8.2 Preform injection moulding

A preform injection moulding system comprises five essential process steps that are the core elements of preform making:

- resin drying
- feeding and plasticizing
- injection and melt distribution
- moulding and cooling
- preform removal and post cooling

This chapter deals in detail with the plasticizing and moulding processes and related equipment. However, other parts of the process have similar importance to the overall preform making and must not be neglected.

Resin drying is essential to PET processing, since a humidity level above 0.003% of RH in the material results in inconsistent processing and components that will not produce the required preform quality. Typical drying requirements are around 5 h at 160°C. Proper material flow must be maintained in the dryer and the hopper systems must be set up appropriately.

Post cooling of preforms is carried out in preform removal systems (robots with takeout plates mirroring the cavity pattern of the mould) or any subsequent handling system. This helps to shorten the overall cycle time by minimizing the cavity cooling time and reduces the preform heat to a level at which deformation will no longer occur. Cooling is effected either by water-cooled tubes that are in contact with the outside shape of the preform or by forced-air convection. A combination of both produces the best results. Patented systems apply conical cooling tubes that suck the tapered preform further into the tube, once it has cooled and shrunk away from the surface. The longer post-cooling can be applied to the preform the better the effect. A robot arm (as shown in Figure 8.12) allows three positions, thus leaving the preform in the cooling tube for at least 2.5 cycles of the moulding system.

8.2.1 Injection machine concepts

8.2.1.1 Plasticizing
While clamping mechanisms and mould concepts vary for different machine suppliers, plasticizing of PET is always carried out in reciprocating screw

Figure 8.12 Robot arm. (Source: Husky Injection Molding systems.)

extruders. As the resin is introduced to the screw from the dryer at 160°C, the feeding of the material must be well controlled with appropriate sectional barrel temperatures. The material must be melted gently and low compression screws used since the 'bulk to melt density ratio' is comparatively high (at 70–80%) for PET. A considerable enthalpy increase must be provided by the plasticizing process, through screw torque and heater bands, to transform the solid into the melt stage. The objective is to control the melt temperature mainly through the heater band energy, thus keeping the mechanically induced portion of the energy increase slightly below 100% of the total required enthalpy mark, which can be described according to the first law of thermodynamics by:

$$P + Q^\circ = m^\circ * \Delta h \tag{2}$$

where P is electrical power, Q is heat flow rate, m is mass throughput and h is enthalpy.

The temperature-sensitive nature of PET demands that the melting process is well controlled; thus, the mechanically induced energy increase resulting in melting must never be above 100% of the total required enthalpy increase. This provides the opportunity to control the critical melt temperature by adjusting the barrel temperature. Barrel temperature should be adjusted with a negative profile from 275–290°C between the feeding section and compression zone down to 260–275°C towards the screw tip. Combined with the appropriate installed power at the operation point, this provides control of the melt temperature by the barrel temperature setting.

The melt temperature must be as uniform as possible and at the lowest achievable level in the flow channel. This assures that cycles are short and that no local overheating can occur. AA levels are maintained and material homogeneity is provided. Many of the requirements for PET processing in injection screw machines conflict with each other and often compromises must be made to control the process optimally.

Pressure gradients in the melt conveying zones are typically low due to the low viscosity of the material, which makes all the above tasks more challenging. Screw design is crucial to obtain efficient and high quality melt conditions targeted at consistent throughput, uniform melt temperature and controlled shear mixing homogeneity. Figure 8.13 shows some typical output figures achieved in PET machines as a function of the screw diameter. Use of recommended screw speeds, as indicated, leads to the desired output rates without the risk of shear degrading of the material.

Additional specific features are required to operate competitive cycle times for PET processing. One principal feature is the shooting pot arrangement (see Figure 8.14) of the two-stage injection unit, which gently plasticizes the PET in the screw at low pressures, with smaller diameter screws, operating at lower rpm (revolutions per minute) and allowing better process control. The shooting

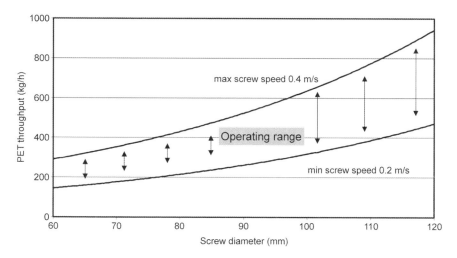

Figure 8.13 Output rates of PET plasticizing systems.

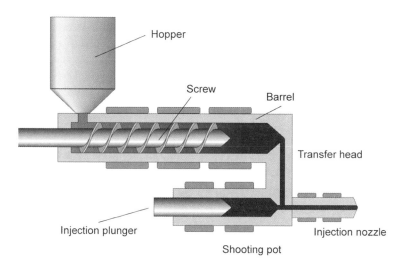

Figure 8.14 Shooting pot arrangement.

pot is used to fill the cavities by fast application of the injection pressure. The plasticizing extruder can run 90% of the total cycle time and must not stop during the injection and hold pressure phase. Packing of the mould cavity, under high pressure, is another feature provided by the shooting pot with better pressure control.

Another way of taking advantage of process differentiation between plasticizing, injection and application of hold pressure is the so-called packing pot arrangement. The packing pot is filled during the injection phase. After this period, the shut-off valve closes and the screw starts plasticizing again, while the packing pot applies the hold pressure during the cooling of the preforms in the cavities. Both of these systems are commonly used. A conventional injection-moulding set-up with just a reciprocating screw cannot provide competitive cycles.

8.2.1.2 Clamping

Injection clamps are either toggle based or directly cylinder actuated. Both versions can be operated hydraulically or hydromechanically, and toggle systems can also be actuated electrically. Tiebarless systems have proved to be unsuitable for PET preform moulding simply because of platen parallelism. Even toggle machines have been shown to overload the mould, resulting in excessive wear and alignment problems once injection pressures higher than those actually needed are applied. However, this should no longer be an issue, since the actual required clamp force in PET injection machines is only about 50–70% of that provided by machines in use today. This provides high strength and stability in the platen design, which is important for the large moulding faces, and also goes hand in hand with the tiebar spacing required for the multicavity moulds. Rigidity is essential for long mould life and moulding accuracy. The surface faces of the multicavity moulds must remain parallel due to the considerable number of precision moving parts. In light of these requirements, hydromechanically operated clamps have proved successful for long mould life and good preform quality.

Large, multicavity preform moulds also require a complex ejector system that must be tailored to the mould and well aligned on the platen. This needs to be considered when choosing the clamp and platen type of the machine.

The clamp force requirement can be derived from the specific injection pressure that PET preforms are operating with, which ranges 2.5–4.0 N/mm^2. The projected area of a preform is given by the maximum diameter of the preform, thus the total required clamp force (F_{clamp}) can be calculated as follows:

$$F_{clamp} = P * \pi * d_{max^2}/4 \qquad (3)$$

where, P is the injection pressure and d is a preform diameter. Assuming typical support ring and preform diameters, the resulting range of required clamp forces can be displayed as a function of the number of cavities.

In 1997, a revolutionary new moulding system was introduced to achieve faster moulding cycles, with improved preform quality and reduced operational cost: the Index moulding system, as shown in Figure 8.15 While the concept of plasticizing and related steps is maintained as in conventional systems, the clamping and cooling side of the machine is fundamentally different. A rotating block with either two or four faces allows one set of cores and neck rings at a

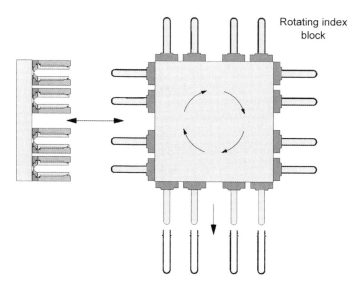

Figure 8.15 Index moulding system.

time to engage into the cavity portion of the mould, which is mounted together with the hot runner plate onto the stationary platen. Injection is effected in the conventional way, until the rotating block, acting as the moving platen of the injection machine, allows the injected preforms to be removed, still in a relatively warm condition. The final cooling takes place on the core set of the face of the indexing block. Following the designated number of cooling positions (either one for a 2-face set-up or three for a 4-face arrangement), the preforms are simply dropped off onto a conveyor belt or taken off by an additional handling system for further processing. This avoids the use of robots, which are known to cause two thirds of all stops in conventional systems. For relatively thick-walled preforms, improved cycle times in excess of 30% are achievable (see Figure 8.16).

Table 8.6 presents the advantages of the Index moulding system, which lead to excellent results in many aspects relative to preform quality (i.e. less gate crystallinity as a result of better cooling, lower AA levels as a result of more frequent flushing of the hot runner, and shorter residence time in the melt flow channels with better melt temperature control). An overall increase of output per capital cost has been achieved with this new moulding technology. After a few years of use, the Index systems have shown reliable machine operation in the field, with small 90 ton clamps and 8–16 cavity moulds, 250 ton clamps with 32 and 48 cavity moulds up to a large size 400 ton clamp with 48 or 72 cavity moulds. Better utilization of energy and floorspace mark another operational advantage of these systems. All the advantages are driven by the cycle reduction potential.

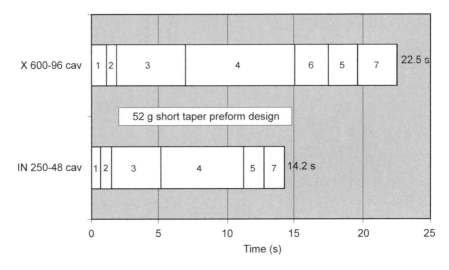

Figure 8.16 Index system cycle time advantages. 1, mould close; 2, clamp up; 3, inject; 4, hold; 5, mould open; 6, cool; 7, robot i/o. (Source: Husky Injection Molding Systems.)

Table 8.6 Advantages of the Index moulding system

Shorter cycle time—higher output per cavity
Higher output per capital
Higher output per floorspace
Simpler operation with less downtime—no robot
Improved preform quality
    –lower acetaldehyde, less gate crystallinity
Reduced specific energy consumption

The downside of the Index moulding system is the higher capital cost for the multiple core plates of the mould. Being cost intensive on these mould parts, high volume requirements of preform of one particular design are best suited to deliver the benefits of this machine concept, through which the highest possible output per capital can be achieved.

8.2.2 Mould design

Injection moulds are designed from four basic components: the mould shoe, the hot runner system, a cavity plate and the core plate with core pins and neck splits. There are various approaches to the design of such moulds but these are not discussed further here. Essential components of the mould that produce the shape of the perform are: the cavity, end-cap, core and neck splits. These all require intensive cooling and elaborate technologies are applied to facilitate this as effectively as possible. They are typically exchangeable components, and it is recommended that key length and diameter parameters are kept within limits. A typical mould lifetime is about 3 years but refurbishment is possible. Smooth and gentle operation of the machinery and process can considerably extend the lifetime of the mould. The most critical components are the neck split and the taper section of the lock rings, which ensure that the multiple cavities are clamped and properly sealed. Most of the clamping force is applied through the tapers, thus they are exposed to the greatest wear in preform moulds.

The neck splits have to be operated to release the moulded preforms from the core. Cam or cylinder actuated systems operate the opening and closing of the neck splits. Ejector systems are an integrated part of the mould and need to be controlled to gently strip the still warm preforms off the slightly tapered cores into the robot arm take-off tubes without distorting them.

Preform moulds are made with multiple cavities, ranging from single cavities for product development purposes up to 96. A pattern of 4, 8, 12, 16, 24, 32, 48, 72 and 96 has developed based on the balance for the hot runner distribution system. The choice of pitch pattern between the individual cavities is product dependent, while certain standards for the two principal neck sizes (28 and 38 mm threaded preforms) are commonly used. With the introduction of lightweight plastic closures and a lightweight finish section of preform and bottle, the neck support rings have been reduced to diameters of 30–31 mm, whereas they had previously been in the 33–37 mm range. As a result, preform moulds could also be designed in smaller pitch, thus reducing the required projected area, hence the tiebar spacing. The new generation moulds dedicated to these finish designs could be used either in smaller machines or the number of cavities per mould could be increased. This also meets the potential in terms of clamp force, as described above.

8.2.2.1 Hot runner systems

The hot runner plate distributes the melt to the multiple cavities of the mould. Flow channels need to be balanced in terms of the cross-section of the flow and also in the pressure drop from the injection point to preform cavity. The cross-section is even less important as long as the pressure drop and cavity fill rates are maintained equally over the number of cavities. In principle, two hot runner design concepts are viable:

- a naturally balanced system (equal flow length from injection point to gate drop, with flow channel diameter balance): $\Delta p \sim L$
- a rheologically balanced system (shortest flow path for every cavity, with the diameter of the individual channels adjusted for equal pressure drop): $\Delta p \sim D^2$

Although the rheologically balanced system has some advantages, it is more sensitive to changes in material grade and temperature and, as a result, less stable in its operating window. The naturally balanced system is used much more widely and can be viewed as the current industry standard. Ideally, in both systems, it is crucial to maintain equal temperature distribution so that the melt flow in each part of the distribution system is equal and well balanced. This results in a consistent melt temperature going to each cavity, producing uniform AA levels.

In fact, this is not achievable in most systems, thus temperature control has been a key issue in hot runner and melt-distribution systems for preform moulds. The appropriate balance is vital between good thermal control and stable control circuits that do not make the adjustment of the temperature profile an art but are rather better understood; the joining of 4–8 channels in one control circuit has proved to be a practical arrangement. Nozzle tips, however, are often individually temperature controlled. Ideally, flow channels must be smooth to achieve plug flow and the shortest possible residence time in the flow channel. At times, it may advisable to introduce a static mixing element into the hot runner channel to support more uniform temperature distribution by mixing the flow. The increase in shear energy and therewith temperature is offset by the effect of a more uniform temperature profile in the melt running through the hot runner. The system needs to be carefully adjusted and placed appropriately in the mould sprue or the machine nozzle.

The injection process must not lead to an increase in melt temperature, which might result in the introduction of too much shear and degradation of the material. Wherever possible flow channels have to be designed to avoid zones with major changes in the flow cross-section (so as not to accelerate or decelerate the flow). This will give rise to consistent results, with highest quality parameters reached in AA values and retention of IV levels.

8.2.2.2 Gates and cavities

Injection gates are an area of knowledge connected to flow as well as to mechanical stability and thermal control. Behind the gate, the aim is to cool the melt. In front of the gate, correct high temperature control is required to give good flow and minimise the residual time that causes clogging, increased shear, degradation or high AA in the melt. As a result, gate systems are closely linked to the cavity design. There are two concepts of gate arrangement, one valve controlled and the other valveless. The valveless systems have advantages (less flow restriction, reduced wear) but their operational stability is questionable. Smaller gate diameters and difficulties in thermal separation of the hot runner and cavity make them difficult to operate for all preforms; smaller preforms with low wall thickness are preferred for this gate system. The conventional gate shut-off valve is more stable, easier to operate and allows larger gate diameters (less shear), but requires more maintenance.

A typical cavity set-up is shown in Figure 8.17; the key elements are the core, the body cavity, end-cap/gate insert and the neck split, the latter being the most expensive part to manufacture. The interface between cavity and neck splits is determined by the principal shape of the perform. If the body diameter of the preform is larger than the preform diameter (measured just below the

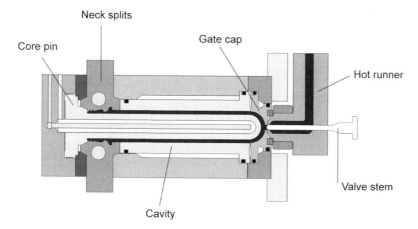

Figure 8.17 Typical cavity set-up.

neck support ring), then the join between neck splits and body cavity must lie at the end of the tapered preform section joining the full diameter body. If the largest diameter of the preform body is that below the neck support ring, then the interface between cavity and neck split can be as high as the edge of the support ring, as long as the preform can be ejected from the cavity without the limitation of an undercut section. The shape and dimensions of a preform mould must deliver the best possible cooling for the preform; thus, great attention is given to the cavity, core and end-cap designs to increase water flow and heat transfer, and produce the best possible cooling effects. A critical portion is the end-cap close to the hot runner gate, which is hot, while the gate insert must provide optimal cooling after the material has been transferred to the cavity. A steep temperature gradient must be accomplished in a non-stationary temperature environment, eventually determining the cycle time limits of a moulding system.

8.2.3   Productivity parameters

8.2.3.1   Cycle time
The cycle time of the preform moulding system has the following essential components:

- dry cycle of the machine (closing, clamping and opening of the mould)
- injection
- hold pressure application
- cooling
- preform take out

All other functions are operated in parallel to these process steps. The machine dry cycle is supplier dependent. Tuning some of the hydraulic components may help to reduce time; however, this is already done by experienced manufacturers. Smaller machines run faster (less mass to move) and have less oil to move, provided they are hydraulically actuated.

The injection process must be gentle. Fast injection rates of 12 g/s, subject to the preform design, with injection pressures at the screw tip of up to 500 bar, are sufficient to fill L/D (flow length over minimum wall thickness of the preform) ratios of up to 50. As a result, shot rates are a function of the number of cavities.

The hold time is a function of the part weight and can be plotted as shown in Figure 8.18. As this time is also used for cooling, however, with the gate still open and mass pushed into the cavity, it is sometimes difficult to differentiate between hold time and cooling time. For lightweight preforms in particular, the process can be set up to minimize the cooling time and apply hold pressure for the whole packing and cooling time.

Preform removal times are again machine dependent. However, it is essential to reduce the interface time between machine and robot control systems by integrating the robot control into the machine control.

The cycle time is critical to the performance of a preform moulding system. Comparisons between different machine systems must be made under the

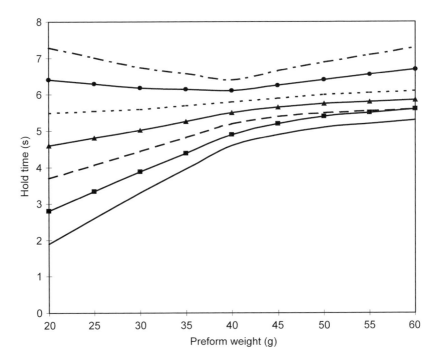

Figure 8.18 Hold time and part weight. Preform wall thickness: ——, 3 mm; —■—, 3.25 mm; — —, 3.5 mm; —▲—, 3.75 mm; - - -, 4 mm; —●—, 4.25 mm; — ., 4.5 mm.

Table 8.7 Output per capital comparison.

Preform injection moulding systems: machine, mould, take-out configuration					
	System A	System B	System C	System D	System E
Number of cavities	48	48	48	32	96
Cycle time (s)	16.1	16.3	21.0	15.6	17.5
Yield (%)	98.0	97.0	96.0	98.5	97.0
Total preforms produced (pph)	10,518	10,283	7,899	7,274	19,156
Capital cost (1'000 €)	1,200	1,250	1,100	850	2,000
Output per captial (pph/1'000 €)	8.77	8.23	7.18	8.56	9.58

System A: 250 ton press, two-stage injection process, accelerated take-out device.
System B: 350 ton press, two-stage injection process, accelerated take-out device.
System C: 300 ton press, reciprocating screw injection, standard take-out device.
System D: 175 ton press, two-stage injection process, accelerated take-out device.
System E: 500 ton press, two-stage injection process, accelerated take-out device.

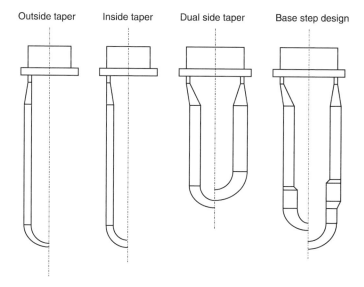

Outside taper   Inside taper   Dual side taper   Base step design

Figure 8.19 Basic preform shapes.

assumption that equal preform quality is achieved and an equal yield of preforms produced over longer running times. Table 8.7 shows a typical comparison of systems that are measured in terms of output per capital.

8.2.3.2   *Preform design and key related parameters*
Preform design itself is dealt with in the context of bottle blowing but it should be understood that there are designs of preform shape that impact on productivity (i.e. cycle time). The shapes shown in Figure 8.19 indicate the possibilities

available where they tie into either neck dimensions or diameters and shapes given through the container design. A critical element is the portion between neck and body section; it is in effect the bottleneck for cavity filling. It determines the amount of material that can be applied to the neck portion during the hold pressure phase, and is thus crucial to the dimensional stability of the neck. The mould temperatures in that section may not be as cold as in the thicker walled body section, thus leaving hold pressure to be applied to the neck portion.

Cycle times are also influenced by the ratio of preform gate to body wall thickness. A thinner wall in the end-cap of the preform is beneficial for cooling but, at the same time, may restrict the application of hold pressure to compensate for shrinkage of PET volume in the cavity. A similar situation applies to long preforms with thin walls ($l/s \gg 50$), which at the same time are difficult to produce due to flow in the cavity, as discussed above.

8.2.3.3 Preform quality and key related parameters

Preform quality can be determined by dimensional stability. In addition, there are a number of other measures which are summarized in Table 8.8, together with some key dimensional criteria.

To be able to blow a consistently good bottle, the IV of the preform material should not differ by more than 3% from that of the virgin resin. A greater IV drop indicates the likelihood of a broad molecular weight distribution that negatively influences the adjustment possibilities of the blowing parameters.

Crystallinity and colour are further factors that affect preform quality. Crystallinity levels must clearly be below 2–3%, and ideally in the main portions of the preform below 1%; all colour defects which are also crystallinity related and can impact quality must be avoided. Brown colouring of the material is a result of degradation, indicating severe damage to the material. Crystallinity can be seen as white effects in the material, either stripes or clouds. If solid white portions are visible, the preforms are clearly unusable.

AA levels in preforms need to be kept below the limits that food and drink companies specify—typically, for 48-cavity preform moulds this is around

Table 8.8 Preform quality measures

Weight
Dimensions
Wall thickness consistency
Density
Intrinsic viscosity (IV)
Acetaldehyde (AA)
Transparency
Colour tint
Sink marks
Gate crystallinity
Stripes or clouds of crystallinity

4–5 µg per litre for CSD preforms and below 5–6 parts per million (ppm) for mineral water applications. Different size moulds and injection systems can give other results that may or may not be acceptable to the preform user. These values are measured differently and are not necessarily correlated; CSD qualities are measured with headspace technology, while mineral water companies specify the more complex ground preform technique. When these two measurement techniques were established, the results of different laboratories differed and certain standards in the industry have been developed accordingly. The IV drop from resin to preform is an indication of the potential level of AA generated during the moulding process but again the correlation is rather vague and only indicative.

AA levels are influenced by material residence time in the melt channel and the temperature profile over the flow channel cross-section in the mould, as well as the temperature level of the melt. Processing and machine design must focus on controlling these parameters and make every effort to further reduce the potential for AA increases and poor processing.

8.3   Stretch blow moulding

About 80% of all bottle-grade PET material is processed using two-stage equipment. This demonstrates the importance and breadth of application of this process and not only for high volume production. Bottle producers and fillers have the flexibility to optimize the distinct advantages of the two-stage process to their production needs. Bottle weights are low, thus optimizing container cost and providing very high container performance properties. High volumes are easily produced with low floorspace requirements and little energy input, and the operator skills required do not demand plastic-processing trained engineers. Overall production costs are lower for medium to large volumes and the capital utilization is superior to single-stage machines once critical volumes are exceeded.

The application flexibility in smaller two-stage machines is higher than in single-stage machines, since the preform market has developed in such a way that injection mould investment can often be avoided. Alternative container shapes and sizes are easily implemented with new blow moulds and product-dependent parts required only for the blow moulding machine.

8.3.1   Principles of the two stage stretch blow moulding process

The two-stage stretch blow moulding process involves reheating the preform and then stretch blow moulding the reheated preform. Both processes overlap and the result of reheating the preform is often underestimated in its impact on the stretch blow portion of the process.

8.3.1.1 Preform reheating

The reheating of preforms determines process consistency and final container properties. The target of reheating is to apply an energy level, combined with a temperature profile, over the length of the preform and through its wall thickness. This allows it to stretch in such a way that the desired material distribution and properties in the final container can be achieved during the subsequent stretch blow moulding stage. Thus, the reheating conditions link preform and container design, in preparation for stretch blow deformation as the final stage. Whatever is achieved in the reheating stage will be just fine-tuned by the process parameters of the stretch blow deformation.

The reheating process has three key objectives:

- increasing the energy level of the preform to the stretch blow moulding temperature range (90–115°C)
- temperature profiling along the preform axis to support the control of the wall thickness distribution (slight gradients)
- temperature profiling through the sidewall section of the preform, in order to achieve a higher temperature on the inside of the preform wall than the outside (gradient subject to predesigned stretch ratio)

These three functions are usually achieved by the infrared radiation oven of stretch blow moulding machines, in which an array of lamps is adjusted in such a way that the longitudinal axis of the preform is fully covered in ideal conditions for the shape of the preform (see Figure 8.19). The preforms are moved along this array of lamps, while being rotated, to achieve a uniform temperature profile on the circumference. In addition to the radiation, convection is used—typically applied through forced airflow. This is also useful so as to not burn the outside surface of the preform but allow the radiation to penetrate the sidewall so that, eventually, a uniform energy level can be achieved throughout the preform wall.

After heating, the preform is allowed to 'equilibrate', in ambient conditions with natural convection, to produce a uniform energy level over the length of the preform wall in the 'stretching temperature' range. This will eventually cause a temperature gradient to develop through the sidewall of the preform. This gradient will give a higher temperature on the inside of the preform wall than on the outside because the heat transfer coefficient is higher on the outside than on the inside of the closed tube of the preform body. This assumes that the predesigned stretch ratios, developed from the inside of the preform wall to the inside of the bottle wall, are larger than those from the outside of the preform wall to the outside of the bottle (see Figure 8.20). These need to be achieved during the stretch blow deformation without negative effects from the differences in additional internal stresses developed between the inside and the outside layer of the material. The deformation force required is higher on the inside of the preform wall than on the outside. The larger the overall stretch

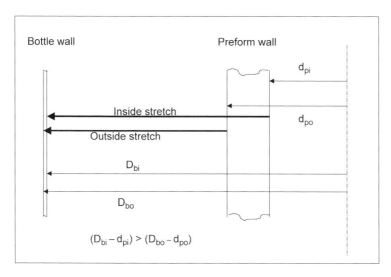

Figure 8.20 Stretch ratios. Variables: D, d, distance, as shown. Subscripts: b, bottle; p, preform; i, inside; o, outside.

ratio of the preform, the larger the difference needed between the inside and outside temperatures.

Reheating of preforms is accomplished by radiation in combination with convection. Reflection of radiation within a machine is an additional factor but will have the impact of 'fuzzy' radiation on the preform reheating. Surface reflection, transmittance and heat absorption affect the internal heat conduction occurring during the process within the machine. Figure 8.21 shows the impact of various factors on preform reheating. It is crucial to operate in the appropriate range of radiation wavelength to ensure that the heat can penetrate into the sidewall of the preform. While resin suppliers have made efforts to enhance the absorption of radiation into the sidewall of a preform through modified grades, there are limits to that effect. It is imperative to operate in a temperature range that allows high heat flows in the right wave length sector. Once that is accomplished, the heat level in a preform can be raised to the right temperature range in less than 10 s.

Axial temperature profiling is accomplished by the heat flow emitted from the combined length of individual heater lamps, or the distance of the individual lamps from the surface of the preform. The latter is typically accomplished by mechanical adjustment of the lamps; although this adds another facet to the potential adjustment variables, it is not always desirable, particularly when there are many product changes and flexibility is required in a machine. A predetermined distance is therefore useful for certain preform types. The lower level lamps (see Figure 8.22) are usually as close as possible to the preform surface, not only because of the shape of larger preforms that have a taper

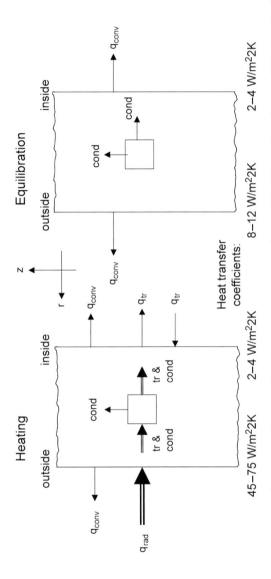

Figure 8.21 Impact of various factors on preform reheating: thermal balance for the preform wall. Variables: q, heat flow; r, radial axis; z, longitudinal axis. Subscripts: rad, radiation; conv, convection; cond, conduction; tr, transmission.

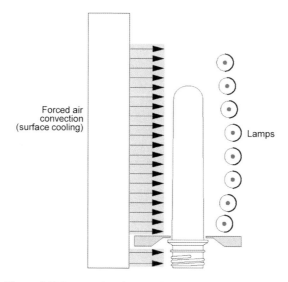

Figure 8.22 Lamp and preform configuration in a reheating oven.

in that section but also because there is little (second lamp) or no (first lamp) neighbour-lamp impact to get the heat flow to high levels. This makes it difficult to apply much heat in this section of the preform. Heat is required in this section of the preform to start the stretch blow process and predetermines the overall material distribution in the bottle.

The middle lamps are usually moved a little further away from the preform (larger distance to the surface) to ensure that they can be operated at high power with good heat penetration into the preform wall (wave length). This is because the close array of lamps cause heating overlap effects from the individual lamps, which result in high heat flows to the middle section of the preform. Here, the effect of surface cooling on the outside of the preform wall is most important.

The top lamps can return to a reduced distance from the preform surface; ideally they are adjustable in height position and proximity with flexibility that makes the end-cap of the preforms easier to reheat. While the middle lamps typically have a distance of 15–20 mm, depending on the length of the preform, the lower-level lamps are at the low end of this range. The upper lamps, with the target heating area in the end-cap, can be at the high end of the range.

The geometrical arrangement of the lamps relative to the surface of the preforms allows pre-profiling of the heat along the preform length, while the power of the heater lamps gives the additional variable of adjusting the final heat profile. Ideally, as mentioned above, the lamps should be operated in a power range that gives the maximum lamp temperature allowing the best possible heat penetration. A typical temperature profile along the preform axis is shown in

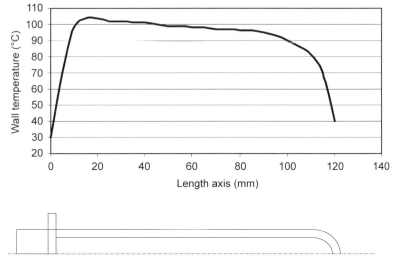

Figure 8.23 Temperature profile along the preform axis.

Figure 8.23, indicating that the higher temperature is necessary between the support ring level and the body of the preform (i.e. in the tapered section). This section is first pre-stretched and subsequently blow-expanded to form the base bubble, which then propagates from the neck of the preform over the whole length into the final container.

Heating and equilibration times may have a considerable impact on the effectiveness of the reheating process. While it is not necessarily advantageous to have long heating times—since this adds to the complexity caused by the influence of heat conduction within the preform wall—there is a relationship between the wall thickness to be reheated and the time needed to reach the required temperature range. Practical experience shows that times between 15 and 25 s are appropriate to reheat preforms to their required temperature range and accomplish the profiling during this time period. Good results have been achieved, in conventional high-performance heating systems, with equilibration times of 40–60% of the heating times. These equilibration times can be split between two heating sections and prior to stretch blowing. However, most machines do not accommodate this favourable feature. The typical effect of having an equilibration period following a heating period is shown in Figure 8.24. It is obvious that the temperature level on the inside wall of the preform lags behind that on the outside during the heating phase, although it recovers and exceeds the level of the outside temperature during the equilibration period. The length of the equilibration period will determine the eventual difference between the inside and outside temperatures of the preform wall. Key factors affecting the gradient of the cooling during equilibration time are the temperature of the

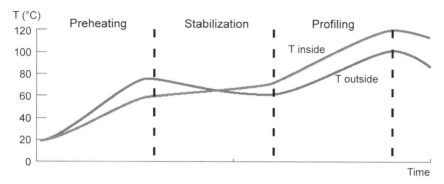

Figure 8.24 Time *vs* T inside/T outside (Ti/To) for heating and equilibration.

environment and heat transfer coefficient assumed for the surface of the preform. It must be supposed that in preform sections with uniform wall thickness this effect will also be uniform over the length of the preform. This assumption may not be true because of variation in lamp radiation temperature (i.e. aging of lamps), outside cooling effects during reheating (forced air convection or sudden changes in room temperature) and the potential conductive heat input from thicker sections at the end of the cylindrical (equal wall thickness) portion of a preform.

8.3.1.2   Stretch blow moulding

Stretch blow moulding is carried out in four subsequent phases, which occur in the blow mould once a reheated (and temperature equilibrated) preform has been delivered to the blowing stations. These four key phases are:

- axial pre-stretching
- pre-blowing and bubble propagation up to typically 90–95% of the final shape of the container
- final blowing to the residual shape of the container
- cooling of the stretch blown container in the closed mould

These four phases are shown in Figure 8.25. The axial pre-stretching (steps 2 and 3) overlap with the start of pre-blowing (step 4) in a way that will help to determine the final material distribution in the container. The longer the pre-stretching occurs, the more material will be transferred to the base end of the container. The earlier the pre-blowing starts, the more material will be expanded early enough to be pressed to the chilled mould wall which prohibits any further material distribution, thus keeping it in that very section. The propagation of the bubble proceeds so that the material makes contact with the wall in the shoulder, body and base section of the container mould (steps 5–8). The high-pressure final blow is applied (step 9) to shape the ultimate contour of the container,

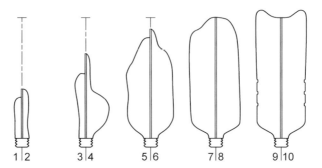

Figure 8.25 Steps 1–10 in the stretch blow-moulding process.

which is then frozen through contact cooling (step 10) on the wall of the water-cooled mould (ideally 4–12°C). Eventually, the pressure is released from the container mould when the container can maintain its shape and dimensions without any impact on the desired properties. This requirement determines the so-called mould-close-time that, for a standard container, is typically in the range 1.8–2.3 s. The most critical section of the container to cool down is the base section, since the wall thickness of the base remains relatively high. Having been heated to the deformation temperature in order to allow the forming of the base shape, it now needs all this heat removing to maintain base dimensional stability.

This time is broken down into the phases displayed in Figure 8.26. The cooling time is required mostly from a physical point of view so that a process optimization often starts with a time for pre-stretch (0.1–0.15 s) and pre-blow (0.3–0.6 s). Depending on the container being blown and the available exhaust

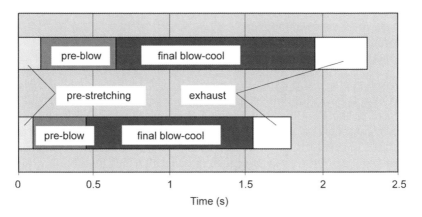

Figure 8.26 Mould close time breakdown.

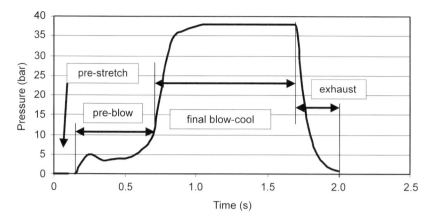

Figure 8.27 Pressure curve.

section of a blow system the exhaust time of the high-pressure blow air takes a relatively fixed amount of time (0.2–0.5 s), leaving a total of 1.2–1.5 s for the final blow and cooling. Blow channel cross-sections and valve configuration then determine how the pressure (which has built up during the final blow stage) will be released. This can be achieved very quickly, leaving most of the time for cooling, or slightly more slowly which can cost about 25% of the cooling effect in the mould. A typical pressure curve is shown in Figure 8.27. Pre-blow pressures are usually set in the range of 5–15 bar, while final blow pressures are operated above 35 to a maximum of 40 bar (limited through readily available compressors). Pre-stretching speeds can range 0.5–1.5 m/s; it may sometimes be necessary to adjust the speed of the stretch rod independently from any other process parameter. This will assist control of material distribution during the blowing process and will allow the stretch rod to advance at maximum speed to the correct position within the preform before actually commencing the stretch.

It is obvious that heavier preform/bottle combinations will require a longer period of reheating as well as stretch blowing, especially when additional processes are added, for example thermal treatments such as heat-setting. Very light preform/bottle combinations do not cause the opposite effect to any considerable extent since timing is already squeezed to a minimum for most standard containers within the limits cited above.

Speciality containers designed with rectangular or other fancy features in the base, shoulder or other sections will require the whole process to be sequenced from the other end. This means stretching the end portion of the preform first and blowing the bottle starting from the base. It requires skill in temperature profiling and process control but may at times lead to good results with very fast blowing processes, almost omitting the pre-blow stage; however, these are complex process conditions that standard commercial equipment does not usually allow.

8.3.1.3 Technologies for thermally stable containers

Thermal treatment of PET has an impact on the properties of the material. The aim is always to increase the thermal stability of the package and its performance relative to heat exposure. Once a container is blown under the standard stretch blow moulding process, any kind of thermal treatment could be carried out in the blow mould or in any other device that will maintain the shape of the container. If this cannot be accomplished, any treatment above the glass transition temperature will release part of the orientation that has been induced during the stretch blow moulding process and the container will lose its shape.

The temperature under which the thermal treatment is carried out determines the level of change in properties of the container. There are four different levels of thermal treatment, as summarized in Table 8.9. These different technologies overlap. Most of the treatment requirements are defined by the filling process of a container subsequent to the blowing process. It is imperative to know the exact filling temperature the container will have to stand and for how long, and under which conditions the container will then be chilled down to room or storage temperature. PET has little thermal stability above the glass transition temperature.

Thermal relaxation and pre-shrinkage. Thermal relaxation is carried out in warm moulds (30–70°C) and is a process that pre-shrinks the container when it is being produced. Machine speed and container cost are only marginally

Table 8.9 Thermal treatment: methods to thermally stabilize PET bottles

		Standard	Relax	Hotfill	Heatset one mould	Heatset dual mould
Typical bottle weight 1.5 ltr	g	40	42	67	58	62
Filling temperature	°C	5–20	5–40	40–88	80–95	88–95
Duration to quench (*)	min	n.a.	n.a.	<1 min	1–4 min	1–4 min
Output range	bph/ station	–1500	1000–1200	700–1000	500–800	1000–1100 400–600
Mould temperature	°C	5–15	30–40	80–110	110–160	10–185
Add'l technique		mould only	post cooling	internal air cooling	internal gas/ air cooling	internal water spray cooling
Crystallinity	%	20–30	20–30	25–35	35–40	35–45
Cost increment		100	105	125	150	200

(*) with less than 2% shrinkage on the container.

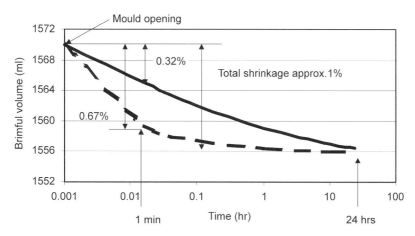

Figure 8.28 Relaxation of bottles over time. Key: —, cold mould; - - -, warm mould.

affected. The implementation of this technology is relatively simple in that the body portion of the blow mould is operated at the elevated temperature while base and neck portion need to stay chilled.

This phenomenon can, however, be accelerated through thermally conditioning the container in the blow mould. Pre-shrinkage by thermal conditioning is often chosen when freshly blown bottles are to be fed directly into a filling line and maybe mixed with older bottles. Once the mould temperature is elevated to 30–40°C, the relaxation occurs in the first few moments after moulding. If the container is kept for more than 6–12 hours after moulding in a cold mould (6–12°C), it reaches the same extent of relaxation that it would reach after moulding from the warmer (30–40°C) mould, typically being 0.5 to 1% smaller than the instantly moulded shape. Figure 8.28 illustrates the basic behaviour, showing that after about 24 hours there is little residual shrinkage left in a container. Obviously, the amount of relaxation can additionally be affected by storage or filling conditions, which are recommended to be in a controlled environment of 20°C and 60% RH. In designing a container and defining process, mould and storage conditions, this relaxation should be considered.

Hot-fill. Hot-fill containers allow filling temperatures of up to a maximum of 87°C with a standard range of 82–85°C. The filling time is kept to a minimum (and should never exceed one minute) before the bottle is quenched to room temperature. It is usually kept between 20–30 seconds, when the container is capped and conveyed to a chilling tunnel. In order to make PET bottles that withstand these kind of conditions, there is an advanced thermal relaxation process applied with mould temperatures ranging as high as 100°C. To ensure that this process works, additional aspects of the package and process design are considered:

- reduced stretch ratios from preform to bottle to limit the amount of residual stress that may be released under the thermal exposure
- addition of thickness to container and preform in critical sections to gain stability
- nucleation of the PET side wall to an extent that it gives further thermal stability which is accomplished by choice of resin grade, mould close time at the elevated temperature and extended heat load during the reheating process.

Production rates on standard machines for this technology are at 600–800 bph per cavity, depending on the extent and the effectiveness of the applied features. The body portion of the mould is heated, while the neck and base must be kept cold enough to deliver the required shape in these areas. Good process control is required.

Container design features, like vacuum panels, are added to retain the container shape after it has been filled with hot product and then compensate for the volume reduction of the product after this has been cooled down. Additional stabilizing rings above and below the panel section help to assert that non-uniform panelling and a collapse of the shape is prevented. The bell-shaped shoulder section provides stability and helps to direct any deformation towards the panel section. The base is ribbed and retains extra thickness to withstand the elevated temperatures. Figure 8.29 illustrates these typical shape features of a hot-fill container, which are almost identical to those used for the heat-set technology. Since there is confusion in the industry about differences between hot-fill capability of a container and true heat-set processes for a container, distinct differences between these two technologies are outlined.

The neck needs to be handled carefully at high temperatures since, without either additional thickness or neck crystallization, the risk of distortion jeopardizes product integrity on the closed container. Neck crystallization has to be applied between the injection-moulding and the blow-moulding stage and is brought about in a separate machine by heating the neck to the crystallization temperature range with aids to secure the key dimensions of the finish while it crystallizes and, as a result, changes density. The amorphous sections, like the neck finish, will turn white when crystallized, while the oriented portions stay transparent. Crystallinity levels in the container reach typically 20–30% and it is predominantly mechanically-induced orientation contributing to this since, at the applied temperatures, there is little thermal crystallization developing in the body.

Heat-set. Heat-set technologies advance the thermal stability and filling temperatures over the 85°C temperature mark and allow longer sterilization

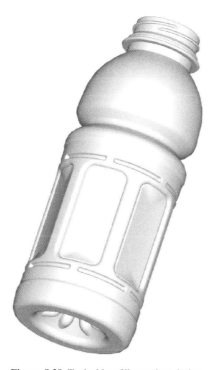

Figure 8.29 Typical hot-fill container design.

times prior to quenching the container to room temperature. Potential fill temperatures as high as 95°C can be achieved, while machine productivity is heavily impacted downwards to 400–700 bph per mould at these higher temperatures. The heat-set effect is achieved through mould temperatures as high as 160°C or in the range of 100–160°C. The container must be cooled prior to releasing from the mould surface. A combination of all the specific features applied in the hot-fill technology leads to thermal crystallization in the sidewall of the container up to 40% crystallinity. Moulds are either heated with oil or through the use of electrical heating segments installed in the mould. Both mould-heating technologies are being applied; lower temperatures are easier to control with oil while electrical heating keeps the machine environment safer.

At these mould temperatures it would be impossible to retain the shape of a container when demoulding. Cooling air is provided through the stretch rod that freezes the inside layer of the container just prior to opening the mould. It reduces the average side-wall temperature of the container for the moulding process and leaves a residual pressure inside the container once it is released from the mould. Shape is retained through the support of this inside pressure. The stretch rod is therefore connected to the high-pressure air system. It is hollow and the allocation of the air outlet holes needs to be adjusted to the container shape and the process conditions.

Base mould temperatures can also be elevated to levels as high as 125°C to gain crystallinity in the base; however, these temperature levels can cause white sections in the amorphous part of the base and have the detrimental effect of the material tending to stick to the mould surface, thus making demoulding complicated. A fine balance between base mould temperature profile, preform and bottle design, and base orientation has to be found to optimize conditions. Base mould temperatures in the range of 70–90°C already contribute to increased base stability and support the lower fill temperatures well.

Super heat-set. Super heat-set methods are applied to exceed crystallinity levels over 40% and create a highly crystalline structure in the side-wall of the container. This thermofixation is either achieved through very long moulding cycles in the blow mould at temperatures above the peak crystallization temperature (160–170°C) or in combination with a secondary moulding process where one of the blowing steps is carried out in a high temperature mould. This secondary process could

- either use an overblown container that has been exposed to a very hot mould in a first step, which is then reheated as a container and then blown into a cold and final mould,
- or use a standard preblown container and then heat-set it in a very hot mould, allowing time to fully stabilize the container.

Cooling the container after such treatment is critical and various methods have been developed, such as cooling with liquid nitrogen or water spray systems, thus adding a level of complexity in equipment additional to the heat-set features of a machine. Commercial success has been limited because of the technological complexity of such machine systems and the related capital and operational costs adding considerably to the overall cost competitiveness of these technologies.

8.3.2 Machinery concepts

Stretch blow moulding machines must be designed to produce the above process criteria with high repeatability, while ensuring good productivity, reliability, and low maintenance and service requirements. In principle, there are two basic concepts available, linear and rotary machines, as shown in Figure 8.30. Linear machines typically operate in intermittent cycles with parallel processes and have been developed particularly for the lower output range. Rotary machines best serve the medium and high output ranges with a continuous material and process flow. Key differences for each version of the

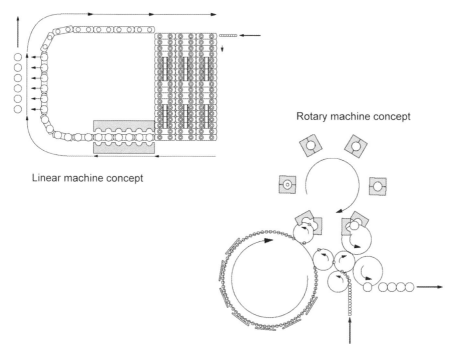

Rotary machine concept

Linear machine concept

Figure 8.30 Linear and rotary machines.

machines are the methods of material handling and preform transport, with the design of the mould station and heating oven configuration being of secondary importance.

Typically, preforms are handled in a neck-down position for the heating process. This position is preferred to avoid any convection heating of the neck portion that may result in deformation during the blowing operation. The neck finish is an important feature of the final container (particularly for lightweight or short-neck finishes). To help reduce this potential neck deformation the blow moulds or mechanisms completely enclose the neck during the blowing operation. It is preferable to leave the preform in a neck-down position for the blowing operation, to avoid turning the heated preform into a neck-up position, where there is the risk of preform bending which may increase scrap rates.

There are currently three different oven design configurations available in commercial machines, as shown schematically in Figure 8.31:

- single lane linear oven configuration
- rotary wheel configuration
- parallel linear oven configuration (multilane)

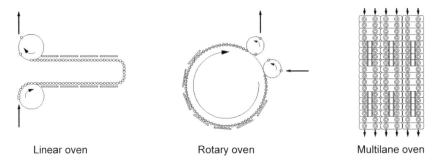

Linear oven   Rotary oven   Multilane oven

Figure 8.31 Three oven design configurations.

The linear oven concept is preferred for many machines since it provides good thermal characteristics for the heating process, the distance between lamps and preforms is constant and floor space requirements are low. The major downside for larger machines, in particular, is the high wear of such systems. Wear and operational consistency are optimal on the rotary oven configuration, which allows the best mechanical synchronization with the blowing wheel of rotary machines, while offering nominally less flexibility to add or subtract heating capacity. This, however, is more of a theoretical argument. The benefits of changing the preform surface-to-lamp distance on a larger wheel configuration are negligible. Small heating wheels may have a detrimental effect on the preform rotation in front of the oven, as their use may result in non-uniform temperature profiles on the circumference of the preform. The benefit of changing the distance may be better heat penetration with built-in equilibration, while it also carries the risk of less energy utilization.

The concept of parallel oven lanes was one of the first in the industry but, over the years, has not been successful. Furthermore, the heating profile differences between multiple lanes result in non-uniform material distribution and properties from cavity to cavity. In recent years, protocols have been developed to improve the control of these differences in oven lanes and to support the control mechanisms. These include better insulation of the oven section. However, differences can still be observed from lane to lane, even if the multiple lanes are supported with individual lane control loops.

Machines, with linear oven configurations, blowing several bottles in parallel per cycle, have the disadvantage that the configuration is either very long, as a result of the intermittent operation, or provides preforms from a continuous flow that then has to be transferred to the multiple moulding station, with the first coming from the oven at a lower temperature than those coming last. This is a disadvantage of many machines that are commercially available today.

As a result of these heating challenges, simple and cost-effective linear machines do have major constraints in output and process consistency; thus they have gradually become limited to low output ranges below 5000 bph.

Preforms can be transported by mandrels, on which they are positioned throughout the process, or by gripping systems that carry the preform through the individual stages of the process. The transfers between the different process steps are critical points for the preforms. Mandrel systems have the advantage that controlled mechanical parts are safely guided through the machine, while the preforms are not touched throughout the whole process.

In-feed for preforms and out-feed configurations for bottles are the most critical components when it comes to machine operational availability. Often, this underestimated component of material handling is crucial to the success of a machine model. Experience has shown that PET preforms and bottles have a limit of around 35,000 pieces per hour in a single lane. Above this, a consistent material flow is very difficult to maintain. This also applies to conveyors and palletizing systems. Multilane systems and lines have helped to overcome this issue in many operational environments.

Linear machines are less expensive with a lower number of mould stations; although scaling up would multiply the cost of the machine with the number of moulding stations. Rotary machines provide much better economies of scale, where the costs for the larger machines, with increased numbers of mould stations, result in lower unit costs (see Figure 8.32). At the same time, this means that rotary machines larger than a certain size bring little capital cost advantage, combined with the fact that efficiency levels are more critical to maintain and more at risk.

Machine controls are typically PLC-based and require high speeds to control the fast processes outlined in the previous sections. Comprehensive

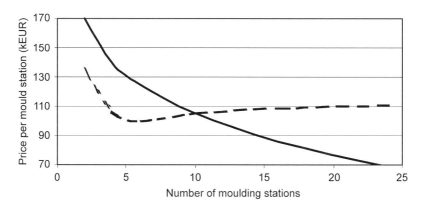

Figure 8.32 Machine cost characteristics. Key: ▬ ▬, linear; ▬▬▬▬, rotary.

machine-status supervision systems are state-of-the-art for the leading equipment models and crucial to allow operators to resolve potential machine stoppages quickly. Process control is mainly limited to temperature control measures, while the concept of in-line process and quality control has been limited primarily to downstream feedback devices that check container quality. Additional special features, such as blow curve display and supervision, stretching consistency measures and preform temperature profile scan are rarely found and mainly serve the experienced machine operator. Incoming preform quality is crucial to ensure high machine utilisation, and thus preform testing devices for high-output machines are commonly in use and mounted as an integral part of the infeed system.

Machine lifetime and stability, drive mechanism and the moulding stations are determined by the design features of the base frame components. The more rigid the base frame design, the lower the impact of the high deformation forces caused by the continuously moving parts on the stability and reliability of the machine. Gear drive mechanisms are preferred, or highly accurate linear drives that ensure repeatability and reduce the need for realignment. Servo-controlled mechanisms are only slightly more expensive and are preferred for accuracy and control in linear machines over any hydraulic or pneumatic drive mechanisms.

Shape-controlled, mechanical drive mechanisms (cams, curves, toggles, levers, etc.) are most popular giving high repeatability and good life-cycles. These are used in all leading equipment supplier's configurations for moulding stations (see Figure 8.33). These high-loaded components must be carefully designed at the lowest possible weight (speed) with the highest available stiffness (blow impulse). Annual operating times typically range 4–8,000 h, thus exceeding, in one year, more than the operational lifetime of an automotive engine. This illustrates how essential serious and competent machine engineering is for these machines.

High pressure blowing air is the main auxiliary requirement for a stretch blow moulding machine, together with electricity and cooling water.

8.3.3 Mould technology

In mould technology terms, a clear differentiation between the neck, body and base sections of the container should be reflected in the design of the moulds. This is not only because the potential thermal control requirements are different in each section but primarily, for reasons of preform selection flexibility and container design, these three basic sections of a mould should be separate.

Moulds are usually manufactured out of aluminium. This is cost competitive and provides good thermal conductivity for fast mould cycle times. The alignment of the cooling channels relative to the moulding surface is important

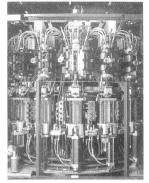

Source: Krupp

Source: Sidel

Source: Tetra Pak

Figure 8.33 Moulding station.

to obtain effective and uniform cooling. This is particularly the case for thicker wall sections in the container. Aluminium moulds are also light, which improves handleability at mould changeovers and keeps the overall weight of the moulding wheel down. Therefore, steel moulds are not recommended, or shell moulds (the inserts give higher flexibility for shape changes but do not prove advantageous for high production runs of containers of the same shape). Quick mould change features are, however, becoming more and more a standard requirement in the industry and mould change times of around 5–10 min are typical in state-of-the-art machines.

It is important to vent areas in the mould, where air can be trapped between the blown container and mould surface, especially for more complex shapes, such as footed-base container designs, contour features, and so on. Venting holes can range in diameter up to 2 mm but are less desirable if no visual impact is required. The surface finish of the mould has an impact on the cooling efficiency and affects the heat transfer between container and mould. In some instances, it may be desirable to retain a rougher surface in some sections of the mould wall to reduce the cooling effect. This increases the temperature of the container wall at mould opening, producing pre-shrink effects that are often considered important when mixing freshly blown bottles with stored bottles for on-line production.

As the container shrinks upon cooling in the mould, the dimensions of the container design are converted to those required in the mould design. 'Shrinkage factors' are applied to establish the difference between the mould dimensions and the final container dimension. The shrinkage factor is a function of the container temperature at mould opening and the release of frozen-in stresses relative to biaxial stretch and shape features. For standard container applications, these shrinkage factors are around 1% in the length and diameter dimensions, while the length shrinkage can be smaller than the diameter shrinkage in the range 10–20%. Variability of this value can be around $+/- 30\%$ and is subject to the resin grade used, applied stretch ratios, shape complexity and features of the container.

8.4   Preform and container design

8.4.1   Container design

In developing container designs consideration must be given to the ability to blow the shape and the potential to design the preform or use an existing preform. Container design is best accomplished via three dimensional computer aided design (3D CAD) systems that can solid model the complete container with all its features. In this modelling, there is a need to predetermine container wall thickness and accurately calculate the container volumes from the solid shape. In some cases, finite element analysis (FEA) might be used to determine optimum material distribution in the more complex shapes required in performance containers, as might be used for carbonated beverages.

Container weights are primarily a function of the volume but also the required application. Figure 8.34 gives some indicative (as requirements can vary considerably) weight ranges as a function of the container volume for different applications.

The container neck finish makes an essential contribution to the container weight. Sizes are focused on the two finish dimensions of 28 and 38 mm, with various design variants. More variability has developed in the neck design and dimension over recent years as a popular means of providing marketing differentiation. The plastic closures only (PCO) finish is a lightweight neck that has been developed specifically to reduce overall container cost and is designed for use only with plastic closures (see Figure 8.35).

Two factors are important for the stretch blow process when it comes to the neck finish dimensions. The first is that the inside diameter of the finish cannot be smaller than the maximum diameter of the container divided by 90% of the maximum allowable stretch ratio. The second is a clearly defined functional support ledge (or neck support ring) for the preform to be conveyed through the machine and to be retained during the blow process. Typically, the diameter of this support ring should exceed the diameter of the preform, below the support

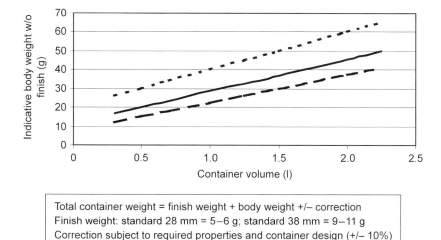

Figure 8.34 Container weight. Key: ———, carbonated soft drinks; ▬ ▬, mineral water/edible oil and sauces; ▪ ▪ ▪ ▪, hot-fill juices, nectars and soft drinks.

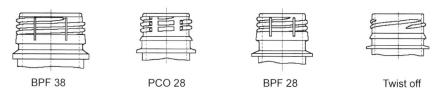

Figure 8.35 Finish design BPF, Aloca, PCO.

ring, by at least 3 mm and should have a minimum thickness of 1 mm for reasons of stability.

Rectangular containers have a wide and a narrow side; the key is to determine an effective diameter that allows a round preform to be designed for such a shape. The ratio of wide and narrow side should not exceed a factor of 1.5, unless it can be made with large radii in the corners of the shape as indicated in Figure 8.36. This then becomes more of an oval-shaped container cross-section, where the wide and flat side can have up to a maximum ratio of 2.5 in ideal scenarios. These figures relate to preform design and stretch limitation that can be analytically derived or found experimentally. In each of the borderline cases, it is necessary to meet a number of very carefully adjusted features. Namely, ideal preform reheat profiling, which can be supported by so-called preferential heating that may be necessary mainly to help compensate for insufficient heat penetration and across the preform wall temperature profiling.

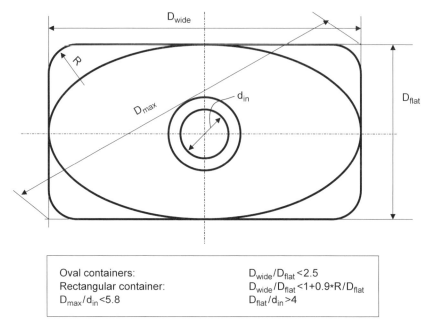

Oval containers:	$D_{wide}/D_{flat} < 2.5$
Rectangular container:	$D_{wide}/D_{flat} < 1 + 0.9*R/D_{flat}$
$D_{max}/d_{in} < 5.8$	$D_{flat}/d_{in} > 4$

Figure 8.36 Rectangular and oval-shaped cross-section.

Shape features such as ribs and panels are often designed to stabilize the blown container and at the same time reduce the weight. Shape forming is then essential to achieve a good contour profile as part of the container design, which may also be required for product differentiation and marketing purposes. The ability to blow such shape features repeatedly and with consistency should drive the design details, particularly such dimensions as corner radii and contour depth. In principle, the depth of a shape feature should never exceed the width of the gap through which the material has to be blown. Although this is achievable in some cases, the resulting wall thickness in that section will be too low and will not provide the desired rigidity enhancement. Shape and contour features are naturally easier to form in container sections with lower wall thickness. Particularly high temperatures can be used in thicker sections of the preform to allow for the desired shape definition. Sharp radii of less than 1 mm in a contoured section should be avoided, while for decent shape definition radii should also not exceed 2 mm.

Container bases are considered a science in themselves. There are many different configurations (a few of which are shown in Figure 8.37 that are either driven by the application requirements or shape and differentiation considerations. Some are also driven by intellectual property rights. Footed bases are used for pressurized containers, although in some cases champagne-base type

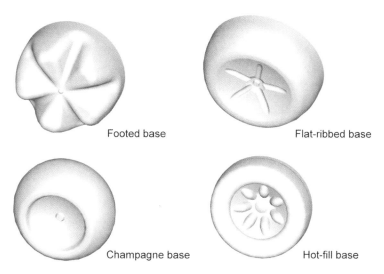

Footed base

Flat-ribbed base

Champagne base

Hot-fill base

Figure 8.37 Base designs.

designs can also be used. In this case, reinforcement is required to reduce the risk of base roll-out and weight is added to the base. Non-carbonated drinks usually have a flat base that can have an increasingly concave shape towards a champagne-type base. This again adds weight but will bring added stability so that the weight of the product in the container will not deform the base. Stability, in general, is enhanced by rib features. Hot-fill and heat-set containers usually have more complex dome designs in the base, with additional rib structures for stability when filling at higher temperature.

The design of the feet is as critical as the preform design, as is quality and the appropriate processing to obtain a stable base configuration. Many different concepts have been pursued while a five footed-base configuration is certainly the most frequent solution globally for stability and processing reasons. Base designs are different in the radii around the feet and in the surrounding transition zones, and also in the stretch bands that go up from the centre of the base to the sidewall of the container. The concept of the stretch band is to provide the strength, while the feet provide the stability and standing capability. Both functions must be carefully synchronized in the design of the base radii. The middle plate around the centre of the base must have specific design criteria to maintain base stability under pressure and leave sufficient base clearance, although this can also be provided by a variety of other workable solutions.

It is essential to all footed-based container designs that the material between the amorphous centre of the base and the highly stretched and oriented section of the feet and the sidewall has uniform transitions, with high strength and

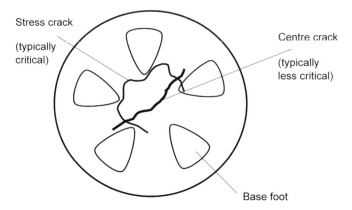

Figure 8.38 Two forms of base cracks.

stability. To achieve this target, the preform as well as the bottle-base design and processing conditions must be considered, otherwise critical base cracks, so-called base stress cracking, may occur during the application of pressure in the filler or on the retail shelves. These cracks develop particularly in the transition zones of oriented and non-oriented sections with significant changes in wall thickness. It is essential to be able to strengthen these areas so that pressure deformation cannot cause surface scratches and subsequent cracking. The transition zones display a molecular instability that can be highly susceptible to pressure or other environmental influences—in particular, certain filling line lubricants on table-top conveyors or caustic material in the supply chain can easily cause considerable damage in footed-based containers. Figure 8.38 shows two typical forms of base cracks. The centre crack that runs through or close to the centre of the gate and base, is clearly related to either low preform gate quality, an incorrect adjustment of the stretch rod stoke (too close to the base mould) or too much residual material that had not been heated sufficiently to be deformed properly from the preform to the bottle. This base crack is easier to cure than the typical stress cracking described above, which occurs in the transition from the centre plate to the feet, where the crack develops circumferentially.

8.4.2 Preform design

The principal preform shapes are summarized in Figure 8.19 . The individual preform design is widely viewed more as an art than a science. In reality, it is a complex matter since the preform quality, the container shape, the reheating conditions and applicable temperature profile, as well as the stretching and blowing parameters, all have a final impact on the container. Once these factors are kept within the limits that have been shown to give an appropriate processing window, it is relatively simple to define preform design criteria that give good results in terms of process performance, with wide processing windows and good container performance. However, it is important to understand the complexity of this field of multiple variables.

The principal relationship between container and preform is demonstrated in Figure 8.39. The hoop stretch ratio is the ratio between the inside diameter of the preform and the outside diameter of the bottle in any given location of the container. Considering that the highest number is the most difficult to blow, a very practical approach refers to the inside diameter of the preform at the end of the cylindrical section in relation to the maximum container diameter, or for rectangular bottles the diagonal and for oval containers the major axis dimension of the bottle.

The axial stretch ratio is the active length of the preform (i.e. total length minus neck height minus non-stretched portion below the support ring [typically 3–5 mm] minus the wall thickness of the preform in the base) versus the stretched length of the container (i.e. the total length minus the neck height minus the non-stretched portion under the neck).

With reference to the stretching behaviour of PET as described in Section 8.1, PET reaches strain hardening in a uniaxial stretching situation at values clearly above 2. This is the value that least needs to be exceeded clearly in axial stretch to obtain containers with uniform wall-thickness distribution and consistent processing.

To blow a preform into a container, a two-dimensional deformation occurs through the axial and the hoop stretch of the preform to the container shape. Since one is a radial and the other is an axial deformation process, these two factors are connected through the relationship of the generic vessel formula by a factor of 2 relative to the potential stretch ratio, in order to reach the same strain hardening effect and the desired equilibrium of stresses for uniform material distribution. This ratio can vary depending on the overlap of axial and hoop stretch in their sequence of occurrence but should not be outside a window of $+/-15\%$ of this value.

These considerations lead to the limits for the stretch ratios as outlined in Table 8.10. Given particular container dimensions, these basic principles lead to relatively effective results. Locally, on any given point in the container, stretch ratios and resulting wall thickness may lie outside the given envelope, either to the very high end or the low end. This is a result of shape features of the container that will not allow the PET to stretch to the desired extent or may make the PET in other portions of the container stretch to much higher levels. This becomes true particularly in the base sections and the shoulder and neck portion of a container. Careful consideration should be given to avoiding situations of overstretching, while at the same time low integral stretch ratios will not allow

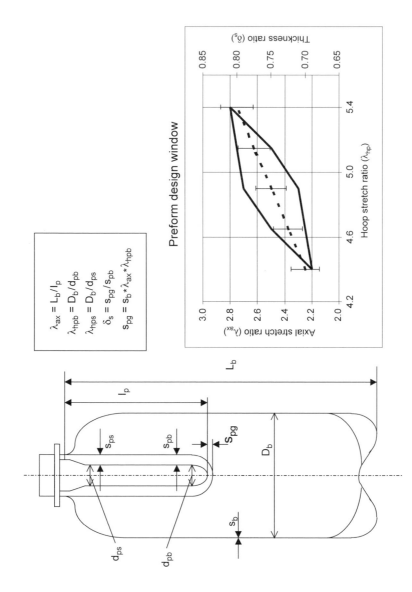

Figure 8.39 Preform and bottle together: stretch ratios. Variables: d, diameter; l, length; s, thickness. Subscripts: b, bottle; p, preform; pb, preform body; pg, preform base; ps, preform shoulder.

Table 8.10 Stretch ratio ranges

	Hoop stretch-ratio
Carbonated soft drinks	4.8–5.2
Flat mineral water	4.6–5.0
Edible oil and sauces	4.6–5.2
Hot-fill/heat-set	4.4–4.8
Flat drinks	4.4–5.2
Oval or special shape containers	4.4–5.6

consistent material distribution to be maintained. Overstretching in a bottle can be seen as white pearlescence on the inside surface of a container, which, when cut open, can be scratched with the fingernail. This is significantly different from the white colouring effect caused by crystallization, resulting from overheating of the preform during the moulding stage. This would prevent the material being stretched to its natural limits.

Once the basic dimensions, such as preform length and inside diameter, of the preform are established, it becomes relatively easy to predetermine the residual parameters of the preform design. The preform wall thickness can be chosen on the basis of the area stretch ratio range that is given in Table 8.10, by multiplication of the targeted container wall thickness under consideration and the overall targeted preform weight as outlined in Figure 8.39 (i.e. higher perform weight results in thicker walls). Consideration should be given to the fact that higher orientation (higher area stretch ratio) results in higher strength of the sidewall of the container, thus allowing a container to be weight reduced to a certain extent; this should result in thicker preform walls. The sidewalls of preforms in standard designs are equally thick throughout the length over the preform, although it is recommended that the wall thickness is reduced towards the gate. Good results have been achieved by using 75–85% of the sidewall thickness for the gate wall thickness. Typically, a taper is applied in the cylindrical section to allow easy stripping off from the injection mould core; as an indication, 1 mm in diameter reduction over 100 mm in length is functional.

The wall thickness of the portion below the support ring, which does not get stretched much, if at all, is typically chosen to be 2–2.5 mm for standard preforms. The length of this section may vary subject to the container design and the processing capabilities of the targeted blow moulding machines; ideally, it should be as short as 4–5 mm but can be as long as 8–15 mm. Heating this portion of the preform is a delicate task, since it is so close to the neck finish section which should not be heated; longer sections make it easier to heat but usually cause an increase in unused material in the neck/shoulder transition.

The tapered transitions are the most difficult to assess when applying simple rules for preform design. The length is determined in the range 5–20 mm, subject to the size of container and length of the preform, overall area stretch ratio and shoulder shape features of the container.

As a result, every container size has a particular range of preforms that is suited to match optimal blowing conditions. Very small variations can make these blowing conditions so unstable that the results in the container are no longer acceptable. For example, 2 litre preforms are not suited for 1.5 litre bottles and vice versa, as some of the principal dimensions may not match at all. Designers often try to compromise the use of injection tooling such that the preform design is adjusted in a few features that are relatively inexpensive to accomplish. However, in most cases this gives unsatisfactory results on the process window and the final container performance.

References

Throughout this chapter, reference is made to much public knowledge in the industry either known to the author through experience or through patents and projects made public in conferences and individual meetings. It should be noted that the information contained in this chapter is generally in the public domain and so the author has not specifically outlined all detailed sources that may have been referenced. If there are specific requests, these should be addressed to the author in writing so that reference can be provided.

Abbreviations

AA	acetaldehyde
BPF	British Plastics Federation
CAD	computer-aided design
CSD	carbonated soft drinks
DMT	dimethyl terephthalate
DSC	differential scanning calorimetry
EG	ethylene glycol
EOS	edible oils and sauces
FEA	finite element analysis
HDPE	high-density polyethylene
IPA	isophthalic acid
IV	intrinsic viscosity
JNSD	juices, nectars and still drinks
LDP	liquid dairy products
MCA	Metal Closures Association
MW	mineral water
OCR	output per capital ratio

PA	polyamide
PAN	polyacrylonitrile
PCO	plastic closures only
PE	polyethylene
PEN	polyethylene naphthalate
PET	polyethylene terephthalate
PP	polypropylene
PVC	polyvinyl chloride
RH	relative humidity
TPA	terephthalic acid

9   Injection blow moulding

Mike Wortley

In the injection blow moulding (IBM) process, which has developed from conventional injection moulding, a moulded preform is blow moulded into a bottle or container. During this process the neck of the preform is kept cold to maintain its moulded dimensions and prevent distortion. The IBM process is used for the manufacture of containers that have close-tolerance threaded necks, wide mouth finishes and highly styled shapes. Good material distribution is possible and finished containers do not need to be trimmed. IBM is used worldwide in the production of millions of containers and technical mouldings, ranging in volume from 1 ml to 1 litre, in a variety of plastics, for packaging a wide range of products: cosmetics, pharmaceuticals, foods, household and automotive products and chemicals. Figure 9.1 presents a typical range of containers produced by IBM.

9.1   Basic principles

Stage 1 of the IBM process is the injection moulding of the preform. In this stage, molten plastic is intruded or injected around the core rod (see Figure 9.2).

The core rod produces the internal finish of the neck of the bottle, while the neck mould produces the external thread form. This area of the mould is cooled to set up the neck finish. The test tube, or body part of the preform, is kept hot and conditioned, so that it will develop a skin and can be removed from the cavity without damage. It is then transferred on the core rod, which also acts

Figure 9.1 Typical injection blow-moulded containers. Courtesy of Jomar Corporation.

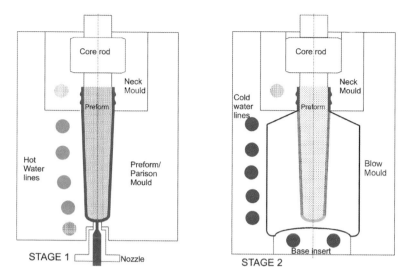

Figure 9.2 In stage 1, the preform is injection moulded. In stage 2, the preform inflates to the shape of the blow mould.

as the blowing mandrel, to the blow mould (stage 2). Here, air is introduced through the core rod and the preform inflates to the shape of the blow mould, where it is cooled for ejection.

9.2 History

The original patents for IBM go back more than 50 years to W. H. Kopitke, who developed the process in a standard injection moulding machine. The early IBM two-station process used adaptations of standard injection moulding equipment, fitted with special tooling. Similar systems using specialised moulds were developed in the late 1950s by Farkus, Moslo and Piotrowsky. The Piotrowsky process used a 180° rotating table, with two sets of core rods and one set of preform and bottle cavities. A commercial vertical rotary turret system, similar to the Piotrowsky process, was developed by Procrea SRL in Italy in the mid 1970s.

The rotary table process (see Figure 9.3) was developed in the early 1960s by Gussoni in Italy and by Wheaton in the US. Gussoni developed the three-station process, which used a horizontal 120° indexing head with split moulds for the preform and bottle cavities, and three sets of core rods. The third station was used to eject the container, while the preform and blow moulding phases were completed simultaneously. Today, development on this horizontal indexing system is continued by companies such as Jomar Corporation and Uniloy Milacron in the US.

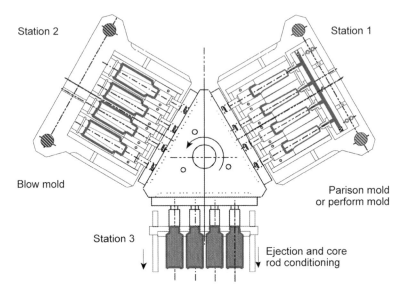

Figure 9.3 Rotary table machine.

9.3   Process identification

Mouldings made by extrusion blow moulding (EBM) will show a horizontal scar across the base split line, generated from the parison weld (see Figure 9.4). Bottles made by the initial injection moulding of a preform can be identified by the injection point or circular scar on the bottom of the container around the central axis (see Figure 9.5). Today, with many bottles produced in polyethylene

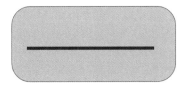

Figure 9.4 Extrusion blow moulding: web or scar of material from clamping of parison prior to blowing.

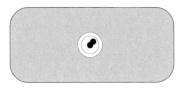

Figure 9.5 Injection blow moulding: injection mark from preform mould, or annular rings where the feedpoint has spread out after blowing.

terephthalate (PET) by injection stretch blow moulding (ISBM), identification becomes more difficult. Generally, larger bottles and containers that require higher orientation, biaxial stretching will be produced on ISBM machines.

9.4 Commercial processes

The design of purpose built, fully integrated, injection blow moulding machines are generally divided into two concepts as defined by the plane of rotation for the core rods mounted to the machine table or turret; the vertically rotating turret of Procrea (see Section 9.6) and the horizontally rotating table developed by companies such as Jomar, Uniloy and Wheaton (see Figure 9.3). The machines with horizontal tables have been further developed and are available in three-station (Jomar, Uniloy and others) and four-station (Uniloy, Wheaton) options. The fourth station can be used for enhanced conditioning of the core pins after bottle stripping. The additional station can be positioned after injection of the preform to allow for pre-blow or temperature conditioning of the preform prior to blowing. Alternatively, the additional station can be used to treat, coat or decorate the bottle after blowing. For reasons of clarity, the more commonly used three-station horizontal-turret and two-station vertical-turret concepts will be used to describe the basic injection blow moulding process.

9.4.1 Rotary table machines: Jomar, Uniloy and similar

The heart of the injection blow process is the triangular table (turret) which rotates in accurate $120°$ steps via a CAMCO or Ferguson indexing unit. A set of individual core rods are mounted on each face of the table and are located accurately in the preform mould neck inserts to form a shut-off at the neck finish of the container. Melted plastic from the plastifier is injected via a hot manifold with individual nozzles into the preform cavities at station 1 at a preset filling and packing pressure. The hot manifold can be one of three main types (see Figure 9.6):

- solid block, with either heating rods or hot oil to maintain melt temperature—even filling is obtained by adjusting the orifice through each nozzle (see Figure 9.2)

Fig 9.6 Plan view of manifolds. ———, Melt; ⌁⌁⌁⌁, Heater rod or hot oil.

- two-plate 'vertical split' manifold, heated by rods or oil, with internal flow path profiling to obtain an even output at the nozzles, which are drilled to one size
- 'coat hanger' type, where flow paths reduce in diameter to maintain even flow

Each individual cavity must be filled at exactly the same instant to ensure that the heating/cooling history within the preform is identical, so that neck shrinkage versus time is consistent and close tolerances can be held in the injection moulded neck finish. Also, when the bottle part of the preform is blown into the finished bottle it will inflate identically with no variation in wall-thickness across the cavities. The preform and blow moulds (see Figure 9.2) are mounted permanently onto diesets for rapid, simple mould changes.

Hot liquid (oil or water), usually above 100°C from a temperature-controlled source, is piped through the preform mould to remove heat from the injected plastic and maintain the preform within the thermoelastic range of the polymer being processed. When the preform has been made and conditioned, the head will lift the core rods vertically from the preform mould and rotate them 120° to the blow mould. Typically, it takes 1.5–2.0 s for the moulds to open, the turret to rotate and the moulds to close. At station 2, the moulds will close around the injected neck finish, with minimal clearance to avoid damage and prevent air loss. Compressed air from the core rod (at 6–15 bar) inflates the preform so that the plastic contacts the wall of the blow mould. The blow mould has similar drillings to the preform mould to allow chilled water (at 8–12°C) to cool the mould. Factors such as the type and thickness of plastic used, the air pressure and chilled water temperature, the shape of the bottle and the material that the mould is made from determine the time it takes to produce a stabilised bottle in a blow mould. When this is achieved, the mould will open and the table is transferred 120° to station 3, the ejection station, where the finished bottle can be mechanically stripped from the core rods. The bottles can be bulk packed, straight from the stripper or transferred to a conveyor for further cooling. If they are stood on a conveyor they can be heat treated, labelled or even filled and capped directly from the machine. When the products have been removed from the core rods, internal and external conditioning can be applied to the core rods prior to the start of the next cycle.

9.5 Tooling

As can be seen from the above, injection blow tooling is complex and consists of three main parts: the preform mould, the blow mould and the core rods, holders and stripper attachments.

Injection clamp forces can be considered low (10–200 tonnes depending on machine size), and the force is spread over the large horizontal split face of the preform mould. The moulds are very hot, which will ease material flow and keep the filling pressure low compared to normal injection moulding.

Moulds have to be designed around the material being processed. For PET, moulds have to be fully hardened throughout. The rest of the moulds can be pretoughened steel for the preform mould and aluminium for the blow moulds.

Multicavity individual moulds (up to 24 cavities) are accurately installed on the dieset for handling. Machining tolerances better than ± 0.01 mm are required to stop flashing in the hot preform mould. Allowances are also required on the cavity centres to take account of the heat expansion in the preform mould and cooling in the blow mould.

Tool manufacturing quality similar to an injection unscrewing mould is required. The preform cavity, as it is split lengthwise, can be profile machined to add extra plastic around the shoulder and base area.

Core rods (see Figure 9.7) have to be stepped or tapered to allow release of the preform, even when cold. A minor undercut or raised rib is added just under the shut-off edge of the neck finish to prevent the preform moving during transfer.

Core rods also act as a blow-exhaust pin and are designed with either shoulder or tip blowing, depending on the material being processed and the product being made. Core rods can be temperature controlled using oil; at 60°C, PET can be released from the core rod. Heated core rods can also reduce the time at start-up in getting the core rods to the processing temperature. Controlling the temperature of the core rods trebles their cost and the oil can cause internal contamination of the preforms and has to be monitored continually. To achieve an even release of the preform from the core rods, a high-quality polish and plating is beneficial. Many materials require the addition of a lubricant (zinc stearate) to assist this release at high temperatures.

Blow moulds are usually made with neck inserts, blow cavities and base inserts. For shallow bases, the inserts can be made in two halves and can be fixed solidly to the blow mould cavity. For crystalline materials and deeper base

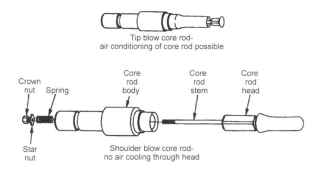

Figure 9.7 Core rods. Courtesy of Jomar Corporation.

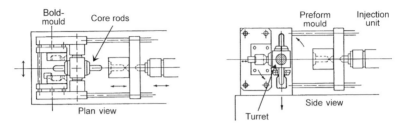

Figure 9.8 The Procrea system. Courtesy of Procrea srl.

punts, base inserts should be retractable as the mould opens, to prevent damage to the product or a distorted neck.

Injection moulding into a temperature-controlled 'hot' mould, under low pressure for most materials, produces a highly homogenous structure that will transfer strength into the even wall thickness of the final bottle and give increased stress crack resistance. The polished finish of the preform mould will also translate into a glossy, smooth surface finish on the container.

9.6   Procrea

The same basic principles of IBM apply to the Procrea system (see Figure 9.8). The injection mould closing system operates in-line with the core rods (preform mandrels). This offers the choice of either having a solid, compact and seamless preform moulding without split line or flash, or a split preform mould similar to the rotary table machines. The clamp force operates on the diameter of the preform. With a solid preform mould, all plastic has to be removable from the preform mould without deformation. Therefore, it is not possible to add external material on the preform to improve the product. The core rods are temperature controlled through the machine turret. The product is ejected from the core rods during head rotation using stripper rings. Tooling tends to be fully hardened.

9.7   Materials

PET is now widely converted by the ISBM process into bottles, containers and jars. PET can also be converted by the IBM process using resins with high/medium intrinsic viscosity (IV) resins. For processing PET it is critical that the core pin is cooled to allow easy ejection of the bottle. Because the IBM process allows less orientation of the preform compared to the ISBM process, less improvement in physical properties is realised.

PET and other polyesters require drying to reduce moisture in the pellets prior to injection moulding. Hopper dryers can be mounted directly on the machine. PET copolyesters can also be converted by IBM, where they are particularly suitable for moulding thick-walled containers with high clarity. These amorphous copolyesters and specially formulated polymers, such as 'Elegante' from Eastman Chemicals, are designed for containers for packaging cosmetics and personal care products, in thin- and thick-wall sections with high clarity and gloss. It is also possible to IBM PEN and PEN copolymers but these polymers are generally more difficult to process than PET.

9.8   Applications

The advantage of IBM is the quality of neck, surface gloss finish and ease of manufacture. Containers with small precision necks can be easily produced and there is no scrap. IBM is particularly suitable for the manufacture of cosmetics and personal care containers, for shampoo, lotions, facial and skin care products, bath and body care products and mascara bottles, and small bottles and containers for beverages and foods, including liquor miniatures.

9.9   Machine and process capabilities

The number of cavities that can be installed on a given size of machine depends on two main criteria: the ability to clamp the preform mould during packing without flashing the preform; and the number of blown bottles that can be fitted within the operational centre distances on the transfer table.

$$\text{Number of cavities} = \frac{\text{Preform clamp tonnage of injection blow machine}}{\text{Cross-sectional area of preform} \times \text{packing pressure of preform}}$$

If the preform is split along the length, this area would equate to the cross-hatched area in Figure 9.9, or the outside diameter of the preform for a solid preform. For larger necked containers (i.e. jars), this area can be reduced (see inset Figure 9.9) by the shaping of the core rod to increase the number of cavities that can be held.

Machines are designed to be highly versatile with the ability to change materials without changing the screw. The limitations of the process are as follows (see Figure 9.9):

1.   Ratio 1: if the core rod length (L) to diameter (d) ratio is excessive, the core rod can deflect during injection, causing uneven wall thickness in the final product.
2.   Ratio 2: similarly, the material can start to stretch unevenly if the outside bottle diameter (D) to core rod diameter (d) becomes excessive, giving an uneven wall.

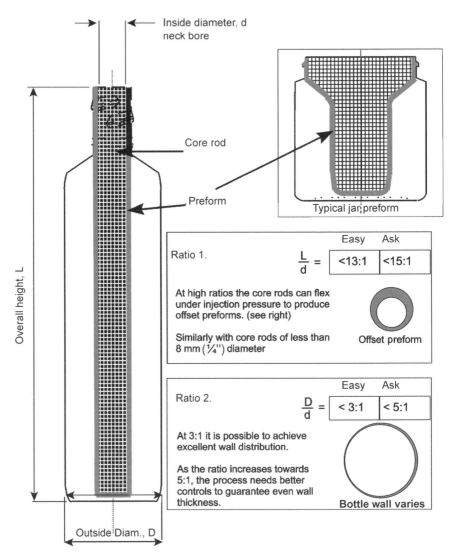

Figure 9.9 Ratios for round containers.

3. Ratio 3: additionally, for oval bottles (see Figure 9.10) it is necessary to compare the major to minor dimensions of the blown container. In extrusion blow moulding, it is possible to compression mould the neck finish from the parison, and get additional plastic at the major axis of the blown container. In injection blow, all material has to be designed within the preform (see oval preform, Figure 9.10). A normal preform thickness would be approximately 2.5–4 mm in order to retain sufficient

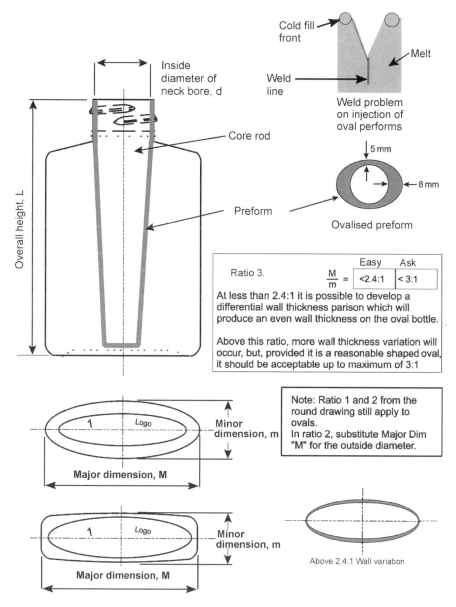

Figure 9.10 Ratio for oval bottles.

heat in the preform between the injection and blow phase of the process. In oval containers this can be increased to 5–8 mm before parison sag will make the process unstable. Handled containers with cut-outs are not possible in IBM because the core rod gets in the way; solid handles are possible.

Thin containers are also difficult. The preform must have an approximately 2.5 mm wall in order to retain heat for blowing. With heated core rods this can be reduced to 2 mm but the container must be an easily blown shape with no sharp shoulders and corners.

Table 9.1 summarises the differences between IBM and EBM.

Table 9.1 Comparison of injection blow-moulding and extrusion blow-moulding.

Injection blow-moulding	Extrusion blow-moulding

1. Bottle weight variation
  Approximately 1%
	1. Bottle weight variation
	  Approximately 3%

2. Wall-thickness distribution
 	
  Uniform wall thickness
  Thick-wall capabilities,
    but thin walls difficult to control
	2. Wall thickness distribution
	
	  Wall-thickness variation
	    approximately 10–20%
	  Both very thick and very thin wall
	    capabilities

3. Bottle finish, shape, size and outputs
 	
  Superior surface finish
  High clarity from fully polished
    mould surfaces
 	
  Ovality ratio limited to 5:1
  L/D ratio limited to 15:1
  Bottle sizes limited to 1 litre for
    commercial reasons
  Hollow handleware not possible
	3. Bottle finish, shape, size and outputs
	
	  Poor surface finish and score lines from die
	    and pin gap
	  Matt surface finish from die gap shear
	
	  Unlimited ovality ratio
	  Unlimited L/D ratio
	  Bottle size unlimited
	  Hollow handleware possible

4. Neck finish
 	
  Controlled external and internal
    neck finishes with injection
    moulding tolerances
  Fully finished neck forms
	4. Neck finish
	
	  Uncontrolled external and internal
	    neck finishes with blow-
	    moulding tolerances
	  Neck finishes may require post operation

5. Waste or scrap
 	
  No scrap during production
	5. Waste or scrap
	
	  20%–40% scrap during production,
	    requiring grinding and recycling equipment

6. Number of cavities
 	
  Greater number of cavities
  Slower cycle time
	6. Number of cavities
	
	  Lesser number of cavities
	  Cycle times slightly quicker

7. Automation
 	
  Process controlled and repeatable
  One operator for 4–6 machines
  Minimal ancillary equipment
    required for product handling
	7. Automation
	
	  Process variable
	  One operator for 1–2 machines
	  Extensive ancillary equipment required for
	    product handling

8. Floor-space efficiency
 	
  Machines are compact and require
    minimal ancillaries
	8. Floor-space efficiency
	
	  Machine is compact but requires
	    extensive ancillaries

Table 9.1 (continued)

Injection blow-moulding	Extrusion blow-moulding
9. Technician input	9. Technician input
Simple mould changes with fast setting times	Complex mould changes with longer setting time
Predetermined and repeatable process settings requiring no adjustment	Predetermined setting conditions, but requiring constant adjustment
10. Resin and materials	10. Resin and materials
Wide range of materials and resins can be processed	Limited materials and resins can be processed
11. Capital costs	11. Capital costs
Mould costs are high but with low maintenance and long life-span	Mould cost low requiring high maintenance and shorter life-span
Lower combined machine and mould costs for long production runs	Higher combined machine and mould cost for long production runs
Higher mould costs for short production runs	Lower mould costs for short production runs
Minimal ancillary equipment costs	High ancillary equipment costs

10 Hot-fill, heat-set, pasteurization and retort technologies

Bora Tekkanat

10.1 The hot-fill process

Hot filling is a well-proven and recognized method for filling high-acid foods (pH < 4.6) that will be shelf stable at ambient temperatures. At the present time, the method is used extensively in the food industry for filling in glass and plastic containers and in paperboard cartons. It relies on heat treatment in tubular or plate-type heat exchangers to temperatures in the region of 90–95°C for at least 15 s (typically 15–30 s). This process produces a 'commercially sterile product' by killing all microorganisms capable of growing in it. The product is then cooled and filled at temperatures ranging 82–85°C into containers, sealed immediately with closures, and then held at this temperature for approximately 2–3 min. The hot filling will sterilize the inner surface of the container. The filled containers are usually placed on their sides so that the neck-finish and closure are also sterilized. The containers are then cooled in a cooling tunnel in order to minimize thermal degradation of the product.

For some foods, the hot-fill process is used to reduce the microbiological content of the food so that a relatively short shelf-life is obtained in cold distribution. Another benefit may be to reduce the viscosity of the food when filled at high temperature. However, the main purpose of hot filling is to provide a product free from microorganisms capable of growing in it at ambient storage; that is, a commercially sterile product. Commercial sterility of thermally processed food refers to the absence of disease-causing microorganisms, absence of toxic substances and of spoilage-causing microorganisms capable of multiplication under a number of non-refrigerated storage and distribution conditions.

The properties of the food will largely determine the sensitivity to spoilage by microorganisms. Temperature and water activity are important factors that affect the microbiological spoilage of foods. Another factor of major importance for the growth and survival of microorganisms in foods is the pH or acidity. Low pH (high acidity) will restrict a number of microorganisms from growing and spoiling the food. It is, therefore, very important to know the acidity of the food when deciding the process conditions and the possible shelf-life of the product. Foods are often divided into two main groups in terms of their acidity: high-acid foods with pH 4.6 or lower, and low-acid foods with pH higher than 4.6.

There are some important reasons for the acidity classification:

- most spores that may survive heat treatment and chemicals will not grow or germinate at pH lower than 4.6
- pathogenic bacteria will not grow at pH lower than 4.6
- the sensitivity to heat treatment is increased at lower pH.

Consequently, microbial spores generally do not have to be inactivated in high-acid foods and these products do not represent a public health risk since pathogenic bacteria will not grow and multiply in such an environment.

However, there are some foods, such as tomato juice, with a relatively high pH (normally 4.1–4.4), that belong to the high-acid food group that can still grow specific spores. These products require a higher heat treatment and hot-fill temperature in order to become commercially sterile. Normal heat treatment is 115–120°C for 15–45 s. The product is then cooled to a hot-fill temperature of about 91–93°C. This treatment is necessary to kill acid-tolerant spores.

Generally, a hot-fill process can be safely performed with liquid foods with a pH of 4.0 or lower. Therefore, the foods that can be processed by a conventional hot-fill process without any modification (acidification) are fruits, fruit drinks such as various nectars, ice-tea (provided that the pH is lower than 4.0) and isotonics.

10.2 The heat-set process

Polyethylene terephthalate (PET) containers were first introduced in 1976 for carbonated soft drinks (CSDs) and were immediately favored over glass because of their numerous advantages, including their lighter weight and shatter resistance. These PET containers are generally made by a blow molding process, in which a previously injection-molded preform is reheated to a temperature above the material's glass transition temperature (T_g), then stretched and blown in a mold cavity to form the final container shape. Containers formed from PET in the shape of bottles, cans, jars and the like are used increasingly for other products, which are traditionally placed in the container at low or moderate temperatures. PET containers produced by the conventional molding process, however, exhibit extremely high thermal distortion, which makes them unsuitable for the packaging of products that require elevated fill temperatures.

One could easily make a quick demonstration of this phenomenon by filling any PET soft drink or water container with boiling water and observing how suddenly the container shrinks and distorts. It is fundamentally important to understand why this happens if one particularly wishes to stabilize a PET container for hot-fill applications. Why did this container shrink and distort?

When a preform is injection molded and cooled quickly in the mold below the T_g of PET, it is in its amorphous state. The morphology of the glassy polymers in their amorphous state (such as the injection-molded PET preform) is described as a random coil in which the molecules are intertwined, like a plate full of spaghetti. The molecules in this glassy state do not have enough energy to slip past one another and they are relatively immobile. Therefore, glassy polymers are heated above their T_g, to what is called the rubbery state to impart orientation. In this rubbery state, molecules have enough energy to slip past one another and have a greater degree of freedom to move under applied force. In its amorphous state, PET is a clear and rigid but brittle material; a proper orientation improves its toughness, impact and barrier properties.

During the blow molding process, PET preforms are first reheated to this rubbery state (above T_g), then stretched and blown in a mold cavity to form the final container shape. The type of orientation a container experiences during blow molding is called sequential biaxial orientation. A typical stress-strain curve, which represents this blow process at a particular reheat temperature, can be depicted as shown in Figure 10.1. The shape of this curve may vary somewhat, depending on the stretch temperature and rate, molecular weight of PET, and mode of extension. However, in general, one could divide these types of curves into three regions to explain the quite complex deformation phenomena taking place that will ultimately determine the final morphology of the blown container.

The elastic region extends up to the yield point, where most of the deformation is bond-stretching type elastic deformation, which is completely recoverable. Past the yield point, permanent deformation starts. In this region, molecules exhibit large viscoelastic movement that results from molecules slipping past one another. This leads to molecular alignment in the direction of applied force. During the stretching process, the weakest point of the preform yields first,

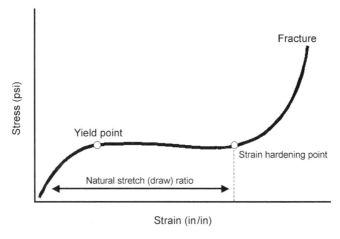

Figure 10.1 A typical stress-strain curve for amorphous PET drawn at 85°C.

which usually happens to be the hottest point of the preform. This plateau region extends up to the strain-hardening point. The level of orientation achieved in the plateau region is very temperature, rate and molecular weight sensitive. The strain-hardening point is reached when molecules can no longer pass one another due to established entanglements, without further increasing the applied load. Strain-hardening gives PET a self-leveling property where the preform is uniformly stretched along the entire length and the container ends up with a fairly constant thickness in the oriented parts.

For orientation to extend in this plateau region, it has to compete against a thermal relaxation phenomenon due to rotation and uncoiling of molecules. This implies that the strain rate has to be faster than the relaxation rate. The alignment of molecules in the plateau region gives rise to another phenomenon, called strain-induced crystallization (SIC). PET has a tendency to crystallize rapidly under the influence of stress. The reheat temperature of a preform is close to the crystallization temperature (T_c) of PET, and this temperature is further raised (sometimes 10–15°C) by adiabatic heating effects due to rapid straining. This creates an ideal condition for the formation of tiny crystallites towards the end of the plateau region. The start of the crystallite formation and molecular entanglements put an end to the plateau region and gives rise to force for further deformation to take place.

The stress-strain curve shows a steady increase in this third region where strain-hardening takes place. Strain-hardening is induced partially by crystallization and partially by the alignment of the polymer chain in the orientation direction. Further crystallization is enabled by trans to gauche isomerization in the amorphous regions. Further stretching and alignment create highly extended chains or 'taut-tie' molecules in the amorphous phase. These molecules are largely responsible for the improved mechanical properties of the highly oriented materials. Further stretching will lead to catastrophic failure as the covalent bonds in the taut-tie molecules fail.

The paragraphs above give a general description of what takes place during biaxial orientation of PET. This may change slightly, or sometimes drastically, depending on the following factors: temperature, strain rate, mode and amount of extension, type of PET and PET molecular weight. Therefore, a plastics engineer has these parameters or variables (among others) at his or her disposal to design and manufacture a functional container for the intended application.

When a blown PET article hits the cold surface of the mold, it starts cooling below its T_g rapidly while still being kept inflated with high air pressure against the mold. This rapid cooling stops all the previously discussed orientation and crystallization phenomena. It locks or freezes the article in the oriented structure but leaves the PET in a state of non-equilibrium.

The blown container is now clear and transparent and demonstrates improved physical and barrier properties. The crystalline and strain-hardened structure is largely responsible for these improvements. A container prepared in this way

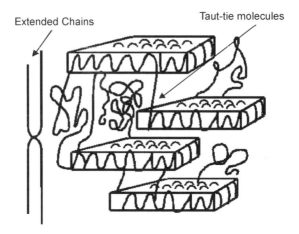

Figure 10.2 A schematic representation of oriented PET morphology.

may contain about 25% crystallinity. Depending on the above factors, the size of these tiny crystallites ranges 10–200 Å but is usually around 50–75 Å. The crystallites, which form in an oriented matrix, are referred to as 'microparacrystallites' (mpc) if one subscribes to the paracrystalline theory of polymers [1]. Since they are small, they do not scatter light and therefore the sidewall of the PET container remains transparent.

The final morphology and microstructure of the blown container is quite complex to model or represent schematically in a simplified form but Figure 10.2 depicts some of the attributes discussed above.

Addressing the question of the shrinkage and distortion of the soft drink container upon filling with boiling water: boiling water rapidly elevates the temperature of PET above its T_g and returns it to the rubbery state where molecules have enough energy and freedom to slip past one another. As this temperature is reached, the molecules are unlocked from that frozen-in structure. These molecules now demonstrate a memory effect. As soon as they have the freedom to move, all the highly oriented molecules in the amorphous region tend to coil back to their original random-coil configuration, which is a lower energy state. If we did not induce any crystallinity during the blowing process, it would be plausible to assume that most of the orientation in the amorphous phase would be recoverable and the container would shrink back almost to a preform in a true rubbery behavior. The presence of crystallites prevents this much shrinkage from happening; but, nonetheless, the container shows significant shrinkage and distortion. This shrinkage makes the PET containers produced by the conventional molding process (cold-mold process) unsuitable for the packaging of products that require elevated fill temperatures.

The success of PET packaging in CSD applications in the late 1970s and early 1980s already had plastics engineers considering how to stabilize the PET containers so that they could handle the hot-fill process. By then, the juice companies realized the advantages of PET containers and wanted to convert from heavy glass to PET, particularly in large, multiserve sizes. They wanted converters to develop a hot-fillable PET container and the converters accepted the challenge. This was not a task just for converters—the major equipment manufacturers also had to rise to the challenge.

The first converter to develop and commercialize a hot-fillable PET container was Yoshino Kogyosho Co. Ltd of Japan. In 1982, they commercialized the first hot-fillable PET container, a 51 g, 1 liter container for an orange juice application (see Figure 10.3).

Yoshino had a long history of technical interaction with Monsanto on blow molding. Monsanto, with Yoshino's collaboration, developed the hot-fillable PET container called 'Monopet'. The first commercial application of Monopet was a 64 oz container for the Ocean Spray Co. in early 1985 (see Figure 10.4). This technology and all Monsanto's PET assets were transferred to Johnson Controls Inc. Plastic Container Business in 1988. The strong technical relationship with Yoshino continued with Johnson Controls and grew even stronger with the acquisition of Johnson Control's Plastic Container business by Schmalbach-Lubeca in early 1997. As a result of Yoshino-Monsanto technical partnership, a new technology was developed which made PET capable of handling hot-fill applications. The process of making PET capable of hot filling is called the 'heat-set process' and the resulting container is called the 'heat-set container'.

Figure 10.3 First commercial heat-set container by Yoshino.

Figure 10.4 First commercial PET heat-set bottle in the US.

Heat-set technology is not actually new. Polyester fiber and film manufacturers have been using heat-set technology to stabilize their products for years. Container technologists simply had to reinvent it for container applications. The biggest challenge they faced was to overcome the uncontrolled shrinkage upon exposure of PET containers to hot-fill conditions. The PET containers had to be thermally stabilized and this was accomplished by the heat-set process.

The heat-set process requires the use of hot molds. The mold halves are usually set to 125–145°C and the mold base is set to 90–95°C. Heat setting takes place as the blown container encounters the surface of the hot mold while being constrained by high pressure air against the mold. The mold is hot and the container needs to be cooled down below its T_g before it can be removed from the mold and handled. Therefore, blow molding equipment suppliers for the heat-set process cleverly incorporated a circulating (balayage) chilled air feature by bringing it through the stretch rod. As the chilled, pressurized air expands into the container from the restricted holes on the stretch rod, it creates adiabatic cooling, which is quite effective. Heat setting requires a little longer in-mold residence time than conventional blow molding. The blow molding speed is usually about 850–1000 bottles per hour per mold for the heat-set process. A conventional blow process can be run at a speed of about 1000–1200 bottles per hour per mold or even faster, depending on the mechanical limitation of the blow equipment.

A fundamental understanding of the heat-set process is important. With a couple of exceptions, the complex deformation/orientation phenomena and the resulting microstructure and morphology described earlier in this section for the conventional blow molding process also holds true for the heat-set process. In the heat-set process, preforms are usually reheated to 100–110°C, about 10°C hotter than in the conventional blow molding process. This increased preform reheat temperature helps raise the crystallinity level to about 30% in a heat-set container as opposed to about 25% in a container produced by the conventional blow molding process. The higher reheat temperatures also move the strain-hardening point of PET to a higher draw ratio and, therefore, the level of strain-hardening experienced in the heat-set process is not as high as in the conventional blow molding process. The containers for heat-set applications do not have to be strain-hardened as much as CSD containers, as they are not used for pressure applications. It is important to design the heat-set preform such that it will yield a lower draw ratio and strain-hardening and therefore lower tendency for shrinkage compared to a CSD application. Therefore, when a properly designed heat-set preform is stretched and blown into a mold, it will have a very similar morphology to that described for the conventional blow molded preform but with perhaps fewer extended chains or taut-tie molecules.

When the blown container encounters the hot mold surface, it is still being constrained by the blow air. This short encounter of the container with the hot mold is long enough to modify the morphology in such a fashion that it can handle hot-fill conditions, at a later stage. Since the container is constrained, it cannot shrink but the extended chains and taut-tie molecules will still have a chance to relax in the amorphous phase. This relaxation and the increased level of crystallinity are the key reasons why a heat-set container will not shrink later as it encounters the hot-fill conditions. The level of relaxation (and therefore resistance to shrinkage) is absolutely time and temperature dependent. As the heat-set temperature increases, the onset of shrinkage temperature also increases. The temperature of heat setting and the extent of molecular relaxation (sometimes also referred to as 'annealing') determine PET's ability to withstand hotter fill conditions and higher temperature processes, such as pasteurization and retort.

Molecular relaxation and the ability of molecules to move around in the amorphous phase will lead to reordering of crystallites into smectic-like structures (layers or planes) and result in a higher level of crystallinity in a heat-set container. The increase in crystallinity is accomplished by refinement and growth of the existing crystallites, perhaps by formation of more new crystallites and by reordering of crystallites [2]. When the chilled air lowers the temperature of the container below its T_g, it stops molecular movement and freezes-in this modified morphology. This heat setting process can be considered as setting a new 'memory' effect to the material/container, so that when it encounters the hot-fill conditions it will 'realize' that it had already been through a more severe thermal exposure.

As important as the thermal stability of the container body is, it is even more important for the finish of the heat-set container to be thermally (and dimensionally) stable during the hot-fill conditions. The finish is the threaded neck section of a container or preform, where the closure is placed. When a PET preform is injection molded it will have some flow-induced orientation. One can observe this flow-induced orientation under cross-polarized light as colorful strain patterns. It is a recommended practice to minimize the flow-induced orientation as much as possible during the injection molding process but it is not possible to eliminate it entirely. Since the hot-fill temperature is above the T_g of PET, an amorphous finish may shrink, distort or ovalize upon hot filling. Depending on the amount of residual flow-induced orientation, this may cause capping problems or result in a higher than desired removal torque. Yoshino realized this when the first heat-set containers were being developed and patented the 'crystallized finish' concept.

Upon crystallization, the finish of the container becomes thermally and dimensionally more stable and can withstand higher fill temperatures and process conditions. Finish crystallization is a secondary operation performed after preforms are injection molded and before they enter the blow molder. During finish crystallization, the finish of the preform is exposed to heat to quickly bring the temperature of the material up to the optimum T_c while the body of the preform is protected from heat exposure. The finish is exposed to heat long enough to create the desired level of crystallinity and then cooled to an ambient temperature for the blow molding operation.

A fundamental understanding of the crystallization process is important. PET will crystallize in its crystallization temperature range, which is above its T_g and below its melting temperature (T_m). This can be accomplished by either heating it from the glassy state or cooling it slowly from the melt state to this temperature range. Crystallization rate and crystalline morphology will depend on how crystallization is carried out. The crystallization half-time curve as a function of temperature is presented in Figure 10.5. The shape of this curve and the magnitude of crystallization half-time will be affected by PET's intrinsic viscosity (IV) or molecular weight, co-monomer content and type, residual catalyst type and level, and moisture content. However, PET generally demonstrates the fastest crystallization rate at about 180°C.

The crystallization now being described is called the 'quiescent' crystallization and should not be confused with the SIC occurring during the stretch blow molding process, which takes place under an applied force. Quiescent crystallization takes place by either heating an amorphous PET, which has a random coil-type structure, from a glassy state or cooling it from melt state into the crystallization range. In this temperature range, the molecules have enough energy and freedom to rearrange themselves into superstructures called the 'spherulites'. Molecules begin to crystallize at sites of heterogeneity (nucleation site), where there is a catalyst or impurity for example. The polymer chains nucleate on the surface of this heterogeneity and lamellar structures grow outward by chain folding. The lamellae radiate out from the nucleus,

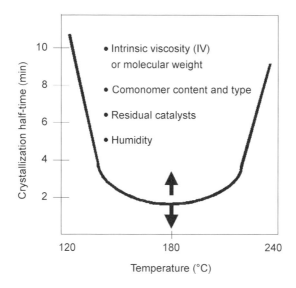

Figure 10.5 The crystallization half-time curve of an amorphous PET as a function of temperature.

Figure 10.6 A single spherulite of PET captured by transmission electron microscopy (TEM).

branching as they grow, to create the spherulitic superstructures. Spherulites grow until they impinge on the neighboring spherulites. A single spherulite of PET captured by transmission electron microscopy (TEM) is shown in Figure 10.6 [3].

The size of the spherulites can vary depending on whether the molecules are heated from the glassy state or cooled from the melt. On the left-hand side of the crystallization curve, when they are heated from the glassy state, the molecules have plenty of nucleation sites; however, they are limited by growth as spherulites rapidly impinge on their neighbors. These spherulites tend to be smaller. On the right-hand side of the crystallization curve, when they are cooled from the melt state, the molecules have much higher energy; however, they are limited by nucleation sites. As they nucleate, they grow into much larger spherulites that are up to several micrometers in size. Since the spherulites are quite large in size, they scatter the incident light. Therefore, the crystallized finish turns opaque and is recognized by its white appearance. Although the body of a heat-set container might have as high a crystallinity level as the finish (about 30%), because of the huge size difference between microparacrystallites and spherulites, the body stays transparent and the finish turns opaque in the same material.

The microstructural details of spherulites, inter-spherulitic interactions and crystallinity level are key factors in controlling the thermal (and dimensional) stability of a crystallized finish. Table 10.1 demonstrates that as the isothermal crystallization temperature of an amorphous PET increases, this increases the amount of crystallinity that can be achieved. A higher crystallinity level, in turn, increases the 'apparent' T_g of PET up to a certain temperature. This increase translates into the thermal stability of a crystallized finish.

The crystallized finish is therefore much more thermally and dimensionally stable compared to the amorphous finish and allows higher temperature applications. An amorphous finish can marginally function in the range 82–85°C for hot filling but requires well-controlled filling lines to eliminate higher temperature spikes. During the early years of conversion from glass to PET, PET

Table 10.1 Effect of degree of crystallinity on the 'apparent' glass transition temperature (T_g) of an amorphous PET crystallized isothermally at different temperatures

Crystallization temperature (T_c) (°C)	Degree of crystallinity (%)	Glass transition temperature (T_g) (°C)
–	Amorphous	75
88	0.07	76.5
89	0.14	79
90	0.24	85
100	0.29	88
105	0.30	89
120	0.34	92
150	0.39	91
200	0.54	85
230	0.61	84

containers were run on existing glass lines that lacked sophisticated controls. Crystallized finishes provided that added safety margin. Since the crystallized finish was, and still is, a patented feature by Yoshino, others have developed injection molded finishes using polymers with T_g higher than 100°C, such as polyarylate (U polymer) or polycarbonate (PC). These finishes would be used as inserts in PET preform injection molding. However, this approach did not become popular, probably due to the higher cost of the insert materials.

The importance of thermal stability of the body and finish of a heat-set PET container for hot-fill applications has been discussed. The process conditions are also important and determine the morphology and, in turn, the morphology determines the expected properties. Design is another important aspect of making a hot-fillable PET container, as discussed next.

When a 64 oz (1893 ml) liquid product (a typical multiserve size for juices) is cooled from a hot-fill temperature of 85°C to room temperature (25°C), it shrinks about 57 ml. If this liquid is hot-filled into a glass container at 85°C, capped immediately and then allowed to cool down to room temperature, it would generate about 17 inches of mercury (0.58 atmosphere of vacuum). This amount of vacuum would be fine in a glass container but it would definitely buckle or collapse a plastic container to an unacceptable extent, even if it was thermally stable. Therefore, the design engineers had to devise solutions to prevent this. They innovated a base design and panels at the sidewalls of the container that move inwards with vacuum as the product cools. As the panels and base move in, they provide a controlled deformation that allows the container to maintain an acceptable final shape. This movement also minimizes the final vacuum level in the container. The base is also thermally stabilized to prevent distortion. Most of these features that manage the vacuum and thermal stability in a heat-set container are patented.

Heat-set containers may still demonstrate some acceptable volume shrinkage upon hot filling. As a rule of thumb, this shrinkage should be limited to less than 1%. The shrinkage can be determined easily by measuring the overflow volume of the container before and after hot filling. Designing a volumetrically correct container is a challenging task [4]. Design engineers need to take into consideration the amount of expansion and shrinkage that both the container and product will display upon hot filling and cooling. By experimentation, the starting overflow capacity of the container at room temperature is determined prior to filling. When this container is hot filled, capped and cooled down to room temperature, it should hold at least the minimum amount of product stated on the label.

The containers produced need to be consistent, repeatable and able to withstand the rigors of the fill lines, transportation, storage and handling. The containers need to provide protection to the product for the intended shelf-life. The shelf-life and barrier properties of PET containers are extremely important. The author will not elaborate here on the subject with the understanding that the other chapters in the present volume will do so. Usually however a multiserve-size heat-set PET container will provide adequate shelf-life for most fruit-based products.

Another important component of the package is the closure. The finish of the heat-set container must be stable enough during hot filling so that when the container is opened for the first time, removal torque is within specified ranges. The reason for a high removal torque is usually the ovalization of finish. The closure needs to be functional for several openings and closings, and it must have a safety mechanism to indicate if the container has been opened or the seal tampered. It must also provide adequate barrier to gas permeation, especially at the top-sealing surface. The finish sizes of PET hot-fill applications have been limited up to 43 mm until recently. The 28, 33 and 38 mm finish sizes have also been used, and the 43 mm finish has been the most common in the US.

Heat-set PET containers are now well accepted by consumers for their clarity, light weight, convenience, recyclability and shatter resistance. Today, almost 80% of multiserve juice is packaged in PET. Historically, heat-set PET containers have been in cylindrical shapes with vacuum panels to reduce deformation. Recently, design engineers have been able to incorporate a grip feature that also functions as a vacuum panel. This is a convenient feature for multiserve containers (Figure 10.7). The growth in the heat-set arena is now focused on single-serve sizes, as more economical barrier technologies are becoming available.

Figure 10.7 A 64 oz heat-set PET container with a pinch-grip feature.

Figure 10.8 First tomato-based juice container in the US.

As mentioned previously, tomato-based juices require higher hot-fill temperatures than 85°C. Schmalbach-Lubeca has developed a container for tomato-based juices that can withstand filling temperatures up to 91°C. Campbell's Soup Co. commercialized this container for their V8 tomato juices in 1995 (Figure 10.8). In the US, this truly marks the first hot-fill application beyond 85°C, which was previously considered the limit. The heat-set technology that allows hot-fill applications beyond 85°C is referred to here as 'enhanced heat-set' technology. Tomato-based juice products also require improved barrier for shelf-life. Therefore, enhanced heat-set containers are slightly heavier than standard heat-set containers. The heat-setting process is also optimized to provide a higher crystallinity level. The combination of improved heat setting, slightly heavier containers and crystallized finishes provides a container that meets the better barrier and hotter fill requirements of tomato-based juices and other high-end hot-filled products.

Since the mid-1980s, Yoshino in Japan have been producing mineral water containers that are filled at 88°C, using a 'double blow' process. In the mid-1990s, again using the double blow process, they commercialized carrot and tomato-based applications filled at 93°C. The double blow process, as the name implies, uses two blow molding wheels joined by a shrink (or annealing) oven. In the first blow wheel a slightly over-sized article is blown. This article is then allowed to shrink to a size slightly smaller than the final container size in the shrink/annealing oven. Finally, it is blown to the container shape in the second blow wheel. The double blow process produces containers with excellent thermal stability. Most of the thermal stability is introduced in the shrink oven, where the article is allowed to relax and increase its crystallinity. Since the size of this article is just slightly smaller than the final container, the second blow operation does not introduce much orientation; and therefore the container stays thermally stable. However, as one might imagine, the capital cost of this equipment is more than twice the cost of standard heat-set equipment. The output of double blow equipment is limited by the output of the slower of the two wheels.

As soon as the 85°C barrier was broken, the next logical step after beverages for 'enhanced heat-set' technology was to target hot-fill applications up to 96°C for fruit and vegetable sauces. These products are traditionally packed in glass jars and cans. PET's attack on these products began with apple sauce. Apple sauce was an ideal product with to which begin the wide-mouth PET revolution. Because apple sauce is frequently positioned as a children's product, converting from glass to PET was a natural choice because of container weight, breakage and safety issues. Commercially launched in 1998, Knouse Food's 48 oz, Lucky Leaf, Musselman's, and Apple Time apple sauce brands were among the first commercial, wide-mouth PET packages on US shelves.

The conversion to PET brought many additional value-added benefits, such as handling and stackability. Before this container could be commercialized, however, preform design and mechanical process enhancements were necessary to address bottle geometry, pinch-grip characteristics and a higher base design. The bottle design also played a key role in filling and labeling speeds. The diamond pattern that appeared on the shoulder of the glass version was incorporated into the shoulder of the PET container. This not only helped the Knouse brands keep their aesthetic identity but also provided structural benefits. With the experience of apple sauce as a foundation, it was time to explore other food categories.

Campbell's Soup then launched a 32 oz wide-mouth PET container with a pinch-grip, containing two different tomato-based soups. One of Campbell's goals was to give consumers an alternative to the typical, last minute search for a meal. A common premise is that consumers have a tendency to go to the refrigerator first and the pantry second when making meal selections. The introduction of this container enabled soup to be readily visible and accessible in the refrigerator. Because of its resealability—a feature not shared by metal cans—it enables the consumer to pour out the desired amount and reseal the rest for later use. For soup, the biggest challenge was to design a bottle that could appropriately handle the amount of vacuum pulled as the soup cooled from its hot-fill temperature down to ambient temperature. Mold and preform design and process enhancements were implemented to achieve a desired solution.

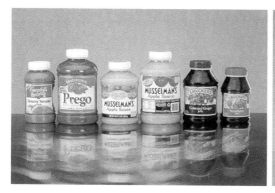

Figure 10.9 Food categories that are packaged in wide-mouth, heat-set PET containers.

Next in the marketplace was a club-store (or hypermarket) size jar—3 lb of strawberry jam or grape jelly. Although the hot-fill temperature requirements were not as stringent with jam and jelly, this category presented challenges of its own. Jams, jellies and preserves are more viscous; therefore, the product retained heat much longer. These products are also light sensitive. Another critical challenge was to get even material distribution throughout the sidewall, so that the container did not collapse under a heavy topload. Topload is very important for products that are going to be stacked several cases high in warehouses, through distribution and at the retail level. Again, the preform design, process conditions and design elements had to be optimized to produce a container that would meet the performance parameters.

Spaghetti sauces and salsas were the next categories to convert. All the conversions so far had 63 mm finishes and were in large size containers. Figure 10.9 presents all the categories that have converted to wide-mouth, heat-set PET so far.

What does the future hold for wide-mouth, heat-set PET containers? We will see more conversions for other product categories in large sizes but market growth will come from smaller sizes, such as 24 oz and smaller, as more affordable barrier technologies become available. We will also see larger sizes, such as number 10 can convert in the food service category. These containers might require wider mouths, such as 89 mm or 110 mm.

10.3 The pasteurization process

Pasteurization was invented in 1809 by a French chemist, Nicolas Appert, who realized that bacteria could be killed by heat without adversely affecting the taste and nutrients of food. Louis Pasteur, who gave it his name, later explained this process. Pasteurization is a general sterilization term for a process that utilizes the application of heat at atmospheric pressure, after the food or beverage product is filled and sealed in a container. The process is commonly used for high-acid foods, such as juices and pickles, and for carbonated beverages, such as beer and juice drinks. High-acid foods are filled into the container at about 60°C and carbonated beverages filled cold (>4°C). The containers are then capped or sealed and subjected to high temperature heat, usually at around 60–100°C depending on the application. The heat is usually applied in a 'pasteurization tunnel' by steam, hot water or a mixture of the two. The product in the center of the container has to reach a desired temperature (usually about 60–75°C), for a desired amount of time (usually about 5–10 min). After the heat treatment, it is essential that the product be cooled down promptly. The entire pasteurization process may take about 45 min to one hour.

The pasteurization requirements of a carbonated beverage and a non-carbonated food application should be discussed separately. In the case of a carbonated beverage, the container remains under pressure and therefore no vacuum panels are needed. The container becomes more pliable at the highest pasteurization temperature, at which point the internal pressure is highest. Containers that can withstand the rigors of pasteurization of a carbonated product are called 'heat and pressure resistant' containers. These containers have been produced by the double blow process. However, the heat and pressure resistant container applications have been limited to only a hemispherical shape with a base cup or footed base designs. Pasteurization of a container with a champagne base design, usually the base style beer producers prefer, still remains a challenge for the future.

Pasteurization of a high-acid food, such as pickles, in a plastic container poses somewhat different challenges. Cucumbers are placed in jars and the jars are filled with brine solution at about 74–82°C and capped, usually with steam flushing of the headspace. Therefore, the containers need to be vacuum resistant, as the cooling of the product to room temperature will generate considerable vacuum. The containers also need to be pressure resistant; when the container is exposed to high pasteurization temperature, the product will expand considerably. This expansion must be accommodated by the container, as the internal pressure will test the integrity of the seal. The cold spot temperature of cucumbers (pickles) needs to be raised to about 75°C for about 5–15 min, depending on brine viscosity and the type of pickle. To achieve this pasteurization at the cold spot, the containers are exposed to much higher temperatures, usually showered by a steam and hot water mixture for an extended period of time.

To accommodate the pressure and subsequent vacuum, the design of the containers becomes very important. The bottle geometry is also important, to achieve the desired cold spot temperature in the shortest time. However, nothing is more important than the thermal stability of the container. If thermal stability is not achieved, no matter how clever the design it will not work.

Schmalbach-Lubeca has developed a process called 'True heat-set™'. The True heat-set™ process, for which several patent applications are being pursued, is capable of providing containers with a thermal stability that can withstand the rigors of pasteurization and retort. The True heat-set™ process does not rely on expensive, double blow equipment to impart thermal stability.

As discussed in Section 10.2, the molecular relaxation and the increased level of crystallinity during heat-set processing are the key reasons why a heat-set container will not shrink subsequently as it encounters the thermal process conditions. The level of relaxation (and therefore resistance to shrinkage) is absolutely time and temperature dependent. As the heat-set temperature increases, the onset of shrinkage temperature also increases. The temperature of heat setting and the extent of molecular relaxation determine PET's ability to withstand higher temperature processes. True heat-set™ utilizes optimized process conditions to impart such a thermal stability. A 40–45% or even higher level of crystallinity can easily be achieved with the True heat-set™ process.

The thermal stability of a container can be quantified by hot filling it at different temperatures and determining the amount of shrinkage measured by its overflow volume before and after hot filling. Again, as a rule of thumb, this shrinkage has to be less than 1% for the container to be considered thermally stable. The thermal stability of containers produced by the 'standard heat-set', 'enhanced heat-set' and 'True heat-set™' processes are compared in Figure 10.10.

The hot-fill trials for the True heat-set™ containers were carried out with hot oil. Obviously, demonstrating a hot-fill capability does not make the container pasteurization or retort capable. However, it is a first and compulsory step towards achieving it. The thermal stability combined with unique design

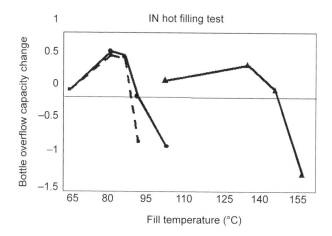

Figure 10.10 The thermal stability of containers produced by different heat-set processes. Key: ▬■▬, heat-set; ▬●▬, enhanced heat-set; ▬▲▬, True heat-set™.

Figure 10.11 First commercial, pasteurizable, wide-mouth PET jar.

features and appropriate closure systems can make PET containers pasteurization and retort capable. Such a pasteurization-capable container was developed by Schmalbach-Lubeca and commercialized by Vlasic Foods International Inc. for a test market of pickles in 2000. This marks the first commercial, pasteurizable, wide-mouth PET jar (Figure 10.11).

This is a 24 oz container with an 82 mm finish, which is a very convenient finish size for easy removal of pickles. It has panels on the sidewall which act to accommodate both pressure and vacuum. As the product expands during pasteurization, the panels move out to accommodate this expansion and therefore the container maintains a lower internal pressure. The panels then move in as the product cools and a vacuum is established. The container has a crystallized finish and champagne base. The champagne base is further stabilized to resist base roll-out during pasteurization. These containers can be run on existing pasteurization tunnels without any modification. The closure is an 82 mm plastic vacuum-holding closure with continuous threads and a thermoplastic elastomer liner.

10.4 The retort process

Retort is a thermal sterilization process for foods that require severe thermal treatment in a container for a period between its production and consumption in order to render the product shelf-life. The thermal process is carried out in a retort vessel that can be over-pressurized. Heating has significant effects on both the microbiological populations and quality attributes of the food. Important quality attributes are color, texture, flavor and nutritive properties that have varying sensitivities to heat. Vegetative microbial cells and spores present a major danger to public health safety in low-acid foods and they need to be destroyed during the thermal process. Obviously, to destroy all potentially harmful spores, such as those of *Clostridium botulinum*, the heat treatment is several minutes long. The destruction of spores can either be achieved at a high temperature for a short period of time or at a lower temperature for a longer time. The quality losses of food would be much less during high temperature-short time processing. Although the emphasis on quality attributes of thermally processed foods is increasing, the first concern must be for safety in processing.

For certain foods, the retort process can take as long as an hour, including the heat-up and cool-down times. Moreover, the product needs to be kept at a temperature, sometimes as high as 121°C, long enough (thermal death time) to obtain commercial sterility. Common reference temperatures are 121°C for low-acid foods and 100°C or 93°C for high-acid and intermediate or acidified products. Researchers refer to a 12D requirement in the retort process for the destruction of *Clostridium botulinum* spores.

Over-pressure conditions within retort systems are commonly achieved using steam and air mixtures, water immersion over-pressure, water spray over-pressure or water cascade over-pressure processes. Agitation of fluid and semi-fluid foods during thermal processing operations can increase the effective heat transfer rate throughout a product and reduce the time required to achieve a safe thermal process. With a shorter period of exposure to sterilizing tempera-tures, heat-labile attributes will degrade to a lesser extent and thereby result in improved product quality. Batch rotary retort systems are the preferred type for processing plastic containers. They are versatile and achieve forced convection heating of foods within containers by rotating entire product cars filled with containers.

An effective retortable container must provide seal integrity to prevent spoilage, and have oxygen- and water-vapor-barrier properties adequate to retard quality degradation. It must withstand the retort processing conditions and abuse during distribution but still be easily opened by the consumer. Because of the requirement for stability and barrier properties at high temperatures, relatively few polymers are suitable for retort applications.

Thermal processing of the retortable PET container poses many challenges for food processors, particularly in establishing appropriate pressure regimes to maintain container integrity during processing. It should be kept in mind that the seal strength of plastic containers is much weaker at high thermal processing temperatures. During high-temperature processing, the container must maintain adequate seal strength to assure protection of package integrity. Furthermore, since plastics become soft and pliable at retort temperatures, the differential pressure between the inside and outside of the container must be minimized to prevent container distortion as well as stress on the seal.

The important consequence of low seal strength during thermal processing is that very close control of pressure is necessary during the thermal process. If external pressures are allowed to drop, particularly at the onset of the cooling phase of the retort process, the high internal pressure due to the vapor pressure of the hot product could cause the package to lose its seal. For this reason, the pressure profile during the heat process for a plastic package must be carefully controlled.

A further consideration when specifying containers for high solids or conduction-heating foods is the container geometry. Product away from the center of the container will experience a more severe time/temperature treatment than the center point or 'cold spot' in the container. In order to design thermal processes on the basis of least lethal temperature history within the container, heat penetration tests should be carried out with thermocouples placed in that 'cold spot' location. As a consequence, most of the food within the package will be overprocessed in relation to the center point, with the degree of overprocessing being greater at larger distances from the center. It follows that slender containers provide an optimal shape for reduction of overprocessing of food within the package, while achieving an adequate thermal process at the center or mid-plane.

Schmalbach-Lubeca is working on the development of retortable PET containers utilizing the thermal stability that the True heat-set™ process provides, combined with unique design features and appropriate closure systems. Two different markets are being targeted with two different styled containers. The first is a 63 mm finish, wide-mouth container that is targeting retorted products, such as chunky fruits, soups with meats, and vegetables (Figure 10.12).

Figure 10.12 Wide-mouth retortable PET container.

Figure 10.13 Narrow-neck, single-serve, retortable PET container.

The second is a narrow-neck (<43 mm), single-serve container that is targeting retorted products, such as milk or cream-based drinks, alcoholic beverages, nutraceuticals, pharmaceuticals, dietary products and infant formula (Figure 10.13).

10.5 Concluding remarks

The value-added packaging that consumers demand is driving packagers, converters, equipment and resin manufacturers to work together to redefine the state-of-the-art PET technology. More and more frequently, today's consumer is asking for convenience, portability, clarity, shatter resistance, reclosability, shelf-life and recyclability. Food—one of the final bastions for glass—is where the battle cry has been heard most recently and most often.

The partnership between suppliers and packagers to create the next generation of PET containers may be even more evident in the future. As this decade continues, there will be further inroads made in food applications. The temperature and performance bar will continue to be set even higher, until such time as the differences will be minimized between PET and glass.

Shape and design will also be the watchwords for the next decade. Researchers will continue to push PET to achieve things it has never done before. And PET will continue to answer the demographics and lifestyle demands consumers place upon it.

References

1. Hoseman, R. (1975) *J. Polymer Sci.*, Symposium No. 50, 265-281.
2. Hoseman, R. (1973) *J. Polymer Sci.*, Symposium No. 42, 563-576.
3. *Polymer Microscopy* (eds. L. Sawyer and D. Grubb), Chapman & Hall, 1996.
4. Weissmann, D. *Proceedings of Bev-Pak 1992*, Sixteenth International Ryder Conference on Beverage Packaging, pp. 117-140.

11 Environmental and recycling considerations

Vince Matthews

11.1 Introduction

Recycling has been practised ever since humans first utilised raw materials to assist and shape their living and working environments. The main purpose was to save resources in terms of raw materials, time and effort. Recycling of industrial scrap/waste is a well-established practice, which generally contributes to reducing the overall costs of manufacturing operations by increasing conversion efficiencies and minimising the generation of disposable waste. This applies to all raw materials, including plastics.

Extending the practice to the recycling of post-consumer waste is a relatively recent innovation and initially concentrated on easy-to-recover materials, such as waste paper, board, glass bottles and large metal artefacts. Recycling schemes were usually organised by commercial groups, who could generate a reasonable financial return, offering incentives to participants who could provide the waste for reprocessing and return to the manufacturing and user chain. If the operation ceased to be profitable, then recycling would cease.

Modern manufacturing operations for commodity materials now make use of relatively large-scale plants and manufacture at low-cost and high-purity levels, which makes recycling a much less attractive financial proposition than in years past. This trend is very apparent with materials used for packaging, especially plastics, which are now extremely cost-effective. These advantages derive from inertness, high strength:weight ratios and judicious development of property combinations to maximise protection of the packaged goods. The growth in use of protective packaging has significantly reduced levels of product damage and wastage of the packaged product. However, this success has raised the disposal of increasing amounts of used packaging as a serious issue for national and regional waste management policy considerations.

The development of PET bottles to replace the heavier and more easily breakable glass variety, coupled with an early lack of facilities to recover and recycle the containers in a similar manner to glass, has undoubtedly contributed to public concern about the apparent waste of resource and rejection of the containers to landfills. Public concern is eventually manifest in political interest and related activity to regulate, control and, if possible, eliminate the cause of concern.

Recycling is seen as a well-established technique that can deliver environmental benefits in terms of reducing waste disposal to landfill, consumption of raw materials—especially from non-renewable fossil fuel resources like oil and gas—and, as a consequence, reductions in other atmospheric emissions. Not surprisingly, most environmental protection policies demand some form of recycling which is included in European Union (EU) and national legislation.

11.2   European environmental policy

The European Union has been particularly active with regard to environmental legislation and has provided guidance in the form of strategy papers and regular Environmental Action Programmes [1]. The broad target of the fifth programme was to promote a concept of sustainable development, which included integration of environmental and social measures in addition to economic factors when developing policy that affects communities. It claims some success in reducing emissions to the environment of toxic metals, acid gases and sewage, which have been directly associated to improvements in the general health of lakes, rivers and the atmosphere [2].

The sixth programme is designed to build on these successes and the priority issues, which are to form the basis of the programme, are selected as:

- limiting climate change
- nature and biodiversity—protecting a unique resource
- health and environment
- ensuring the sustainable management of natural resources and wastes

On the subject of waste management, the strategy will be to focus on waste-prevention initiatives. However, a majority of the waste still being generated will be expected to be recycled back into the economic cycle, helping to reduce society's demand for raw materials. *The aim is to recover and recycle waste that makes sense—that is, to the point where there is still a net environmental benefit and the recovery is practicable both economically and technically.*

11.3   EU Packaging and Packaging Waste Directive

In the 1980s, The European Commission (EC) initiated legislation designed to control the generation of packaging waste, in particular targeting beverage packaging. The directive, 85/339/EEC, covering containers of liquids for human consumption, was the result of these initial moves [3]. This particular legislation was withdrawn when more comprehensive legislation arrived in the form of the Packaging and Packaging Waste Directive, 94/62/EC, in 1994 [4]. It is this directive which is currently driving the initiatives to recycle post-consumer PET bottles. For comprehensive coverage of legislation on packaging, the reader should consult more specialised sources [5].

The Directive covers all packaging marketed in the EU and all household, commercial and industrial packaging waste. The key points from the directive are outlined below.

The aims of the Directive are:

- to harmonise national measures, so as to prevent or reduce the impacts of packaging on the environment of all member states and of third countries, and to remove obstacles to trade and distortion and restriction of competition
- to prevent the production of packaging waste and reduce the amount of waste for final disposal through packaging reuse, recycling and other forms of recovery

Member states must:

- take action to reduce the quantity and the harmfulness to the environment of materials and substances used and in general promote 'clean' products and technology
- ensure implementation of other preventive measures, such as collecting and taking advantage of packaging waste prevention initiatives
- set up systems to recover at least 50% of packaging waste and no more than 65% by July 2001, and recycle at least 25% and no more than 45% of packaging materials, with no material recycled at less than 15%
- notify the EC of measures adopted or to be adopted
- report on progress and set up national databases so that implementation can be monitored
- 'where appropriate', encourage the use of materials recovered from recycled packaging waste in the production of new packaging and other products
- ensure that by January 1998 packaging is allowed on the market only if it complies with certain 'essential requirements', which include minimisation of packaging weight and volume to the amount needed for safety and consumer acceptance of the packed product, and suitability for reuse, material recycling, energy recovery or composting
- limit heavy metals content to 600 ppm by July 1998, 250 ppm by July 1999 and 100 ppm by July 2001
- allow free access to packaging complying with the Directive

Use of recovery capacity outside a member state counts towards achievement of that member state's targets, provided this takes place on the basis of agreements and within EC rules.

Member states may:

- encourage reuse systems for packaging 'which can be reused in an environmentally sound manner', provided they do not conflict with the EC Treaty
- introduce economic instruments to implement the objectives of the Directive, provided they are in accordance with the principles governing EC environmental policy

- set themselves targets higher than 65% recovery, 45% recycling and a minimum of 15% recycling for each material, but only if they have appropriate recycling/recovery capacity and provided the measures taken do not distort the internal market or hinder other member states' ability to comply with the Directive. However, they must prenotify the EC, which must verify that the proposals will not constitute arbitrary discrimination or a disguised restriction to trade

Some countries have been allowed to extend the compliance dates and Greece, Ireland and Portugal may decide to set lower targets than those required for the other member states, but must achieve at least 25% recovery by mid-2001. By the end of 2005, they must however meet the targets laid down for the other member states to achieve by mid-2001.

It is important to understand that the target of 15% by material does not refer specifically to PET but covers all plastics, including PET.

11.3.1   Compliance with the Directive's targets

It is the responsibility of each of the member countries to institute local legislation, which will deliver compliance with the Directive's demands. To assist with the development of the legislation and resolution of issues in application of the Directive to particular types and sections of packaging, a committee of national civil servants, chaired by an EC official—The 'Article 21 Committee'—decides how to deal with any problems. For example, exemptions from the heavy metal limits (e.g. for recycled materials and materials in closed loops) and any necessary adaptations to scientific and technical progress.

The Committee is examining member states' practical experience in implementing the targets with the objective of fixing new targets for the second five-year phase, which are likely to be higher than the present targets.

New European standards on criteria and methodologies for evaluating packaging are required, for example: Life Cycle Assessments (LCAs)—a technique for auditing the environmental benefits of systems that can be applied to recovery and recycling procedures; methods for measuring and verifying the presence and release into the environment of heavy metals and other dangerous substances in packaging and packaging waste; criteria for minimum recycled content in appropriate types of packaging; criteria for recycling methods; criteria for composting methods and produced compost; and criteria for the marking of packaging. More specific definitions are required to encompass all possible recovery methods, for example: Is energy recovery to be classed as recycling? What specific factors or quantities will be the most appropriate to use for the calculations of compliance?

The above paragraph illustrates the extent of the developments required to enable complete implementation of the Directive and achievement of its aims.

Two important appendixes describe an 'Identification system' and the 'Essential requirements' for compliance.

The importance of these appendixes is often lost to many smaller companies, who struggle with understanding the overall aims. Trade associations are an important source of interpretation and should be used to clarify any doubts in the minds of users of packaging products. They all issue guidelines to the essential requirements, which are available to members [6].

Two examples illustrate their importance:

- All packaging entering the EU needs to be coded or marked to identify the material used and whether it is reusable or recoverable
- All packaging should also fulfil the essential requirements—one of which is that the packaging should be constructed in such a manner that it can effectively be recycled in 'large enough quantities' by the current standard 'recycling' systems used in Europe

This last paragraph is an interpretation, which really says non-recoverable packaging is no longer acceptable.

11.3.2 Material identification

The EC Decision on Material Identification, 97/129/EC, lays down an identification system for all packaging. The numbers and abbreviations for material identification should appear in the centre of, or below, the 'recyclable' or 'reusable' markings (e.g. PET can be coded 'PET 1') (see Figure 11.1). This type of coding has been derived from an earlier identification system, developed by the Society of Plastic Industries (SPI) in the US to identify plastics materials, and is often referred to as the SPI code.

Use of the numbering and abbreviations will be voluntary at first but may be come mandatory later if agreed. The marking symbols have been generally accepted and are now used widely. Many composite plastic materials will prove contentious and difficult to classify as one specific material. Suppliers of such

Figure 11.1 European Commission identification of packaging material.

packaging will need to evaluate the recyclability of the product and declare how it will be managed after use.

11.3.3 CEN standards and the essential requirements

CEN (Comité Européen Normalisation) the European Committee for Standardisation [7], has prepared a set of standards to guide users along a route to compliance with the Essential Requirements of the Directive. The CEN proposals are now complete but have not yet been accepted.

In view of the enormous range and complexity of packaging types and the recovery and the disposal situations which have to be taken into account, CEN has opted for a management system approach aimed at ensuring a continuous effort to improve the environmental profile of packaging. A checklist procedure will ensure that decisions take account of the often conflicting social, environmental and economic factors affecting the choice of packaging.

In addition to five mandated standards (on prevention, reuse, material recovery, energy recovery and organic recovery), CEN experts have prepared an additional draft standard on requirements for the use of European standards in the field of packaging and packaging waste, the so-called 'umbrella standard'. This is the key document explaining how the interlocking mandated standards and reports are to be used.

The CEN thinking is that by providing practical guidelines on how the Essential Requirements can be interpreted and implemented, the new standards will ensure that packaging designers and specifiers keep potential environmental improvements under continuous scrutiny, as well as giving added value in developing the European Single Market for packaging and packaged goods.

11.3.4 CEN 'umbrella standard' (pren 13427)

This 'umbrella standard' has been introduced to establish an overall methodology, which guides users through the five main standards indicating which are applicable to each type of pack. Compliance with the Essential Requirements will require the supplier to assess all his products against each standard before placing them on the market. It is possible that some of the standards may conflict with each other: hence, the umbrella standard to ensure harmony. For a full understanding, the reader should consult more specialised descriptive texts or the drafts themselves [5].

A summary of the assessment results must be prepared. The supplier must retain records of the assessments and supporting documents for at least two years after the relevant packaging has been placed on the market for the last time. These records must be available for inspection.

The supplier is recommended to apply these principles as an integral part of his formal management system in order to improve the environmental performance of his operation and to provide the opportunity for continuous improvement, for example by incorporating the procedures into an existing EN ISO 9000 I 14000 scheme [8].

The five standards (drafts) are:

- Prevention (prEN 13428)
- Requirements For Relevant Materials And Types of Reusable Packaging (prEN 13429)
- Material Recycling (prEN 13430)
- Energy Recovery (prEN 13431)
- Organic Recovery (prEN 13432)

The most relevant standard for PET packaging is that dealing with material recycling. It covers all forms of packaging and types of packaging material, and all collection and sorting arrangements and recycling facilities. It formalises a procedure by which design, production and use of packaging can be checked against the requirements of various material recycling systems. For example:

- check recyclability of the materials from which it is produced
- check that the design of packaging makes use of materials or combinations of materials which are compatible with known and relevant recycling systems and technologies—or potentially compatible if in development
- check if the presence of substances or materials is liable to have a negative influence on the quality of the recycled material
- check that the design of the primary packaging (e.g. its shape, design and location of the opening) will enable emptying of the packaging, so that the used packaging is compatible with the recycling process

Material identification should be recognisable so as to facilitate clear and unambiguous identification of the predominant material. This may help the packaging user by indicating a disposal option, or may facilitate collection and sorting, or the aggregation of materials into recycling streams. However, the nature of some materials is clear without the need for further identification, and recognition may also be assisted by means such as colour or container shape.

Considering all these standards, and the complexity of certain types of packaging, there may be good opportunities for specialist recyclers who can sort, separate and recycle the more complex types of packaging. Organisations dedicated to recycling of specific materials have been active in promoting container design principles compatible with current recycling practice and they issue instructive guidelines, for example see Figure 11.2 [9].

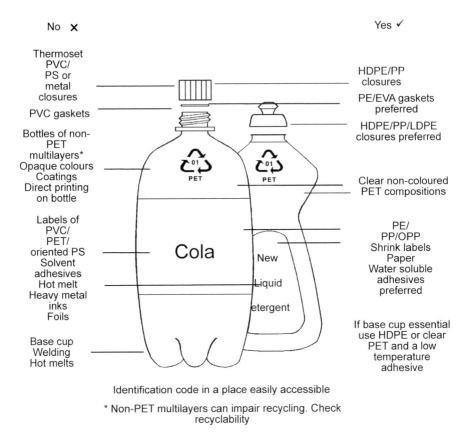

Figure 11.2 Design for recycling: Source PETCORE (PET Container Recycling Europe) [9] (redrawn image). Abbreviations: PVC, polyvinyl chloride; PS, polystyrene; PET, polyethylene terephthalate; HDPE, high-density polyethylene; PP, polypropylene; PE, polyethylene; EVA, ethylene vinyl acetate; LDPE, low-density polyethylene; OPP, oriented polypropylene.

11.3.5 Implementation of the Directive in the member countries

The EU Member States have slowly enacted the required legislation within their own national law. The relative pace of implementation and progress towards achieving the targets has been widely different, depending upon culture and historical structures already in place dealing with waste management issues.

In general, the producers/packers/fillers/importers/distributors of packaging have been mandated with the responsibility of achieving compliance. They have formalised the process by creating national organisations to develop procedures to collect and recover the required amounts of packaging. The same organisations also procure the finance. Funding is primarily via a 'Green Point Fee', a levy per package paid to the recovery organization, which takes up the legal obligation to recycle the package. The funds are then used to build up an infrastructure to ensure collection and recovery over the longer term. However, other structures operate quite successfully without the need for a formal Green Point Fee, for example in the Netherlands and Italy.

A list of the relevant country organisations that can provide more comprehensive details on how they operate is given at the end of this chapter. For reports covering all European systems see reference [5].

Collection and 'valorisation' of PET containers is an integral part of these systems. Valorisation, in this context, means recovering the full value by whatever technique is considered the most acceptable, when evaluating both economic and environmental factors. Recovery of the intrinsic energy content of the plastic material is also given a weighting. Modern 'waste-to-energy' facilities, which utilise heat for industrial use or local heating, can have much higher efficiencies and, equipped with gas scrubbers, much lower emissions than older waste incineration plants. Switzerland, for example, recovers 13.2% of its plastics waste by mechanical/feedstock recycling methods and 77.9% by energy recovery. The Netherlands recovers a similar ratio of 14.3:57% and Denmark 9.7:74.7%. By comparison, the UK ratio is 7.9:10.5% [10].

Use of refill and/or deposit systems in most northern Europe countries and absence of such systems in some of the countries who are large users of PET, such as Italy, France and the UK, creates two different recovery processes adapted for local circumstances. The deposit systems produce high recovery rates (> 80%). However, deposit schemes are costly to maintain and any environmental benefit should be evaluated against other schemes, with economics, free trade restrictions and social benefits included as judgement criteria. Very recent studies [11] indicate that cost savings delivered by the use of non-refillable packs, on balance, outweigh the small environmental gains by a large factor. The Austrian collection agency (ARA) concluded that deposits, or other penalties in support of refillable systems, are becoming irrelevant in the light of this research and there is an increasing trend towards non-refillables; the study is subject to peer review. For a more detailed discussion of the effects of deposits see [12].

11.4 Collection and recovery procedures

A brief description of the collection and recovery procedures is given here but for a more comprehensive review covering all aspects of starting a system, schemes available, promotion and case studies see [9]. The PETCORE guide to collection of plastic bottles is the main source of the data included here.

All the collection systems described fall into one of three main categories: drop-off programmes, kerbside collection systems and deposit refund systems. National preferences for a specific system are again related to historical custom and practice and which one to adopt often depends on local circumstances—for example, potential volume/resident, demographics/population density.

11.4.1   Drop-off schemes

Drop-off schemes are the easiest to set up. Also known as 'bring' or 'bank' schemes, they are mostly run by local authorities, supermarkets and, to a lesser extent, NGO voluntary organisations (non government). The public deliver their recyclables to a central site. Convenience, security and vehicle access/parking are the key success factors. The receptacles must have sufficient capacity for the area, be easy to use and discharge, and substantial/secure enough to discourage vandalism. Typical containers include igloos, front-end loading box-type containers, wheel bins, net cages—useful for self-classification of the waste—and large volume containers or 'skips'. Well-run schemes can achieve recovery levels in the region of 40–50%. Economics are estimated at 140–250 Euro/tonne for efficient plastic bottle collection.

11.4.2   Kerbside collection schemes

Kerbside schemes usually collect directly from the household and are popular because they are convenient. Recovery rates are higher if containers are provided and achieve levels of > 60%. Recyclables are often sorted at the vehicle, which can simplify subsequent operations at a material reclamation facility (MRF) or recycler, in effect saving capital equipment costs. However, moves to larger schemes have a preference for less sorting at the vehicle and devote more resources to larger MRFs with sophisticated automatic sorting facilities. Many combinations—for example, vehicle design, mix of recyclables and frequency of collection—are constantly evaluated but support the trend. Net costs of < 7.5 Euro/household/year are recorded but the variation ranges 3–29 Euro. Some Canadian schemes report net costs lower than equivalent costs of waste disposal.

11.4.3   Reverse vending machines

This technology is specifically designed for beverage containers and the machines are mainly utilised in regions that have used traditional deposit systems, based on consumer returns back to the vendor. In place of the vendor issuing a monetary or token deposit, a machine is installed to complete the transaction automatically 24 h/day. The machine accepts the container and, if the required identification is positive—that is, bar code, material sensor, video image or similar detection device—the container is transported to a storage section for later treatment or collection and the consumer is rewarded, usually with a token to exchange for goods. The costs are notably higher than conventional bring or kerbside schemes, *ca* 310–440 Euro/tonne, but they target more specifically and are claimed to return higher yields.

11.4.4   Other systems

Switzerland operates a very successful used PET container collection scheme by use of a pre-disposal fee per container, paid by the filler/distributor. The money is then used to organise a countrywide placement of collection bins, together with widespread publicity and promotion. Recovery rates of > 80% are claimed [13].

11.4.5   Plastic bottle sorting

The local or regional authority MRFs are usually the first to sort and concentrate specific containers such as PET bottles. However, to enhance quality and avoid specific contamination harmful to the outlets supplied, the eventual recyclers will effect a more careful separation. Separation facilities use similar techniques to those in MRFs but the recyclers often include more sophisticated automated techniques. Manual sorting and separation of bottles is based on colour, shape or product recognition. If bottles made of different materials are the same shape and/or are used for similar products, confusion is inevitable. The materials recognition code (see Section 11.32) is useful for the identification of new bottles on first sorting but 'look alike' bottles will not be subjected to such close scrutiny and will not be sorted effectively; therefore, automatic sorting using some specific property of the material is essential. The chlorine atom in PVC or a spectroscopic response, in addition to colour and opacity will be recognised by a sensor (X-ray, NIR, visible light), which will activate an air jet to remove the identified bottle. Automatic sensing and sorting is far more reliable, faster and more efficient and yields better quality but is more expensive. A number of companies specialise in this type of business and will supply equipment for sorting specific wastestreams. Single devices that identify at least three properties are now replacing multiple single-property sensors; two devices can then sort into nine different streams. The effect on productivity is significant, raising throughputs by a factor of two to three [14].

11.5   Plastics packaging waste

The regulations in force to guide the waste management practice of used plastics packaging do not single out PET materials for specific consideration and PET is included in the overall classification of plastics. The Association of Plastic Manufacturers in Europe (APME) [6] has been the main source of statistics on recovery and recycling of all plastic materials and publishes annual reports on the amounts recovered and recycled. Table 11.1 is adapted from this report to illustrate the recycling of the various plastics and the position of PET. The latest report covers data for 1998 [10].

Table 11.1 Plastics recycling in Western Europe—source Sofres 1998 data

	LDPE/ LLDPE	HDPE	PP	PVC	PS	PET	EPS	Other thermoplastics	Thermo-sets	Total
Total plastics consumption (K tonnes)	6526	4473	5003	5119	2087	1305	767	2679	2422	30381.0
Plastics consumption for packaging (K tonnes)	4210	2826	2389	769	959	1229	210	0	0	12592.0
Actual amounts available for recycling[1] (K tonnes)	3577	2161	1781	605	750	1019	194	0	0	10087.0

By mechanical recycling[2]

	LDPE/ LLDPE	HDPE	PP	PVC	PS	PET	EPS	Mixed plastics	Feedstock recycling[3]	Total
Plastics packaging recycled (K tonnes)	644	285	87	15	42	170	32	55	361	1691.0
% of consumption	15.3	10.1	3.6	2.0	4.4	13.8	15.2			13.4
% of actual waste available	18.0	13.2	4.9	2.5	5.6	16.7	16.5			16.8

Energy recovery[4]	2218.0
Total recovered	3922.4
% of consumption	17.6
% of available waste	38.9

[1] Sofres use a formula based on the life-cycle of the end-use to calculate the amount becoming available as waste in any one year.
[2] Mechanical recycling refers to material returned to the economic cycle without chemical change.
[3] Feedstock recycling refers to material recovered as monomers or used as a raw material source, excluding use as fuel.
[4] Energy recovery excludes simple incineration; it is largely modern waste-to-energy facilities with some recovery in cement kilns.
Abbreviations: LDPE, low-density polyethylene; LLDPE, linear low-density polyethylene; HDPE, high-density polyethylene; PP, polypropylene; PVC, polyvinylchloride; PS, polystyrene; PET, polyethylene terephthalate; EPS, expanded polystyrene.

Table 11.2 Recovery rates for PET containers in Western Europe

	1993	1994	1995	1996	1997	1998	1999	2000
Consumption WE[1] (K tonnes)	560	700	800	840	1050	1200	1500	1700
PETCORE gross collection[2] (K tonnes)	25	33	45	75	110	175	220	280
% of consumption	4.5	4.7	5.6	8.9	10.5	14.6	14.7	16.5

[1] Based on APME data, annual report publications.
[2] PETCORE data, gross collection figures for PET containers.
Abbreviations: APME, Association of Plastic Manufacturers in Europe; PETCORE, PET Container Recycling Europe.

The waste available calculation estimates the amounts of packaging which will have a life-cycle longer than one year and will therefore become available for recycling as waste in later years. The table shows that overall mechanical recycling of plastics amounts to about 13% of consumption; however, assuming the waste available figure is correct, it rises to nearer 17% based on actual available waste, very close to the target set in Directive 94/62/EC. The contribution from the recycling of PET containers, at about 17%, is significant to the overall figure. The low-density polyethylene (LDPE) figure is largely from recycling film used as shrink-wrapping for pallets.

Table 11.2 presents PET container recovery figures in more detail. The figures may be slightly different from those shown in Table 11.1 as they are from different sources. However, Table 11.2 shows that the collection rates are increasing faster than consumption and that PET is likely to continue to make a significant contribution to plastics recovery statistics in the foreseeable future.

Various estimates indicate that consumption figures in the year 2005 could be as high as 2.6–2.7 million tonnes. If recycling rates just keep pace with this growth it would release about 500 K tonnes into the market for PET. With a similar growth in collection volumes, it is not inconceivable that collection volumes could reach 800–900 K tonnes, recording recycling rates of about 30%. Historically, recycling capacity has kept pace with the increased volumes, as illustrated in Figure 11.3. These recycling figures are indicative only because factors such as exports and imports of filled containers are not taken into full consideration when they are compiled. The variation in trade patterns will make it difficult to assess actual consumption and available waste volumes with full certainty; this makes accurate compliance with fixed recycling targets difficult. More information on PET container recycling can be obtained from PETCORE [9]. Similar data for the US are published by the National Association for PET Container Resources (NAPCOR) [15]. NAPCOR figures are calculated a little differently and they make determined attempts to assess the true volume of used PET containers available for recycling, which takes into account imports and exports of containers. Table 11.3 presents NAPCOR data for 1999.

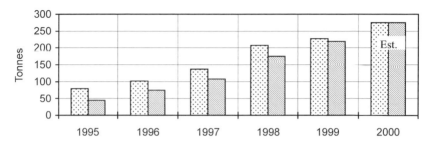

Figure 11.3 Recycling of PET containers in Western Europe, 1999. Source PET Container Recycling Europe (PETCORE). Key: ⊡, total recycling capacity (input); ▨, amount collected (gross).

Table 11.3 Gross PET bottle recycling rate 1999—source National Association for PET Container Resources (NAPCOR)

Year	Total US Material recycled (R-PET) (MM lbs)	PET bottles on US shelves estimated as available for recycling (MM lbs)	Gross recycling rate %
1995	775	1950	39.7
1996	697	2198	31.7
1997	691	2551	27.1
1998	745	3006	24.8
1999	771	3250	23.7

Abbreviation: R-PET, recycled polyethylene terephthalate.

11.6   Recycling of PET

Historically, PET manufacturers have recovered their own process waste because the intrinsic high value of the material provided an economic incentive. Depolymerisation processes, such as methanolysis and glycolysis, were operated alongside simple melt incorporation of clean process scrap. Technology improvements and larger manufacturing plant led to improved conversion efficiencies and the economics of scale gradually reduced the amounts of internal waste and, as learning was reflected in lower virgin prices, the external market price of all PET grades.

It became economically less attractive for the manufacturers to reprocess waste internally. Recovery therefore became more of a non-essential business and attracted the attentions of smaller entrepreneurial companies who created sufficient business by reprocessing waste from several of the mainstream manufacturers. The technology used was based on standard mechanical recovery techniques, such as grinding/washing and extrusion filtration to regenerate pellets.

Improved purity and consistency of intermediates, together with the development of automated process control techniques, moved manufacturing technology for virgin polymer increasingly towards continuous manufacturing capacity and a consequent reduction in the more expensive but versatile batch capacity. These trends supported the external growth in the business of reprocessing on a smaller scale, primarily existing on good quality waste from large manufacturers. Availability, type of waste and purity were usually reliable.

The 1980s was the decade of growth for the grades of PET used in packaging, particularly for containers, and a European consumption of around 500,000 tonnes in 1993 generated a whole new source of high-quality PET for those companies whose businesses were based on material recycling by melt processing. The sheer volume of used PET containers that promises to become available is now sufficient to re-examine those recovery processes which are more volume sensitive, namely glycolysis, methanolysis and hydrolysis.

11.6.1 Principal methods of PET recovery

PET can be recovered and reutilised by a wide variety of methods which fall into one of three categories: melt reprocessing, breaking down by depolymerisation to precursors, purification and repolymerisation, and as a last resort recovery as energy. These methods follow a general hierarchy of plastics recovery systems, simplified for illustration in Table 11.4. A number of factors need to be considered when deciding on recovery of appropriate PET waste.

Table 11.4 Suitable recovery processes for waste PET materials

Degree of contamination	Recovery process	General economics	Process convenience	Typical feedstock
Low	Washing and remelting Partial glycolysis	Satisfactory	Simple	Refillable PET/ One-way PET bottles Uncoated heavy gauge film
Medium	Glycolysis Methanolysis Recovery of TA and EG	Increasing costs	Increasing complexity	Generic fibrous waste 100% Coated PET Coloured PET Mixed 'barrier' PET containers
High	Energy	Well-established costs	Relatively convenient	Laminates, coated thin-gauge films, complex composite PET materials

Abbreviations: TA, terephthalic acid; EG, ethylene glycol; PET, polyethylene terephthalate.

11.6.2 Contamination issues

Physical contamination at the macro-level (i.e. with dirt, glass fragments, stones, grit, soil, paper, glues, product residues and other plastics like polyvinylchloride (PVC) and polyethylene (PE) used as labels, seals and closures) can usually be removed by simple washing techniques. However, embedded, ingrained soil caused by abrasion or grinding (e.g. during baling, transport or handling in poor storage conditions) will not be removed by simple techniques and will need special treatment, such as filtration, to ensure removal. Physical contamination at a micro-level is more difficult to remove, especially if sealed in with glues or embedded by abrasion or impacting. Impurities of this nature create weak spots in the form of 'blebs', 'blobs', 'fish eyes' and black specks, which lead to problems during reprocessing, such as excessive breaks in fibre spinning. The consequence is loss of quality, loss of productivity and excessive costs.

Oils, fats and greases need more detergent and, although leaving no residual quality problems, can necessitate additional treatment with wash waters before disposal.

Chemical contamination occurs by adsorption of contents like flavourings, essential oils or similar ingredients used in the product formulations. Adventitious contamination can be introduced by use of the container for purposes other than the original intention—for example, storage of pesticides, household chemicals or motor and fuel oils. Complete removal will require desorption, which can be a slow process and therefore reduces productivity; however, these occurrences are few and are not known to cause many problems during subsequent reprocessing, i.e. into textile fibres. For other uses, appearance and odour are important. The increasing demand to look at recycled-PET (R-PET) for food contact end-uses raises the importance of the quality issue by several factors.

Developments in PET container technology are now combining materials with the PET to improve the 'barrier' properties in order to prevent loss of carbon dioxide, ingress of oxygen and protection against UV light. These newer container types will complicate conventional recycling operations and demand specific treatments prior to application of conventional procedures or new recovery techniques altogether. The intended use of R-PET often determines the feed purity requirements.

11.6.3 Recycling of PET by mechanical methods

Mechanical recycling is the simplest and currently the most effective method of recycling used PET containers. Mechanical recycling, in this context, is defined as: the reprocessing of the PET waste material for the original purpose or for other purposes without changing the chemical structure of the processed material. The clean material is either used 100% or selectively blended with virgin to produce the desired properties and then processed directly in a similar manner to the virgin resin. The quality of the recycled stream determines end uses and degree of blending required for meeting any specification. Key requirements are molecular weight (intrinsic viscosity, IV), colour and the level of physical contamination.

Equipment for preparing and cleaning the waste PET before returning it back into the user chain is relatively simple in function and is freely available from a number of specialist manufacturers [16]. The description given here is primarily geared to handling of post-consumer PET bottles but with relatively simple adjustments and additions can handle most types of PET waste, including fibres, films and composites.

Table 11.5 illustrates all possible stages that presently occur during the process commonly referred to as 'recycling of PET bottles', in order to mechanically clean the used PET to a level satisfactory for all applications—including return to food-contact end-uses. It is important to include all possible operations if there are to be assessments of the true environmental costs of the overall recycling process. All of the operations could be completed by a single recycler. However, in actual practice, different operators handle various aspects of the process—for example, some local authorities will manage all the sorting and separation of bottles in their own MRF, with the costs supported by 'green point fees'. However, most MRFs operate multi-material sorting (i.e. glass, metals and paper) and the sorted 'plastic' bottles need additional sorting.

11.6.3.1 Prewashing and granulation

Washing before granulation removes loose contamination and prevents 'grinding in' of dirt that can be difficult to remove in subsequent stages. However, dry grinding and air classification to remove paper and some 'light' contamination, such as thin labelling or even micro-thin barrier layer materials, can reduce the load on separation at later stages. Excessive and strong glues affect granulation and process efficiencies by preventing separation and increasing the weight of the 'lights'.

11.6.3.2 Washing

The mainwash is a key part of the process and can be operated at ambient temperature or temperatures up to 80°C. It will almost certainly use a detergent or some caustic alkali promoter, depending upon the degree of contamination or the prior contents of the containers (e.g. oils or salad dressings). Labels, glues, inks, caps, inserts and so on affect efficiencies. Water insoluble glues become soft and pick up soil and grit that stick to the PET reducing the quality, and should be avoided. Poor quality inks can colour wash water, which taints the R-PET. Cheap papers disintegrate easily, blocking filters.

Table 11.5 Process stages during complete mechanical recycling of PET bottles

Process activity	1	2	3	4	5	6	7	8	9	10
	Collect	Transport	Sort	Bale	Transport	Unbale	Sort	Bale	Transport	Unbale
	Drop off, kerbside, deposit	Ship to MRF or sorting centre	Sort PET 'quality' bottles. Reject PVC & HDPE	Pack PET bottles for transport	Ship to recycler	Prepare for recycling	Secondary sort. Removes all PVC bottles	Pack for transport	Ship to recycler	Prepare for recycling

	11	12	13	14	15	16	17	18	19	20
	Prewash	Classify (sort)	Grind (wet/Dry)	Classify (wet/dry)	Mainwash	Main classification (via density)	Rinse	Dry	Final sort (flake)	Product (store)
	Remove loose soil, dust, soluble contents, paper labels, glues	Damaged, excessively soiled, discoloured bottles, etc.	Cut/convert to flake	Air/hydrocyclones (removes papers, labels, caps, seals)	Caustic/detergents (cleaners) hot removes serious soils	Hydrocyclone—removes other 'plastics', caps, labels, papers	Clean water, removes caustics, detergents	Hot or cold (removes surplus water)	Removes black specks, stones, grit, degraded bits, etc. (also grades flake size)	**First commercial product**

	21	22	23	24	25	26	27	28	29	30
	Transport	Dry	Extrusion (vented)	Filtration (melt)	Pelletise	Crystallise (store)	Product (store)	Transport	Polymerise (SSP)	Product (store)
	Ship to customer or for secondary treatment	Serious drying (to avoid hydrolytic degradation on melting)	Removes adsorbed volatile organics	Removes particulate solids, plus degraded particles	Extrude and cut to virgin standard	Simple crystallisation to avoid sintering during subsequent processing	**Second commercial product**	Ship to customer	Raise molecular weight (purify free of volatiles—acetaldehyde, low oligomers)	**Third commercial product**

Abbreviations: MRF, material reclamation facility; PET, polyethylene terephthalate; PVC, polyvinylchloride; HDPE, high-density polyethylene.

11.6.3.3   Classification

A second, 'sink float', separation usually follows directly after the mainwash. These can be large settling tanks or combinations with cyclonic separators. Hydrocyclones centrifuge out lower density materials, such as polypropylene (PP) and polyethylene (PE) (density < 1.0; PET has a density of > 1.33). PVC has a similar density to PET (> 1.33) and will remain in the same fraction. Complete separation of all PVC materials before grinding and washing is essential for good quality. Heavily printed label sections and metallic/metalised materials can also appear in this fraction. Hydrocyclones are much faster, more efficient and easier to adjust for quality than the older settling tank systems, and are a major component of modern plastic recycling plants.

11.6.3.4   Final sorting

This is an optional stage designed to reject discoloured, degraded or non-conforming species. This equipment is expensive to install but essential for high quality R-PET. Excessive black specks from degraded adhesive or dirt can reduce yields significantly. Sorting is usually based on photometric selection according to optical reflection, colour, shape and size. Once the non-conforming flakes are identified, a jet of compressed air removes them. The actual set-up of the selection is critical; if it is too sensitive, the yields of the main product will fall as more good product is rejected with the impurities.

11.6.3.5   Melt filtration, repelletisation

This is an optional stage, often combined with solid state polymerisation (SSP) to produce the highest quality product. Here, non-melting particles are removed and pellet uniformity is achieved, which is essential for good drying and uniform processing. Feedstock from selected used PET containers is treated by this type of technology and the R-PET used in an identical fashion to virgin material.

Recyclers will eliminate as many of the stages as possible to reduce costs and at the same time provide the product quality assurance needed by the end user. The product achieved in stage 20 is relatively pure, clean flake and is suitable for the majority of R-PET end-use applications (e.g. fibre and sheet). The end uses here are only limited by molecular weight and high purity constraints. Strapping tape is one end use that demands a relatively high molecular weight (IV usually > 0.80). This is achieved by SSP of either the washed flake or repelletised chip at about $220°C$, under vacuum or an atmosphere of inert gas, for up to 10 h (stage 29). If flake is used for this additional process, it is important that the chip is of regular size to ensure uniform molecular weight distribution, an essential condition for homogenous melting during subsequent reprocessing. If repelletised chip is used, then the final product is of almost the same consistency as virgin and can be blended and used together with virgin without any major difficulties.

For further detail on types of equipment available for cleaning PET, the reader should contact specialist suppliers [16]. The reader can also consult a comprehensive review of the recycling and recovery of all plastics [17].

11.6.4   R-PET and food-contact quality

Health and safety of the consumer is the prime concern in considering the reuse and recycling of any materials in food-contact applications. In this context, R-PET has to comply with all the rules and regulations that apply to virgin PET. One particular property of PET, when compared to other plastics, is its low diffusion coefficient. This is a factor of 100 lower than some other materials, which means it is more difficult for contaminants to diffuse into or out of PET [18].

There is no specific EU legislation that covers recycled plastics; most regulations neither preclude nor allow such use in food contact applications. However, there are exceptions and some national regulations expressly forbid such applications. Research work to define safe boundaries and possible limitations of use can be consulted [19].

EU legislation covering the recycling of packaging materials has provided incentives to look at all potential outlets for R-PET. As food contact applications are the largest market, this area has received much attention. Some manufacturers of recycling equipment now offer plant that is said to be capable of purifying R-PET to a standard which allows it to comply with the current regulations. One such construction, the Erema 'two-step process' (illustrated in Figure 11.4), claims to be capable of raising the molecular weight sufficiently in the first section to produce R-PET polymer to reuse for manufacture of bottles. The final purification is also claimed to be of satisfactory quality for food contact applications [20].

A number of other inventions seeking similar purity without resorting to the more certain but costlier solvolysis techniques can be studied in the relatively recent patent literature. The references here are not comprehensive; the reader can use them as a good initial source of information and they also provide other 'prior art' references. Starting with washed, clean flake followed by a degassing technique in the solid phase and subsequent melt extrusion/vacuum venting, produces a high purity R-PET [21]; spiking techniques with contaminants, such as toluene, benzophenone, metal arsenates and chloroform, are used to show the effectiveness of the cleaning. A similar technique with two independent extruders was previously patented [22]. Washed, clean R-PET is further decontaminated by leaching with solvents, such as acetone and ethyl acetate, which dissolve in and swell the PET and are easily removed afterwards [23]. Pre-extraction of R-PET flake with methyl ethyl ketone, tetrahydrofuran (THF) or dioxane is claimed to remove PVC and PS, in addition to other impurities. The extracted R-PET is then dissolved in ethyl carbonate or dimethyl phthalate before recovery by precipitation using cool solvent [24]. Steam 'stripping' of

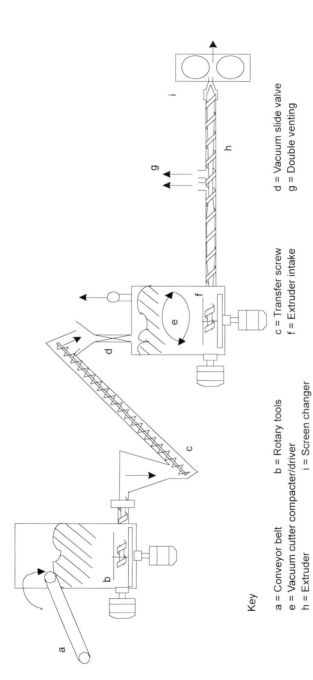

Key

a = Conveyor belt   b = Rotary tools   c = Transfer screw   d = Vacuum slide valve

e = Vacuum cutter compacter/driver   f = Extruder intake   g = Double venting

h = Extruder   i = Screen changer

Figure 11.4 Erema 'two-step process' plant for high purification of recycled PET. Figure redrawn from Erema publications [20].

prewashed R-PET flake at 150–190°C is claimed to achieve high purity [25], as is nitrogen purging at 218°C using washed flake of specified particle size [26].

Treatment of washed PET flake, reclaimed from used bottles, with a calculated amount of 50% sodium hydroxide solution to react with about 3–10% of the PET, at temperatures just below melting (*ca* 250°C), partially saponifies the PET and is claimed to strip free the contaminated surface layer. This process is the technological basis of new plant commissioned in Switzerland in 2000 [27].

11.6.5 Recycling of PET by chemical methods—'solvolysis'

Solvolytic methods are much more complicated than straightforward washing processes and are more energy intensive. They also need specialist plant and are therefore more costly, especially if performed on a small scale (i.e. below 20–30 K tonnes per annum). However, they can utilise lower quality feedstock as the processes allow additional purification to be performed.

PET can be recycled by breaking down the polymer chains by treatment with chemicals, such as: alcohols, e.g. methanol (methanolysis, producing dimethyl terephthalate [DMT] monomer); ethylene glycol (glycolysis, producing bis[hydroxyethyl]terephthalate {BHET} monomer); acids (hydrolysis); or alkalis (saponification). These treatments effectively depolymerise the PET, reform the partially esterified monomers and, finally, the intermediates of terephthalic acid (TA) and ethylene glycol (EG). However, the two major processes that have been used commercially are glycolysis and methanolysis. Polyesterification is an equilibrium reaction and the depolymerisation is carried out with an excess of the solvent to force the reaction backwards and regenerate the intermediates (see Figure 11.5).

Selection of the most appropriate technique is dependent upon the quality of the available feedstock and the quality of the products required. Glycolysis is a compromise between simple washing and reprocessing of the clean flake by melt extrusion and complete recovery of the intermediates by methanolysis and hydrolysis procedures.

To be economically attractive, depolymerisation facilities should achieve a minimum economy of scale dependent upon the cost and purity of the feedstock.

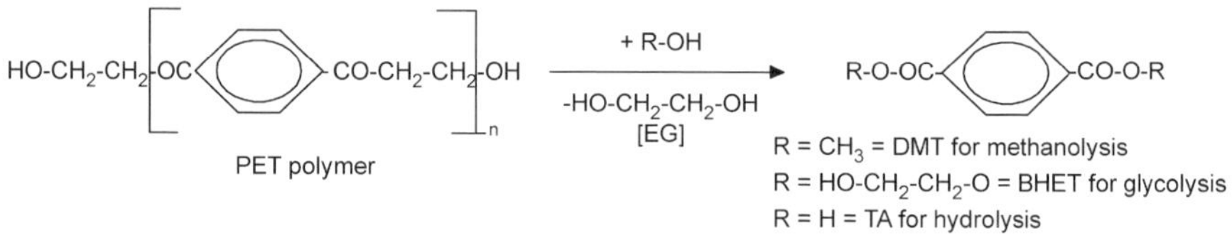
HO-CH2-CH2-OC
CO-CH2-CH2-OH
n
PET polymer
+ R-OH
-HO-CH2-CH2-OH
[EG]
R-O-OC
CO-O-R
R = CH3 = DMT for methanolysis
R = HO-CH2-CH2-O = BHET for glycolysis
R = H = TA for hydrolysis

Figure 11.5 Depolymerisation reactions of PET. Abbreviations: PET, polyethylene terephthalate; EG, ethylene glycol; DMT, dimethyl terephthalate; BHET, bis (hydroxyethyl) terephthalate; TA, terephthalic acid.

Using post-consumer recovered PET containers as a source, the expected economical plant size will exceed 15,000 tonnes per annum, with plant sizes in the region of 40,000–50,000 annual tonnes more advantageous. Finding reliable and continual sources of feedstock to supply plants of this size has not been sustainable so far but with the increasing volumes of PET bottles collected by the national recovery associations this limitation may soon be overcome.

It is possible to convert PET into alternative intermediate compounds, known as polyols, by breaking down the chains with polyglycols and ethylene oxide treatment. These compounds are available for conversion into polyurethane and polyisocyanurate foams and resins. This topic has also been the subject of research trials but as yet no widespread commercial development [28].

11.6.5.1   *Glycolysis* [29, 31]

Glycolysis involves treating the PET with excess of ethylene glycol, driving the reaction backwards and forming BHET with smaller quantities of oligomers (n = 3–6). BHET is the intermediate formed by the reaction of TA with EG in the direct esterification process. This reaction is normally effected under pressure, at temperatures preferably about 210–220°C, to achieve a fast enough reaction rate; an ester-interchange catalyst (e.g. zinc acetate) can be used to speed up the process. At room temperature, BHET is a waxy solid of relatively high melting point, which cannot be purified by distillation (unlike DMT). BHET is not isolated as a generic product and is part of the dynamics of the conventional esterification process. Purification is normally effected by melt filtration under pressure to remove physical impurities, preceded (or followed) by possible treatment with carbon, which by selective adsorption removes chemical impurities that lead to poor colour and oxidative degradation. The filtration processes can be very cumbersome if significant quantities of fine pigments or colours, like titania, are to be removed. In glycolytic processes, non-pigment colours that are intentionally added to PET or copolymer constituents are not generally removed. Acid-based modifiers, such as isophthalic acid (IPA), will also remain and react again in the process. Heating ethylene glycol at high temperatures also forms higher glycols, such as diethylene glycol (DEG), as by-products. This component will also polymerise and has the effect of reducing the melting point of the resultant PET; it therefore needs to be kept well under control. Catalysts (e.g. sodium acetate trihydrate) can be used to suppress DEG formation [30].

The process is therefore preferably carried out with recovered material of known history and high quality. Feedstock will usually be well washed and cleaned before use. It is normally integrated well into conventional plant operations and nearly always used as a careful blend with mainstream BHET, usually less than 25% concentration. The risk of contamination of mainstream operations is high and careful attention to the quality of the recovered feedstock is vital. To place the risk in context, modern direct esterification plants produce *ca* 250,000 tonnes per year in continuous steady-state operation; any disruption will cause a major disturbance. Glycolysis polymer is nearly always inferior in colour and physical properties when compared to virgin and the end-use outlets are carefully selected [30].

Glycolysis has some significant advantages over either methanolysis or hydrolysis, primarily because BHET may be used as a raw material for either a DMT-based or a TPA-based PET production process without major modification of the production facility. Another significant advantage provided by the glycolysis technique is that the removal of glycol from the depolymerisation solvent is not necessary.

Glycolysis principles have been utilized to generate polyester diol precursors for manufacture of polyurethanes (PU) or polyisocyanurates. The mixture obtained from boiling PET with glycol at 200°C is esterified with adipic acid (220°C) and forms a product suitable for manufacture of polyurethane foam insulation [29, 31]. Scrap PET can be treated with diethylene glycol to form a low molecular weight modified aromatic polyol, which is further reacted with an aliphatic polyol (ethoxylated methyl glycoside) to form another suitable precursor (see [32] for useful references to prior art). Figure 11.6 illustrates some examples of the depolymerisation of PET by the glycolysis and methanolysis routes.

11.6.5.2 Methanolysis

Treatment of PET with methanol, again under pressure and at temperatures around 200°C, reverses the polymerisation in a similar manner to glycolysis but the intermediate produced is dimethyl terephthalate (DMT) along with ethylene glycol. The reaction uses a transesterification catalyst, usually the acetate of cobalt, magnesium, manganese or zinc; this intermediate is very well characterised and a commercially traded product with defined properties. DMT is purified by distillation, a process that yields high purity, especially free from physical contaminants. In this respect, it is much easier to purify than BHET and can generally accept an input specification of lower physical quality. However, organic impurities leading to poor colour may still persist and complete removal is not always possible. The product is usually isolated prior to reuse, and can be subject to final analysis followed by appropriate blending before reuse. DMT derived from methanolysis of post-consumer recovered containers generally has a product quality identical to that of DMT from virgin routes and is used for all DMT applications.

Development of the process has concentrated on increasing the yields of DMT, which means moving the reaction away from the equilibrium condition reducing the formation of oligomers and BHET. One improvement describes dissolution of the PET in oligomers of EG and TA or DMT; super-heated methanol is then passed through the solution. The EG and DMT are recovered overhead [33]. In a further refinement, the process is adjusted to produce 'half-esters'

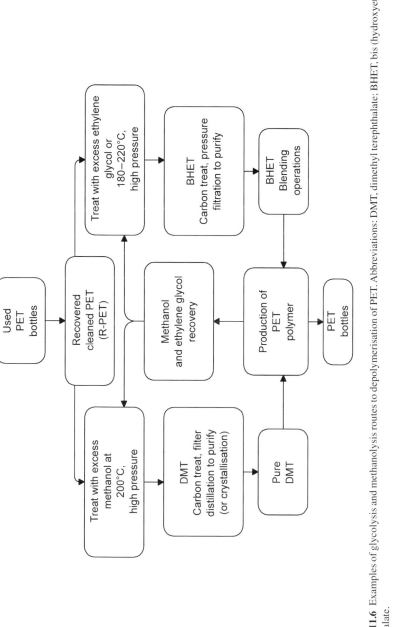

Figure 11.6 Examples of glycolysis and methanolysis routes to depolymerisation of PET. Abbreviations: DMT, dimethyl terephthalate; BHET, bis (hydroxyethyl) terephthalate.

(esters of TA with methanol and EG). The advantage of this development is that these half-esters have the potential to be used in either DMT or TA based polymerisation processes without additional purification. This should reduce the number of stages, use less equipment and consume less energy than previous methanolysis type processes (see [34], which includes a good review of earlier processes).

Both glycolysis and methanolysis use materials that are well established within the PET manufacturing process flowsheets, and their impact on product and process quality is very well understood. Ethylene glycol and methanol are easily recovered and reused within the process, with little solid or other waste. Economics are also very well understood; plant design and achievement of throughputs can be forecast with high accuracy.

11.6.5.3 Hydrolysis

Hydrolysis of PET is conducted by treating the PET with water, mineral acids or caustic soda (see Figure 11.7). The TA obtained is referred to as 'crude' and is then subject to purification by crystallisation, from solvents like acetic acid, to achieve standards equivalent to available commercial grades of TA. Like the other depolymerisation processes, it is also possible to treat with activated carbon to remove the organic impurities. The economics of these processes are highly dependent on scale, degree of purification required and utilisation of the by-products. Chemical purity of the TA is only one aspect of its acceptability; modern direct esterification technology demands TA that forms 'pre-mixtures' with EG and resultant viscosities that allow rapid transfer by pumping methods. Similarly, dispersion rates of the TA within the glycol to form 'slurries' are dependent upon the particle size and shape. The glycol segment is purified by distillation. The TA product is not easy to purify to a standard normally acceptable for direct esterification but this can be done. Multiple recrystallisations of TA improve purity but at extra cost. Following these processes, uses must be found for neutralisation residues, such as sodium sulphate.

Hydrolytic breakdown of PET into its prime intermediates followed by re-polymerisation is likely to be more expensive than either glycolysis or methanolysis, as it involves additional processing stages. However, it allows purification of both intermediates to levels equivalent to that of virgin products.

Although these processes work technically, satisfactory commercial processes are not operating at present. Older technology describes the use of concentrated sulphuric acid (11 acid:2 water) at room temperature to achieve hydrolysis within minutes. The resultant crude TA is precipitated by dilution, filtered, and dissolved in alkali to effect further purification [35]. Recent developments have used solid sodium hydroxide to initiate and speed up the reaction, driving off glycol and then diluting with water. This allows operation at atmospheric pressures. Purification can be effected by extracting impurities with butanol

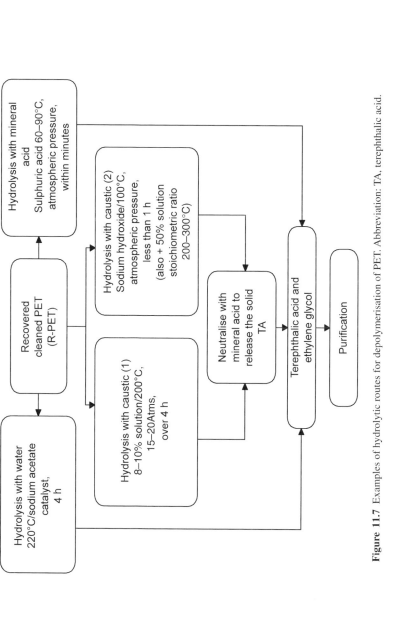

Figure 11.7 Examples of hydrolytic routes for depolymerisation of PET. Abbreviation: TA, terephthalic acid.

before passing the solution down a column of activated charcoal to remove the remaining impurities. The technology is known as the RECOPET process [36]. A similar technique uses a strong caustic solution (*ca* 50%) to form a slurry of disodium terephthalate, which is heated to 300°C to drive off the glycol and destroy other organic contaminants. The resultant mobile slurry is then diluted with water, and further purified by centrifugation and filtration before recovering the TA by neutralisation with mineral acid. The high temperatures used suggest that the process could be used for highly contaminated PET wastes. This technology is known as the 'UNPET' process [37].

In another variation, relatively clean PET bottle scrap is first treated with boiling EG using a countercurrent flow apparatus. The PET is rapidly glycolysed to a brittle composition of BHET and oligomers. This brittle flake is then crushed and separated from other coarse impurities, such as PVC. It is further purified by settling methods before hydrolysis to crude TA. The process, which claimed to use relatively crude PET bottle scrap, was known as the 'RENEW' process [38].

Methods of recovering TA by hydrolysis techniques nearly always fail on the issue of repurification of the TA. The PET waste arises from many sources and contains variable quantities of impurities, which pass into the crude product. Attempting to match virgin quality is very difficult without excessive quality control procedures, which are complex and costly. Improvements to the process technology are still being attempted. PET waste is first hydrolysed with an excess of pure water at *ca* 250°C/autoclave, and the crude TA is then hydrogenated at 280°C using a palladium/carbon catalyst. However, viable and competitive commercial processes are some way off (for a good description of the prior art see [39]).

11.7   Environmental benefits of recycling

Recycling is assumed to deliver environmental benefits in terms of material savings. The appropriate technique for measuring these savings is termed 'life cycle assessment' (LCA). LCA is a technique that catalogues all the environmental aspects associated with a product or, more specifically, the processes used to produce the product or provide the service. The catalogue (inventory) is then used to assess the impact of those aspects on the environment. The procedure is similar to a financial audit but examines the environmental viability of a process.

An LCA is useful because it collects all available environmental factors from the time resources are extracted from nature to the time when the resources are returned back to nature, a so called 'cradle to grave' process. For example, total energy consumed and emissions generated to air, water and earth during product manufacture, use and disposal.

The EU Packaging and Packaging Waste Directive, 94/62/EC, refers to the use of LCA techniques for use in examining achievements towards the environmental targets set in the directive with the clear intention of determining a programme of regulation. Following from these initiatives, an LCA international standard ISO 14040 [8], now exists which defines how LCAs should be completed, verified and used.

In many instances, the environmental benefits are immediately obvious with limited analysis; for example, in the case of aluminium recycling the whole process of recovery and recycling uses about 10% of the energy used to extract virgin aluminium from its ores. A simple energy analysis suffices if the major judgement criteria are energy consumption and emissions to air/water/land— most emissions arise from burning of fuels and extraction of ores. In other cases, the savings are not always so clearly delineated and careful assessment of a wider range of parameters is necessary.

11.7.1 *General principles of recycling* (adapted from Boustead [40])

There are essentially two types of recycling: 'closed loop' recycling classifies those processes which recover and reprocess material and feed it back into the same production system, and 'open loop' recycling classifies processes which recover and reprocess material then feed it to a different production process. PET 'bottle-to-bottle' recycling is an example of closed loop recycling; and using recycled PET from used bottles to manufacture fibres, rigid sheet, film or engineering PET plastics are examples of open loop recycling.

Closed loop recycling is probably the simplest form. It is illustrated schematically in Figure 11.8, which shows a simple linear production sequence consisting of three unit operations, labelled: Production (1), Use (2) and Disposal (3). A recycling loop (4) is included, in which some of the waste materials are extracted from the system, reprocessed and fed back into the same production sequence to displace virgin raw materials.

Material flows. Assume that all of the operations operate without loss of material and that the consumer is located somewhere in operation 2. The saving in raw materials due to recycling can be calculated. Assume also that f is the fraction of the waste leaving operation 2 that is reprocessed and returned to the main production sequence.

If the recycling loop was absent (i.e. $f = 0$), then normal operations would see a mass flow (m) moving through the sequence. With the recycling loop in place, the consumer in operation 2 will still see a product of mass, m, flowing past. However, the raw materials input to operation 1 has been reduced from m to $m(1 - f)$ and the solid waste entering operation 3 has also been reduced from m to $m(1 - f)$. Therefore, as a consequence of introducing the recycling loop, both the demand for raw materials and the solid waste generated have been reduced. Even if an allowance is made for material loss in all of the operations, the same general conclusion will apply.

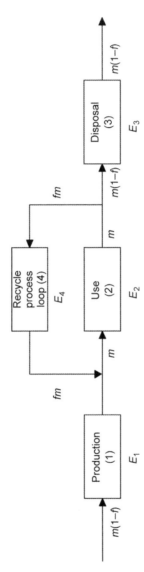

Figure 11.8 Simple production sequence when a closed recycling loop is included. For explanation see Section 11.7.1.

An alternative way of viewing this effect is to suppose that a single mass, m, is fed into operation 1 in Figure 11.8 and then monitor the quantity of product that will pass the consumer as some fraction of this initial mass is repeatedly recycled. On the first pass, the product mass will be m. After the first recycle, the mass will be mf. At the second recycle it will be mf^2 and so on. Thus, after an infinite number of recycles, the consumer will have experienced a total mass, M, of product passing through, where:

$$M = m + mf + mf^2 + mf^3 + mf^4 + \cdots \cdots$$

and summing this series gives:

$$M = m/(1 - f)$$

Thus, if there is no recycling, f will be zero and $M = m$. However, say 50% of the materials leaving operation 2 is recycled, then $f = 0.5$ and $M = 2$, so that for an input of 1 kg of material, the consumer will experience 2 kg of product. Similarly, with 90% recycling, $f = 0.9$ and $M = 10$, so that an input of 1 kg of raw material will lead to the passage of 10 kg of product. In all of these cases, the amount of solid waste generated will remain at 1 kg, irrespective of the mass of product passing the consumer.

Thus, if the aim of recycling is to minimise consumption of raw materials and generation of solid waste, then the goal should be to try and achieve the highest recycling rate possible.

Energy flows. The energy needed to drive the systems shown in Figure 11.8 presents a more complicated picture than the raw materials flows. To evaluate these energies, suppose that the energy requirement per unit output for each of the unit operations 1 to 4 is: E_1, E_2, E_3 and E_4, respectively; thus, E_1 is the energy required by operation 1 to give an output of 1 kg from operation 1, and so on.

For the system of Figure 11.8 where no recycling occurs (i.e. $f = 0$), the system energy, E_s, is given by:

$$E_s = m \cdot E_1 + m \cdot E_2 + m \cdot E_3 = m(E_1 + E_2 + E_3) \qquad (1)$$

The system energy with recycling will be E_s' where:

$$E_s' = (1 - f)mE_1 + mE_2 + (1 - f)mE_3 + fmE_4$$

and this rearranges to:

$$E_s' = m(E_1 + E_2 + E_3) + fm(E_4 - E_1 - E_3) \qquad (2)$$

Using equation (1), this can be rewritten as:

$$E_s' = E_s + fm(E_4 - E_1 - E_3) \qquad (3)$$

If the energy change introduced as a result of incorporating the recycling loop is written as ΔE_s, where:

$$\Delta E_s = E_s - E'_s$$

then, from Equation (3):

$$\Delta E_s = fm(E_1 + E_3 - E_4) \tag{4}$$

If the righthand side of this equation is positive, then there will be energy savings. However, if the term is negative, then the recycling loop will have incurred a net expenditure of additional energy. The critical condition, which determines whether energy will be saved or lost in recycling, is the value of E_4 relative to the sum of E_1 and E_3. Note that that this general conclusion does not change if there are material losses from the recycling loop as would occur in a real recycling operation.

For the recycling of post-consumer waste, operation 1 in Figure 11.8 will represent all processes starting with the extraction of raw materials from the earth through to the production and delivery of a material at the converter. Operation 3 will represent all materials handling operations that occur after the consumer has finished with the product at the end of its useful life, and in most systems represents the collection and disposal of waste. In general, the energy associated with operation 3 will be small compared with the energy associated with operation 1. Consequently, provided that the materials can be collected, reprocessed and delivered to the converter with an energy requirement less than that needed to produce virgin materials via operation 1, then an energy saving will occur.

One important consequence flows from this conclusion. As the production of virgin materials is made increasingly more energy efficient by improved manufacturing techniques, the scope for energy saving by recycling is gradually diminished because the parameter E_1 in Equation (4) is decreased relative to E_4. Conversely, when recycling is made more energy efficient, as for example through improved waste collection, larger amounts of material can be recycled without loss of energy.

11.7.2 Application of the general principles to the recovery and recycling of PET bottles

To simplify the illustration, only the energy parameter is discussed in the following section and emissions to the environment are not included. However, this does not mean that these factors should be ignored; other assessments of the viability of recycling processes may rank emissions higher than energy consumption—for example, global warming gases may take priority.

Figure 11.9 illustrates typical PET-user chains through extraction of oil, manufacture of products (i.e. bottles, fibre or engineering plastics), through the

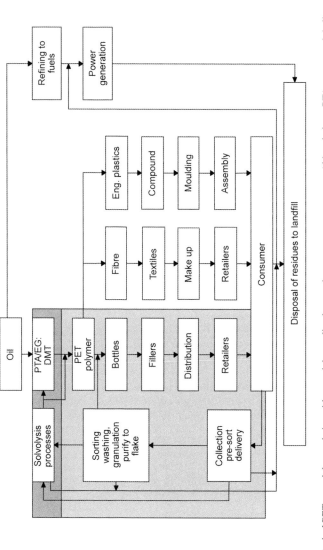

Figure 11.9 Typical PET material user chains with material recycling loop and route to energy recovery. Abbreviations: PTA, pure terephthalic acid; EG, ethylene glycol; PET, polyethylene terephthalate; DMT, dimethyl terephthalate.

use operation and eventually to final disposal. Also included are a recycling loop, and an indication that the used products can be a possible energy source. Considering recycling of PET bottles, the recycling loop collects the used bottles, progresses them through the sequence of sorting, followed by washing and cleaning to a usable flake before returning the recovered PET back into the user chain immediately prior to the manufacture of more bottles. The lighter grey block delineates the boundary of this process. The recycling loop also illustrates the additional or alternative step of recovery by solvolytic processes; in this case the PET is broken down into its intermediates before returning into the user chain prior to manufacture of the PET polymer. This additional step is delineated by the extension of the boundary, including the darker grey area.

The three basic operations of production, use and disposal are easily discerned. The closed loop example would be the 'bottle-to-bottle' sequence and the open loop example the use of R-PET in the other sequences for fibre and engineering plastics manufacture. It can also be seen from Figure 11.9 that the recycling loop is split into three main operations: collection, sorting and delivery to the recycler, washing and cleaning of the flake, and the more complex solvolytic recovery process. It is possible that used bottles can be fed directly into the solvolytic processes without prior cleaning. However, this is not usual at the present stage of developments and some prior washing is assumed.

11.7.2.1 Potential energy savings

Before one can consider the reprocessing options, it is appropriate to look at what potential energy savings are possible. In principle, collection, reprocessing and reintroduction of recycled PET material will directly displace virgin materials from the consumption chain. It is this displacement value that is the driving force for recycling operations. In Figure 11.9 the chain from oil to PET polymer is represented by only three stages. In actual fact, there are about 13–14 stages. The full sequence of operations outlined in Figure 11.10 consumes a total of 77 MJ for every kg of bottle-grade PET produced [41]. This total is made up of 39 MJ as 'feedstock' and 38 MJ as fuels consumed to drive the process. The feedstock is the organic equivalent to the inorganic glass in bottles or the aluminium metal in cans and is carried through the recycling process as material equivalent before re-entering the user chain again as regenerated PET polymer. Put another way, 38 MJ of fuels are consumed to deliver 1 kg of virgin polymer to the bottle preparation process.

Any recovery and recycling operations that deliver R-PET back to the bottle production process should, therefore, consume less than 38 MJ if savings are to be achieved.

It should be noted that these values will continue to fall as manufacturing technology improves. In the early 1980s, intermediates plants usually had operating capacities of 100,000–150,000 tonnes per annum. Such plants, installed today, will have operating capacities of > 250,000 tonnes per annum Similar changes

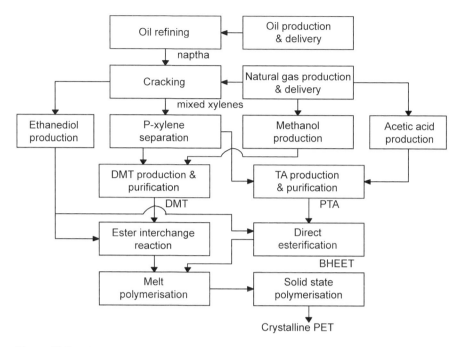

Figure 11.10 Full sequence of operations to produce virgin PET polymer. Abbreviations: DMT, dimethyl terephthalate; TA, terephthalic acid; PTA, pure terephthalic acid; BHET, bis(hydroxyethyl) terephthalate.

are noted with polymerisation capacity; > 150,000 tonnes per annum units are in operation today, an advance on older 50,000 tonnes per annum capacity plants. These unit size changes and technology improvements increase efficiencies enormously, especially energy utilisation. It is these factors that have allowed the price of PET to remain virtually unchanged over the last 15–20 years and reduced the energy consumption to the low figures we see today. Some historical total energy consumption figures for manufacture of PET are presented in Table 11.6. It should also be noted that the figure of 77 MJ/kg represents an average value and in reality there will be a range dependent on a number of factors. Plant size, occupational capacity and efficiencies are factors that are well understood. What is not well understood are the other influencing factors related to use of oil or gas as the principal raw material and the national power supply systems in the country of manufacture. For example, if gas is used as primary feedstock to make the PET polymer, the overall energy contributions will be higher than if oil was used, simply because gas has a higher calorific value. One kilogram of carbon from methane will require a feedstock energy of 72 MJ compared to 1 kg of carbon from oil of feedstock energy 59 MJ (i.e. pentane as an example of naphtha). Similar but smaller differences are seen if all the electrical power is derived from

Table 11.6 Total energy values for manufacture of PET polymer

Reference	Energy (MJ/kg)
Boustead – 1982 [42]	150
Franklin Associates – 1980 [43]	123–127
Boustead – 1989, 1993 [44]	128–135
Franklin Associates – 1992 [43]	89
BUWAL 132 – 1990 [45]	*ca* 70
APME – 1995 [41]	84
APME – 2000 [41]	77

hydro-power compared to power generated from relatively inefficient coal fuels. It is beyond the scope of this article to describe these differences in detail and the reader should refer to the methodology used in compiling the APME eco-profile reports for a more thorough discourse of the subject [41].

11.7.2.2 Fuels and feedstocks

In the above theoretical analysis, materials flows have been illustrated separately from energy flows and the final consequences of recycling are shown to be different for these two different parameters. In general, recycling will always result in a reduction in consumption of raw materials and generation of solid waste per unit product at the consumer. For energy, however, there is no guarantee that savings will occur when recycling is practised. The energy changes resulting from recycling are governed by a number of different parameters and only when certain well-defined conditions are satisfied will energy savings occur.

For products manufactured from inorganic materials, raw materials consumption is invariably kept separate from the consumption of fuels. The units in which they are measured are different and there is therefore no temptation to try and combine the two.

For polymers, and other materials based on combustible feedstocks, such as wood products, the raw materials input, measured as the feedstock energy, can be measured in mass terms but is often expressed in the same units as the fuel energy. Consequently, there is frequently the desire to combine the two contributions into a single parameter for simplicity. This, however, should be avoided. The effect of recycling on feedstock energy follows the treatment given earlier for materials flows, whereas the treatment for fuel energy follows that given for energy and, as noted earlier, they are different. Only if the material is ultimately thrown away as waste, without energy recovery, is it meaningful to add together fuel and feedstock energies as a description of the total energy resource that must be extracted from the earth.

It is also important to distinguish between two different types of recycling processes for plastics: mechanical recycling and energy recovery. Mechanical recycling is the recovery of plastics for further use as materials. This form of recycling is available to all other materials. The environmental benefits arise from the ability to spread the initial material production energy over several uses (shown in the earlier analysis as mass flows through the consumer). Energy recovery is the recovery of the feedstock energy (or a proportion of it) for use as a fuel. In other words, it is the conversion of feedstock to fuel. This option is available to relatively few materials; plastics and wood products are the most obvious examples. These two recovery options are complementary; they are not mutually exclusive. Consequently, recycling should not be regarded as a choice between mechanical recycling or energy recovery, since both are possible for the same material.

11.7.2.3 Collection of bottles

Obviously, before we can reprocess we need a supply of collected bottles. The energy associated with collection is very often accounted for in a social context and as part of the duty carried out by local authorities in connection with waste disposal operations. This figure can be extremely variable; however, if recycling is being operated with the intention to save overall resource, then energy expended on collection must be factored into the calculations.

One source of information on energy consumed during collection comes from a study from the Netherlands on plastic waste management options [46]. Five different collection models were studied, consuming 3–4 MJ/kg for collection, some pretreatment separation and delivery to the processing plant. These models are noted as:

- collect integral with MSW and ship to MSW combustion plant
- kerbside separate collection of 100% of plastics as part of MSW scheme and transport to separation plant
- kerbside collection of 95.9% plastics as in 2; collect 4.1% separately as bottles
- kerbside collection of plastics/metals/drinks cartons (PMD)
- kerbside collection of 54% of all plastics and 46% with MSW; delivery to feedstock recycling

Note: the lowest energy option is not to collect, separate, measure the yields and deliver to the recyclers but to collect and deliver straight to an energy recovery facility (ca 1.5 MJ/kg).

This latter figure could be a realistic representation of a very basic minimum figure, which could be achieved if all that was needed was to collect and prepare 100% PET bottles then delivery to a processing site. These figures are probably representative of the better-developed schemes. There are schemes at earlier stages of development that reflect higher energy figures. Certainly, those with a higher proportion of drop-off collection points and rural areas would expect to consume more energy.

Studies executed for APME [47] indicate collection energies from 0.2 to 1.2 MJ/kg for systems covering collection with MSW to separate green bin collection of all plastic bottles. Rural area collections can be up to 5–6 times more energy intensive.

Other data [48] suggest collection energy figures of up to 15 MJ/kg (including sorting) and also give some indication that volumes from urban collections can be up to seven times more effective in yield when compared to collections from rural areas.

These figures suggest that sensible collection schemes for recyclables should be developed, which are not too ambitious. Simply setting goals of 80–90% recovery when the best available information says less than half of this is achievable is guaranteed to fail. To obtain the correct balance demands careful analysis before compiling a programme. Figure 11.11, which illustrates the general case of total system energy, shows how the balance will change as a programme becomes more successful and recovery of the last remaining material shifts the efficiency of the exercise. Trying to recover the last 5% when the recovery rate is already 95% would require that every last container, including those taken away by overseas holidaymakers, be recovered. Clearly this is impractical but it underlines the fact that at very high recovery rates, the recovery energy would certainly be increasing steeply. As a consequence, the actual relationship between total system energy and recycled fraction is most likely to be a curve, such as that shown schematically. Application of Pareto analysis to target the major effects is recommended and a careful audit of environmental cost, in addition to the normal economic cost, carried out on a regular basis.

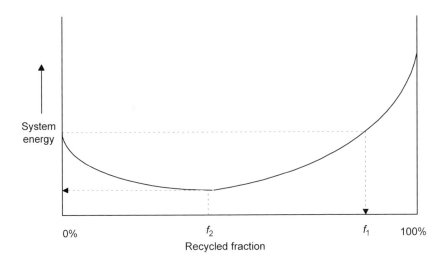

Figure 11.11 System energy as a function of recycled fraction when all of the energy terms are constant.

It is clear (from Figure 11.11) that if the overall recycling process is not to use more energy than the process without recycling, then the recycled fraction should not increase above f_1. It can also be deduced that if the recycling target is to actually save energy, then recycling should be practised only until the total energy curve reaches the minimum of Figure 11.11; that is, until the recycling fraction reaches f_2.

The importance of this is that recycling targets or collection targets should not be set arbitrarily. If energy savings are a target for any recycling process, then the primary aim should be either to establish the form of the curve in Figure 11.11 so that the limits f_1 and f_2 can be determined or to establish the condition set by Equation (4), so that energy is not wasted.

11.7.2.4 *Reprocessing of the collected bottles*

Depending on the type of recycling process used, the reprocessed bottles could be introduced back into the chain either as a clean washed polymer flake, as partly polymerised PET (BHET from glycolysis) or as the recovered interme- diates, DMT (from methanolysis), PTA and EG (from hydrolytic processes).

Reprocessing of the collected containers can be carried out in several ways, but the procedures illustrated in Table 11.7 usually follow the delivery of bottles to the processing plant. For the purposes of this discussion, only the methods of recovery back to a reusable polymer are considered.

Associated with the description 'a collected PET bottle' are labels, caps, residues from the contents and other soil depending on method of collection and storage. This associated detritus can amount to 50% of the total weight. As bottles are engineered lighter, this ratio generally increases. Separation and preparation before final reprocessing can make very variable demands on energy input. This is particularly so with methanolysis and glycolysis. Some hydrolytic processes claim an advantage by being able to use highly contaminated feed- stock needing a minimum of pretreatment. Good quality used PET bottle input is essential for efficient reprocessing.

There is also another option, which should not be ignored for lower quality materials unsuitable for mechanical recycling. PET has an intrinsic energy value equivalent to that of soft coal (i.e. 23 MJ/kg) or approximately half that of fuel oils. Hydrocarbon fuel oils have values of *ca* 40–45 MJ/kg. PET does burn in a clean manner and can be used as an effective fuel substitute. If energy recovery

Table 11.7 Energy consumption for PET recovery processes

Method of recovery	Approximate energy consumption (MJ/kg)	Reference
Reprocessing to clean PET flake	6–15	[42–45]
Hydrolytic processes to PET	20–30	[49]
Methanolysis/glycolysis to PET	24–38	[43, 44, 50]

Table 11.8 Summary of potential energy savings (MJ/kg) from recycling of PET

	Fuel energy to manufacture virgin PET	Energy for collection/ delivery	Energy for reprocessing	Energy for total 'waste management'	Potential savings—from displacement of virgin PET
Washing to flake	38	2–15	6–15	8–30	8–30
Hydrolysis	36*	2–15	20–30	22–45	14–(−5)
Glycolysis/ methanolysis	36*	2–15	24–38	26–53	10–(−13)
Energy	40–45 (fuel oil value only)	2	0	2	19–24 (23 − 2 = 21 MJ reclaimed as fuel)

* This figure is reduced to 36 MJ to allow for the energy for polymerisation of the intermediates to PET (i.e. *ca* 2 MJ/kg).

is considered, the virgin PET will not be replaced in the user chain but fuel oil will be replaced in a corresponding fuel supply chain (see Figure 11.8). After subtracting the energy used to deliver to the waste-to-energy plant (2 MJ/kg), it is possible to recover up to 21 MJ calorific value.

It is important to note that the 39 MJ feedstock value is not related to the calorific value of the PET material and represents a mass value carried through from the initial feedstock. On combustion of PET, only the true calorific value can be recovered. Table 11.8 summarises the potential savings from each of the 'recycling' processes.

The conclusions that can be drawn from this 'broadbrush' analysis are that the most effective method of recycling is the simple washing and regeneration of reusable PET flake. Once the recovery processes include remelting or depolymerisation using high temperatures and pressures, the energy demands rise markedly. However, technology is constantly improving and regular reappraisal of processes on an individual basis is sensible advice. Where there are likely to be special circumstances—for example, large volume sources of very clean used PET or the absence of facilities to use the simple remelt route—then other options can be effective in achieving the environmental gains indicated in energy terms.

The reprocessing of PET has a sound basis when considered in the context of systems energy and provides one driving force for the economic recycling of PET materials. This is likely to be the case for some years to come, especially for reprocessing directly as clean flake. However, for recycling of used PET bottles to be successful and sustainable, financial factors also need to be considered.

11.7.2.5 The consumer factor
The consumer factor is usually ignored in calculating savings but can have a large influence on overall energy savings. The consumer collects the bottles from

the retailer and, after using the contents, delivers the empty bottle back to the collection point; this is also a factor that should be examined and included in the equation. The problem with this sector is that the detailed habits of shopping and returns are so variable that accurate analysis is very difficult. However, Boustead [44] notes that 30% of shoppers make the trip on foot, 32% travel 3 miles (ca 5 km and 90% of these by car) and the remaining 38% travel 7 miles (11 km). The average car consumes approximately 3–3.5 MJ/km. Simple calculation indicates that this is a consumption of approximately 17–20 MJ/trip. This energy would, of course, be averaged over all the contents purchased; ten items purchased each trip is indicated [44]. If the containers are returned on the next shopping trip, the energy consumption is kept to a minimum. However, if the purchase of soft drinks is a singular activity and the return to a drop-off container is also a special trip, the consumer contribution to the energy factor can very rapidly make recycling an irrelevance as far as overall environmental benefit is concerned.

The consumer should be encouraged to combine the purchase and drop-off together with other trips if energy consumption is to be minimised. More research in this sector would help to plan recovery of PET materials in a way that maximises environmental benefits in accordance with the aims of the EU sixth environmental programme.

11.8   Economic factors affecting PET bottle recycling

Calculating actual economics for most processes appears a relatively easy task; however, this is never quite as simple as expected because of the different ways companies have of allocating costs, overheads and profits, and of evaluating what margin they are prepared to accept. It is easier when looking at stand-alone, single-operation plants, which have to allocate all costs and investment directly to the output of a single individual plant, unit or product. However, many large companies can integrate smaller plants into larger or multi-product procedures that can allocate costs over several operations.

What is known is that the selling price of R-PET is directly related to that of virgin PET and this is a deciding factor in determining what the cost structure has to be for an effective operation. In this respect, it should be remembered that wherever R-PET can technically be used so could virgin without any quality risk. This quality/risk factor keeps the price of R-PET at about 50–60% of the virgin price. Long-term monitoring of the ratio is undertaken by special-ist consultants [51], who publish market price data covering a range of PET products on a regular basis. Figure 11.12 illustrates this relationship, with the average price of R-PET about 53/54% of virgin (R-PET clear flake primarily for fibre).

Recycling plants are smaller in size compared to virgin PET plants, with much greater feedstock variation. The whole nature of the recovery process

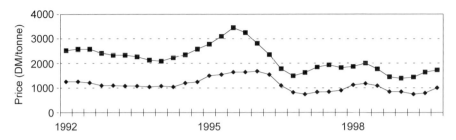

Figure 11.12 European price relationship PET/R-PET. Key: —■—, virgin PET; —◆—, recycled-PET flake. Source—based on PCI data [51].

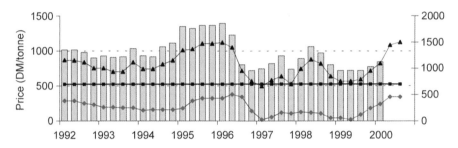

Figure 11.13 European PET price relationships—reclaimers margins. Key: ▨, margin on raw material; —▲—, flake average; —■—, production cost; —◆—, bottles average.

generally demands more supervision, quality monitoring and inspection. It takes little imagination to see that quality defines the success of the whole operation. Figure 11.13 gives some indication of the economics of producing R-PET flake. It is clear from this figure that recycling to PET flake encountered economic difficulties during 1996/97 and early 1999. One major recycler of PET bottles went into liquidation in 2000 but has since commenced trading under new ownership. It is significant that this particular recycler was washing and selling the clean flake with no added value product.

With the cyclical nature of the investment and the need to construct large, high-throughput virgin plant to keep costs low, we are likely to see a continuing cycle in this price relationship with downward pressure on prices and costs.

Another major factor affecting the overall economics of PET recycling in Europe and having a major influence on pricing is imports. The European Commission has recently imposed 'anti-dumping' duties on imports of PET resin from a number of Asian manufacturers, which has had the effect of raising market prices of virgin products and, in tandem, the price of R-PET [52]. Similar duties imposed on imports of PET textile products has also helped PET recyclers because the major market for R-PET is still for manufacture of fibre products.

11.9   Markets for recovered PET

Good quality, high purity, R-PET can be considered virtually equivalent to virgin PET and, as such, should be able to be used as a direct substitute for virgin PET in many markets. Clean, recovered R-PET flake can be converted into many different products. It is used again in bottles for non-food end-uses, such as household chemicals and cleaners, but the major secondary use is for the manufacture of polyester fibres, which are then used to make clothing and household textiles, either directly or as a filling fibre in anoraks and bedding. The modern ultra-lightweight 'fleece' jackets offering warmth and easy care properties are a good example of the very effective use of R-PET. The fibres are also used extensively for carpets (US), and scouring and cleaning pads. The textile industry is well developed and mature in Europe and R-PET finds a ready market. However, it is also highly competitive and low-priced fibre from the Far East has kept margins low from 1997 to 2000. This effect has led to a higher than expected demand for baled European bottles from India and China. It is expected that this market will continue to see similar volatile swings in supply and demand.

Converting R-PET into the form of sheet allows it to be subsequently thermo-formed into protective packaging for delicate articles, such as eggs and plants for despatch through the mail. The relatively lower PET prices in recent years have opened up markets, which have been served by clear PVC and 'crystal clear' polystyrene. More R-PET is now being used in protective blister packaging for point of sale display and merchandising trays. The sheet market is estimated to provide substantial growth but competitive pricing is the key factor.

Application of co-injection technology can encapsulate recovered PET with a layer of virgin material and allow it to be reused again for food and soft drink packaging. This innovation has been operating for a number of years by the Coca Cola® company to produce 'three-layer' bottles (e.g. in Switzerland and Sweden). The technology is relatively expensive and the continuing innovations in standard technology have limited its widespread use.

A number of companies have developed technology that can wash, decontaminate and produce clean R-PET pellets or flake to a hygiene standard which is perfectly acceptable for reuse directly into food contact applications, such as beverage containers. Developments like these bring 'bottle-to-bottle' recycling for PET containers closer to standard commercial practice. Schmalbach-Lubeca regenerates such a grade of R-PET for use internally and Wellman has been offering its Ecoclear, a 25% R-PET/75% virgin blend since 1996. Both these companies are anticipating good growth potential in this area as the products gain acceptance. Improved collection volumes are now sustaining these developments, guaranteeing supplies of feedstock, which in turn allows companies to invest in the technology with confidence. Two recently commissioned plants in Switzerland and Germany [53] support the view that this market sector will

Table 11.9 European end-use markets for recycled PET (R-PET)—absorption capacity (K tonnes)—Source PCI [51]

Market sector	Total market size (includes virgin)		% Penetration with R-PET		Tonnage used	
	1999	2004	1999	2004	1999	2004
Fibre	448	605	20	33	88	199
Food contact containers	1224	1955	<1	16	5	314
Non-food contact containers	208	247	<1	14	1	35
Amorphous sheet (A-PET)	89	134	6	35	16	47
Strapping	27	40	22	75	6	30
Injection moulding	100	122	2	12	2	15
Polyols	30	52	3	33	1	17
Totals	2126	3155	6	21	118	657

use more purified R-PET; the data presented in Table 11.9 allow for expected growth. These figures should be taken as a guide only, especially the textile figures, as the market is subject to change by imposition of duties and tariffs.

PET has been utilised in strapping for many years and competes very effectively with steel and polypropylene. R-PET needs to be free from particulate contamination and its molecular weight increased to provide the correct tensile strength. Price is an important factor but continuity of supply is very important. The world's largest strapping manufacturer (ITW-Signode) has integrated backwards to secure its European R-PET supplies by buying the PET recycling assets of Polyrecycling in Switzerland. The off-take will not be excessively large but the demand will be consistent with steady growth.

Regular, comprehensive independent reviews of the markets for recycled PET flake are obtained from industry specialists. Data supplied by PCI Ltd [51] show the estimated demand for R-PET in the years 2000 and 2004; the projected

Table 11.10 Main markets for melt reprocessing of clean recycled PET flake

Fibres	In staple form for fillings, e.g. anoraks, bedding, cushions and furnishings
	Industrial fibres for belting, webbing, scouring/cleaning pads, filters and cleaning cloths
	Other textiles, e.g. carpets, upholstery fabrics, interlinings, protective clothing and other garments
Sheet	Blister packaging, boxes, trays, shallow pots and cups
Strapping	Binding and strapping tapes, mainly for securing bales or bulky articles on pallets
Blow moulding	Primarily into bottles for non-food applications; co-extrusion technology is likely to allow expansion into selected food applications
Injection moulding	Transparent articles or plates, when reinforced with glass fibre for selected engineering applications

demand exceeding the current and predicted collection volumes of used PET containers in those years. On the basis of these estimates it is expected that there will be sufficient future demand for all the R-PET produced; Table 11.9 and Table 11.10 indicate the market sectors and typical applications.

11.10   PET recycling—the future

If the trend to push costs and prices down is to continue, then the industry has to find ways of ensuring better profitability. There are a number of ways that this can be achieved but the most successful are those that add value to the product. Table 11.11 illustrates the difference between the prices of the major PET products, resin, fibre, sheet and strapping. Conversion direct to sheet can increase the added value by a factor of 2 to 4.

Wellman, one of the most successful PET recyclers, built its fibre business by converting PET waste and used bottles to fibre. Other successful recyclers have produced sheet and have kept themselves profitable during difficult periods. The world's biggest producer of PET strapping, ITW, has now integrated the assets of the Swiss recycler, Polyrecycling, into its own operations and can add margins by direct conversion to strapping to ensure continuing profitability and securing continuity of supply.

Virgin PET has recently become competitive with other materials (PVC and PS) used in the development of packaging that utilises clear or transparent plastic sheet. If this competitive price is held, some European producers can be optimistic about expanding the use of A-PET in selected sheet markets. Penetration at relatively low levels (5%) could expand the demand for R-PET by some 100 K tonnes.

However, the emphasis is not on the suitability of PET for the end use but on what long-term pricing structure will prevail. The clarity, quality, cleanliness and purity of PET from a bottle source is assured. PVC bottle sources are now virtually replaced by PET and a large source of clear used PS is not readily available. However, some suppliers are offering 'take back' facilities to protect some of the major markets. More accurate predictions in this application area

Table 11.11 Prices of PET products (DM/tonne)—Western European source [51] figures rounded

	Q3 1999	Q4 1999	Q1 2000
Virgin PET resin	1500–1800	1540–1900	1900–2200
Fibre chip	1620–1720	1620–1900	1930–2030
Staple fibre	1550–2050	1650–2300	2100–2300
A-PET sheet	2550–3450	2600–3500	2600–3700
Strapping	2450–3300	2450–3350	2500–3500
Post-consumer flake clear	500–1080	700–1200	800–1400
Baled bottles – one-way clear	30–210	60–300	100–400

Abbreviation: A-PET, amorphous polyethylene terephthalate.

Table 11.12 European plastics packaging collection and recovery organisations—contact co-ordinates

Collection agency	Plastics agency	Collection agency	Plastics agency
ARA - Altstoff Recycling Austria Mariahilfestrasse 123 A-1062 Wien, Austria http://www.ara.at tel: +43 1 599 997-0 fax: +43 1 595 3535	ÖKK Handelskai 388/Top841 A-1020 Wien, Austria tel: +43 1 720 7001 fax: +43 1 720 700140	FOST-Plus Rue Martin V 40 B-1200 Brussels, Belgium tel: +32 2 775 0350 fax: +32 2 771 1696 www.fostplus.be	Plarebel VZW Maria Louisa Square, 49 B-1000 Brussels, Belgium tel: +00 32 2 238 9781 fax: +32 2 238 9998
The Finnish Packaging Register - PYR Isoroobertinkatu 1a F-00120 Helsinki, Finland tel: +358 9616 230 fax: +358 9616 23100 http://www.pyr.fi	Finnish Plastics Industries Federation Box 4 F-00131 Helsinki, Finland tel: +358 9 1728 4326 fax: +358 9 171164	Eco-Emballages SA 44 Avenue Georges Pompidou BP306 F-92302 Levallois-Perret Cedex, France tel: +33 1 40 89 99 99 fax: +33 1 40 89 99 88 http://www.ecoemballages.fr	VALORPLAST Le diamant A – 14 rue de la République F-92909 Paris – La Défense Cedex, France tel: +33 1 46 53 10 95 fax: +33 1 46 53 10 90 www.valorplast.com
DSD AG - Duales System Deutschland Frankfurterstrasse 720-726 D-51145 Köln-Porz-Eil, Germany tel: +49 2203 9370 fax: +49 2203 937190 http://www.gruener-punkt.de	DKR Frankfurterstrasse 720-726 D-51145 Köln, Germany tel: +49 2203 93170 fax: +49 2203 931744 www.dkr.de	REPAK 1 Ballymount Road Clondalkin Dublin 22, Ireland tel: +353 1 467 0191 fax: +353 1 467 0197	
Consorzio Nazionale Imballaggi – CONAI Via Donizetti 6 I-20122 Milano, Italy tel: +39 02 54044.1 fax: +39 02 5412 2648 www.CONAI.org	Consorzio Nazionale per il Recupero degli Imballaggi inPlastica (Co.Re.Pla.) Via del Vecchio Politecnico 3 I-20121 Milano, Italy tel: +39 02 760541 fax: +39 02 76054320	Valorlux B P 26 L-3205 Leudelange, Luxembourg tel: +352 370 006 fax: +352 371 137 www.Valorlux.lu	

SVM-PACT
Postbus 11753
NL-2502 AT Den Haag
The Netherlands
tel: +70 382 4042
fax: +70 381 9016

Sociedade Ponto Verde
Edifício Infante D. Henrique
Rua João Chagas,
n° 53-1° Dt°
P-1495-072 Algés, Portugal
tel: +351 21 414 73 00
fax: +351 21 414 52 46
www.pontoverde.pt

Returpack PET
Box 17777
S-11893 Stockholm, Sweden
tel: +46 8702 0880
fax: +46 8643 3985

Plastkretsen
Box 105
S-10122 Stockholm, Sweden
tel: +46 8402 1375
fax: +46 8566 14440
pir.plastkretsen@plastkretsen.se

Other European Groups

Europen
Legerlaan 6
B-1040 – Brussels, Belgium
tel: +32 2 7363600
fax: +32 2 7363521
www.europen.be

Plastretur AS
c/o Den Norske
Emballasjeforening
Sorkedalsveien 6
N-0369 Oslo, Norway
tel: +47 2212 1780
fax: +47 2260 0067

ANEP
Sant Pere Més Alt, 1, pral. Bis
E-08003 Barcelona, Spain
tel: +34 93 3192687
fax: +34 93 3192654

Materialretur A/S
Postboks 252 Skøyen
N-0212 Oslo, Norway
tel: +47 22 12 1500
fax: +47 22 12 1519

Ecoembalajes España SA
Paseo de la Castellana 147
8 Planta
E-28046 Madrid, Spain
tel: +34 91 567 2403
fax: +34 91 567 2410
www.ecoembes.com

PET-Recycling Schweiz
Naglerwiesenstrasse 4
Ch-8049 Zürich, Switzerland
tel: +41 1 342 2960
fax: +41 1 342 2966

Other European Groups

Plastics

APME
Ave E Van Nieuwenhuyse 4
B-1160 – Brussels, Belgium
tel: +32 2 675 3258
fax: +32 2 675 4002
www.apme.org

Plastics

Assure
83, Ave E Mounier
B-1200 Brussels, Belgium
tel: +32 2 772 5252
fax: +32 2 772 5419
www.assure.be

Plastics

EuPC
Ave de Cortenbergh, 66
B-1000 Brussels, Belgium
tel: +00 32 2 732 41 24
fax: +32 2 732 41 18
www.eupc.org

should be developed from a study of the longer-term economics, availability and pricing structure of competitive materials relative to R-PET. This should enable better assessment of likely market penetration.

The uncertainty surrounding the permanence of the EU anti-dumping duties will also be a factor in the future sustainability of PET bottle recycling. These duties are usually in place for a period of 5 yrs, so one could expect the market situation to change if they are removed.

References

1. *A Community Strategy for Waste Management*, Brussels, 18th September 1989, Council Resolution of 7th May 1990. Fifth Environmental Action Programme 1992–2000. OJ C 138, 17.05.1993; also known under the title '*Towards Sustainability*'. A sixth Environmental Action Programme is now in draft, 7th December 2000, which will cover 2000–2009.
2. European Environmental Agency (EEA). Report, *Environmental Signals 2000*, EEA DK-1050, Copenhagen, Denmark.
3. Directive 85/339/EEC, ref. OJ No L 176, 6.7.1985, p. 18. Directive as amended by Directive 91/629/EEC, ref. OJ No L 337, 31.12.1991, p. 48.
4. Directive 94/62/EC ref. OJ No L 365/10 31.12.1994.
5. Perchards; www.perchards.com
6. APME (Association of Plastic Manufacturers in Europe), Ave E. Van Nieuwenhuyse 4, B-1160 Brussels, www.apme.org; EuPC (European Plastics Converters), Ave de Cortenbergh 66, B-1040 Brussels; Europen (European Organisation for Packaging Industries and the Environment), Ave de l'Armée 6, B-1040 Brusssels, Belgium.
7. CEN (Comité Européen Normalisation), Rue de Stassart 36, B-1050 Brussels, Belgium. Tour Europe Cedex 7, 92049 Paris la Defense, France.
8. ISO (International Organisation for Standardisation), Casa Postale 56, CH-1211 Genéve 20, Switzerland.
9. PETCORE publications, *Design for Recycle* and *Collection and Sorting Plastic Bottles. A Comprehensive Guide, 2000 edition*. PETCORE (PET Container Recycling Europe), Ave de Cortenbergh 66, Box 2, B-1000, Brussels, Belgium..
10. Information system on plastic waste management in Europe 1998 data, Jan 2000. Taylor Nelson Sofres Consulting for the Association of Plastic Manufactures in Europe [6].
11. Study by GUA (Gesellschaft für Umweltfreundliche Abfalt Behandlung—Company for Environmentally Friendly Waste Treatment). Sechshauser Strasse. 83, A-1150 Wien, Austria. Commissioned by ARA and the Austrian Beverage Association, 2000.
12. *Economic Instruments in Packaging Waste Policy*, published by Europen [6], October 2000, available from their website, www.europen.com
13. PET Recycling Schweiz, Naglerweisenstrasse 4, CH-8049, Zurich, Switzerland.
14. MSS Inc., 3738 Keystone Ave, Nashville TN 37211, USA; www.magsep.com
15. NAPCOR, 2105 Water Ridge Parkway, Suite 570 Charlotte, NC 28217 USA; www.napcor.com
16. EREMA Plastic Recycling Systems, Freindorf, Unterfeldstrasse 3, PO Box 38, 4502 Ausfelden/ Linz, Austria. www.erema.com. SOREMA Plastic Recycling Systems, Via, Dei Platini 11, 22040 Alzate Brianza, Italy.
17. *Recycling and Recovery of Plastics* (ed. J. Brandrup), Carl Hanser Verlag. ISBN 3-446-18258-6, 1996.
18. ILSI (International Life Sciences Institute) Europe report series. *Packaging Materials. 1. Polyethylene Terephthalate (PET) for Food Packaging Applications*. July 2000. ILSI Europe, Ave E. Mounier, 83. Box 6, B-1200 Brussels, Belgium.

19. (a) EU DG XII Science, Research and Development. AIR2-CT93-1104. Programme to establish criteria to ensure the quality and safety in use of recycled and reused plastics for food packaging. Final Summary Report, December 1997. (b) Recycling Plastics for Food Contact Use – Guidelines. ILSI Europe, Ave E. Mounier 83, B-1200, Brussels, Belgium, May 1998.

20. *EREMA Plastic Recycling Systems*, Freindorf, Unterfeldstrasse 3, PO Box 38, 4502 Ausfelden/Linz, Austria; www.erema.com

21. US Patent 5,876,664. Wellman Inc., March 2, 1999.

22. US Patent 5,102, 594. Stamicarbon. BV, April 7, 1992.

23. US Patent 5,780,520. Eastman Chemical Inc., July 14, 1998.

24. US Patent 5,554,657 Shell Chemicals, September 10, 1996.

25. US Patent 5,824,196 Plastics Technologies, October 20, 1998.

26. US Patent 5,899,392 Plastics Technologies, May 4, 1999.

27. US Patent 5,958,987 Coca Cola Inc., September 28, 1999.

28. Urethanes Technology, March 1999. Repol SA, F-63500 Issoire (subsidiary of Group TBI).

29. Gintis, D. (1992) *Macromol. Chem. Macromol. Symp.*, **57**, 185-190.

30. Klein, P. *Recycling and Recovery of Plastics* (Ed. J. Brandrup), Carl Hanser Verlag. ISBN 3-446-18258-6, 1996, pp. 495-497.

31. Tersac, G. *et al.* Synthesis of insulating foams from polyethylene terephthalate bottles. *Catouch. Plast.*, 1991, **68**, 706 pp. 181-5.

32. US Patent 5,360,900 Oxid Inc., November 1, 1994.

33. US Patent 5,051,528. Eastman Kodak, September 24, 1991.

34. US Patent 6,136,869. Eastman Chemical Inc., October 2000.

35. US Patent 4,355,175. Puztaszeri, October 19, 1982.

36. US Patent 5,254,666. Inst. Francais du Petrole, October 19, 1993.

37. US Patents 5,395,858. URRC December 6, 1995 and 5,580,905 December 3, 1996.

38. International patent No. WO/93/23465. November 25, 1993.

39. US Patent 5,502,247. Amoco Inc., March 6, 1996.

40. Boustead, I. Plastics and the environment, Rapra Review Report No. 75, 1994. *Principles of Recycling*, SPOLD (Society for Promotion of Life Cycle Development), Brussels, 1994. www.Boustead-Consulting.co.uk

41. APME eco-profiles of plastics, Report 8: Polyethylene terephthalate (PET) July 1995. Revised edition published January 2000 on the APME website, www.apme.org

42. Boustead, I. (1982) private communication, Boustead Consulting, 2-4 Black Cottages, West Grinstead, Horsham, West Sussex, RH13 7BD, UK.

43. Franklin Associates, Prairie Village, Kansas. Collective studies for NAPCOR and the US SPI, 1980 and 1989 to 1994.

44. Boustead, I. Studies for Incpen (Industry Council for Packaging and the Environment) (UK) 1989, 1993 and personal communications.

45. SAEFL (Swiss Agency for the Environments Forest and Landscape - BUWAL), Report 132, 1990 *Ecobalances of Packaging Materials.*

46. *Disposal of Plastics Household Waste: Analysis of Environmental Impacts and Costs.* Centre for Energy Conservation and Technology, Delft, The Netherlands, November 1994.

47. Studies for APME by Chem. Systems/TNO (1992) and Research Development and Consulting (RDC) 1995.

48. Maack Business Services, Proceedings of 'Recycling '93', Geneva conference January 1993, Vol ii, p. 201.

49. Boustead, I. unpublished calculations based on data provided by URRC - United Resource Recovery Operation, Spartanburg, SC 29303 USA, process ref [37] and Innovations in PET Pty Ltd, Footscray, Victoria 3012, Australia, process ref [38].

50. Goulin, J.P., private communication, Eastman Chemical, December 2000 in reference to Eastman OPTSYS PET recycling process.

51. PCI Ltd, P.O. Box 319, Derby DE1 2ZZ, UK.
52. EC Anti-dumping duties, OJ L 199, 5.8.2000, p. 48. OJ L 199, 5.8.2000, p. 6.
53. PET Kuststoffrecycling GmbH (PKR), Beislich, *Using the Stehning BtB Process*, October, 1999 and RecyPET AG, Frauenfeld, Switzerland, *Using URRC Technology*, September, 2000.
54. Unkelbach, K.-H., CENSOR® Sorting Centrifuge, Baker process, Cologne. Paper given at APME sponsored Identiplast Conference, Brussels, Belgium, April 1999.
55. Sortex Ltd, London, E15 2PJ, UK; Radix Systems Ltd, Romsey, Hampshire, UK.

Index